Lecture Notes in Mathematics   1926

Editors:
J.-M. Morel, Cachan
F. Takens, Groningen
B. Teissier, Paris

Luis Barreira · Claudia Valls

Stability of Nonautonomous Differential Equations

 Springer

Authors

Luis Barreira
Claudia Valls
Departamento de Matemática
Instituto Superior Técnico
Av. Rovisco Pais
1049-001 Lisboa
Portugal

e-mail: *barreira@math.ist.utl.pt*
        cvalls@math.ist.utl.pt
URL: *http://www.math.ist.utl.pt/~barreira/*

Library of Congress Control Number: 2007934028

Mathematics Subject Classification (2000): 34Dxx, 37Dxx

ISSN print edition: 0075-8434
ISSN electronic edition: 1617-9692
ISBN 978-3-540-74774-1 Springer Berlin Heidelberg New York
DOI 10.1007/978-3-540-74775-8

Springer is a part of Springer Science+Business Media
springer.com
© Springer-Verlag Berlin Heidelberg 2008

Typesetting by the authors and SPi using a Springer LaTeX macro package

Cover design: *design & production* GmbH, Heidelberg

Printed on acid-free paper   SPIN: 12114993   41/SPi   5 4 3 2 1 0

To our parents

Preface

The main theme of this book is the stability of nonautonomous differential equations, with emphasis on the study of the existence and smoothness of invariant manifolds, and the Lyapunov stability of solutions. We always consider a nonuniform exponential behavior of the linear variational equations, given by the existence of a nonuniform exponential contraction or a nonuniform exponential dichotomy. Thus, the results hold for a much larger class of systems than in the "classical" theory of exponential dichotomies.

The departure point of the book is our joint work on the construction of invariant manifolds for nonuniformly hyperbolic trajectories of nonautonomous differential equations in Banach spaces. We then consider several related developments, concerning the existence and regularity of topological conjugacies, the construction of center manifolds, the study of reversible and equivariant equations, and so on. The presentation is self-contained and intends to convey the full extent of our approach as well as its unified character. The book contributes towards a rigorous mathematical foundation for the theory in the infinite-dimensional setting, also with the hope that it may lead to further developments in the field. The exposition is directed to researchers as well as graduate students interested in differential equations and dynamical systems, particularly in stability theory.

The first part of the book serves as an introduction to the other parts. After giving in Chapter 1 a detailed introduction to the main ideas and motivations behind the theory developed in the book, together with an overview of its contents, we introduce in Chapter 2 the concept of nonuniform exponential dichotomy, which is central in our approach, and we discuss some of its basic properties. Chapter 3 considers the problem of the robustness of nonuniform exponential dichotomies.

In the second part of the book we discuss several consequences of local nature for a nonlinear system when the associated linear variational equation admits a nonuniform exponential dichotomy. In particular, we establish in Chapter 4 the existence of Lipschitz stable manifolds for nonautonomous equations in a Banach space. In Chapters 5 and 6 we establish the smoothness of the stable manifolds. We first consider the finite-dimensional case in Chapter 5, with the method of invariant families of cones. This approach uses in a decisive manner the compactness of the closed unit ball in the ambient space, and this is why we consider only finite-dimensional spaces in this chapter. Moreover, the proof strongly relies on the use of Lyapunov norms to control the nonuniformity of the exponential dichotomies. As an outcome of our approach we provide examples of C^1 vector fields with invariant stable manifolds, while in the existing nonuniform hyperbolicity theory one assumes that the vector field is of class $C^{1+\alpha}$. In Chapter 6 we consider differential equations in Banach spaces, although at the expense of slightly stronger assumptions for the vector field. The method of proof is different from the one in Chapter 5, and is based on the application of a lemma of Henry to obtain both the existence and smoothness of the stable manifolds using a single fixed point problem. In addition, we show that not only the trajectories but also their derivatives with respect to the initial condition decay with exponential speed along the stable manifolds. A feature of our approach is that we deal directly with flows or semiflows instead of considering the associated time-1 maps. In Chapter 7 we establish a version of the Grobman–Hartman theorem for nonautonomous differential equations in Banach spaces, assuming that the linear variational equation admits a nonuniform exponential dichotomy. In addition, we show that the conjugacies that we construct are always Hölder continuous.

The third part of the book is dedicated to the study of center manifolds. In Chapter 8 we extend the approach in Chapter 6 to nonuniform exponential trichotomies, and we establish the existence of center manifolds that are as smooth as the vector field. In particular, we obtain simultaneously the existence and smoothness of the center manifolds using a single fixed point problem. In Chapter 9 we show that some symmetries of the differential equations descend to the center manifolds. More precisely, we consider the properties of reversibility and equivariance in time, and we show that the dynamics on the center manifold is reversible or equivariant if the dynamics in the ambient space has the same property.

In the fourth part of the book we study the so-called regularity theory of Lyapunov and its applications to the stability theory of differential equations. We note that this approach is distinct from what is usually called Lyapunov's second method, which is based on the use of Lyapunov functions. In Chapter 10 we provide a detailed exposition of the regularity theory, organized in a pragmatic manner so that it can be used in the last two chapters of the book. In Chapter 11 we extend the regularity theory to the infinite-dimensional setting of Hilbert spaces. Chapter 12 is dedicated to the study of the stability of nonautonomous differential equations using the regularity theory. We note that the notion of Lyapunov regularity is much less restrictive than the notion of uniform stability, and thus we obtain the persistence of the stability of solutions of nonautonomous differential equations under much weaker assumptions.

We are grateful to several people who have helped us in various ways. We particularly would like to thank Jack Hale, Luis Magalhães, Waldyr Oliva, and Carlos Rocha for their support and encouragement along the years as well as their helpful comments on several aspects of our work. We also would like to thank the referees for the careful reading of the manuscript.

We were supported by the Center for Mathematical Analysis, Geometry, and Dynamical Systems, and through Fundação para a Ciência e a Tecnologia by the Programs POCTI/FEDER, POSI and POCI 2010/Fundo Social Europeu, and the grants SFRH/BPD/14404/2003 and SFRH/BPD/26465/2006.

Luis Barreira and Claudia Valls
Lisbon, October 2006

Contents

1

Introduction

In the theory of differential equations, the notion of (uniform) *exponential dichotomy*, introduced by Perron in [69], plays a central role in the study of stable and unstable invariant manifolds. In particular, consider a solution $u(t)$ of the equation $u' = F(u)$ for some differentiable map F in a Banach space. Setting $A(t) = d_{u(t)}F$, the existence of an exponential dichotomy for the linear variational equation

$$v' = A(t)v \tag{1.1}$$

implies the existence of stable and unstable invariant manifolds for the solution $u(t)$, up to mild additional assumptions on the nonlinear part of the vector field. The theory of exponential dichotomies and its applications are well developed. In particular, there exist large classes of linear differential equations with exponential dichotomies. For example, Sacker and Sell [83, 84, 85, 82, 86] discuss sufficient conditions for the existence of exponential dichotomies, also in the infinite-dimensional setting. In a different direction, for geodesic flows on compact smooth Riemannian manifolds with strictly negative sectional curvature, the unit tangent bundle is a hyperbolic set, that is, they are Anosov flows. Furthermore, time changes and small C^1 perturbations of flows with a hyperbolic set also have a hyperbolic set (see for example [49] for details). We refer to the books [24, 41, 46, 88] for details and further references related to exponential dichotomies. We particularly recommend [24] for historical comments. The interested reader may also consult the books [32, 33, 60]. On the other hand, the notion of exponential dichotomy substantially restricts the dynamics and it is important to look for more general types of hyperbolic behavior.

Our main objective is to consider the more general notion of *nonuniform exponential dichotomy* and study in a systematic manner some of its consequences, in particular concerning the existence and smoothness of invariant manifolds for nonautonomous differential equations. Also in the nonuniform setting, we obtain a version of the Grobman–Hartman theorem, the existence of center manifolds, as well as their reversibility and equivariance properties, and an infinite-dimensional version of Lyapunov's regularity theory with applications to the stability of solutions of nonautonomous equations. In comparison with the classical notion of (uniform) exponential dichotomy, the existence of a nonuniform exponential dichotomy is a much weaker hypothesis. In fact, perhaps surprisingly, essentially *any* linear equation as in (1.1), with global solutions and with at least one negative Lyapunov exponent, has a nonuniform exponential dichotomy (see Chapter 10 for details). We emphasize that we always consider nonautonomous differential equations, and with the exception of Chapters 5 and 10 the theory is systematically developed in infinite-dimensional spaces. Another aspect of our approach is that we deal directly with flows or semiflows instead of using their time-1 maps (with the single exception of Chapter 7, where we establish a version of the Grobman–Hartman theorem). Our work is also a contribution to the theory of nonuniformly hyperbolic dynamics (we refer to [1, 2, 3] for detailed expositions of the theory).

We discuss in this chapter the main ideas and motivations behind the theory developed in the book. We also highlight some of the main results and the relations with former work. We mostly follow the order in which the material is presented in the book.

1.1 Exponential contractions

In order to describe the differences between the notions of uniform exponential dichotomy and nonuniform exponential dichotomy, we first consider the case when only contraction is present. We could replace contraction by expansion simply by reversing the time.

Consider a continuous function $t \mapsto A(t)$ with values in the $n \times n$ real matrices for $t \geq 0$. We assume that all solutions of (1.1) are global in the future, that is, are defined for every $t \geq 0$. Let $U(t, s)$ be the evolution operator associated with equation (1.1). This is the operator satisfying

$$U(t, s)v(s) = v(t)$$

for every solution $v(t)$ of (1.1) and every $t \geq s$. We assume in this section that all Lyapunov exponents of solutions of equation (1.1) are negative, that is,

$$\limsup_{t \to +\infty} \frac{1}{t} \log\|v(t)\| < 0 \text{ for each solution } v(t) \text{ of } (1.1). \tag{1.2}$$

We say that $U(t, s)$ is a *(uniform) exponential contraction* if there exist constants $a, c > 0$ such that

$$\|U(t, s)\| \leq ce^{-a(t-s)} \text{ for every } t \geq s.$$

We say that $U(t, s)$ is a *nonuniform exponential contraction* if there exist constants $a, c > 0$ and $b \geq 0$ such that

$$\|U(t,s)\| \leq ce^{-a(t-s)+bs} \text{ for every } t \geq s. \tag{1.3}$$

Thus, a nonuniform exponential contraction allows a "spoiling" of the uniform contraction along each solution as the initial time s increases: while the uniform contraction (given by a) is still present in (1.3), and is independent of the initial time $s \geq 0$, we may have the additional exponential term e^{bs} (and thus the nonuniformity along the solution). This means that even though in both cases we have the exponential stability of solutions (due to (1.2)), in the nonuniform case, in order that a given solution is in a prescribed neighborhood, the size of the initial condition may depend on s (while in the uniform case the size can be chosen independently of s).

The following statement is a simple consequence of Theorem 10.6 (the proof of which is inspired in related work in [1]).

Theorem 1.1. *If the equation* (1.1) *satisfies the condition* (1.2), *then the associated evolution operator* $U(t,s)$ *is a nonuniform exponential contraction, for which the constant* a *is any positive number satisfying*

$$a < -\sup_{v_0 \in \mathbb{R}^n} \limsup_{t \to +\infty} \frac{1}{t} \log\|v(t)\|, \tag{1.4}$$

where $v(t)$ *is the unique solution of the equation* (1.1) *with* $v(0) = v_0$.

We note that the right-hand side of (1.4) is indeed positive (since the lim sup in (1.2) can only take a finite number of values; see Section 10.1). In view of Theorem 1.1, the notion of nonuniform exponential contraction is in fact as weak as possible, since all (exponentially stable) linear equations originate an evolution operator having such a contraction. A similar behavior occurs in the case of nonuniform exponential dichotomies (see Theorem 10.6). Thus, in specific applications we never need to assume the existence of a nonuniform exponential contraction (since this follows from (1.2)) but instead we look for conditions on a and b which ensure the desired results. For example, in general we are only able to establish the stability of the zero solution of (1.1) under sufficiently small perturbations provided that b/a is sufficiently small (see Chapter 12 for related results).

In view of this discussion it is also important to give a sharp estimate for b. We refer to Section 10.3 for details; here, we consider only the case of triangular matrices. The following statement is a simple consequence of Theorems 10.6 and 10.8.

Theorem 1.2. *If the matrix* $A(t)$ *is upper triangular for every* $t \geq 0$, *then the constant* b *can be any number satisfying*

$$b > \sum_{k=1}^{n} \left(\limsup_{t \to +\infty} \frac{1}{t} \int_0^t a_k(\tau)\,d\tau - \liminf_{t \to +\infty} \frac{1}{t} \int_0^t a_k(\tau)\,d\tau \right),$$

where $a_1(t)$, ..., $a_n(t)$ *are the entries in the diagonal of* $A(t)$.

See Chapter 11 for generalizations of Theorems 1.1 and 1.2 to infinite-dimensional spaces.

1.2 Exponential dichotomies and stable manifolds

We now consider the more general case of nonuniform exponential dichotomies. These are composed of nonuniform contractions and nonuniform expansions (see Section 1.1). We also present a first consequence of the existence of an exponential dichotomy, namely the existence of invariant stable manifolds for any sufficiently small perturbation.

Consider a Banach space X and a continuous function $t \mapsto A(t)$ such that $A(t)$ is a bounded linear operator on X for each $t \geq 0$. We assume again that all solutions of (1.1) are global in the future, that is, are defined for every $t \geq 0$. Let $T(t, s)$ be the evolution operator associated with equation (1.1). This is the operator satisfying

$$T(t, s)v(s) = v(t)$$

for every solution $v(t)$ of (1.1) and every $t \geq s$. For simplicity of the exposition, we assume that the evolution operator $T(t, s)$ has a decomposition in block form

$$T(t, s) = (U(t, s), V(t, s))$$

into evolution operators with respect to some invariant decomposition $X = E \oplus F$ (which is independent of the time t). We emphasize that in the remaining chapters we do not assume that $T(t, s)$ has a decomposition in block form.

We say that the equation (1.1) admits a *nonuniform exponential dichotomy* if there exist constants $\lambda < 0 \leq \mu$ and a, b, $K > 0$, such that for every $t \geq s \geq 0$,

$$\|U(t, s)\| \leq Ke^{\lambda(t-s)+as} \quad \text{and} \quad \|V(t, s)^{-1}\| \leq Ke^{-\mu(t-s)+bt}. \tag{1.5}$$

The constants λ and μ play the role of Lyapunov exponents, while a and b measure the nonuniformity of the dichotomy. The assumption $\lambda < 0$ means that there is at least one negative Lyapunov exponent.

We now consider the equation

$$v' = A(t)v + f(t, v), \tag{1.6}$$

where the perturbation $f(t, v)$ is a continuous function defined for $t \geq 0$ and $v \in X$, such that $f(t, 0) = 0$ for every $t \geq 0$ (and thus the origin is also a solution of (1.6)).

The following is one of our main results on the existence of stable manifolds for a nonautonomous differential equation, and is an immediate consequence of Theorem 4.1.

Theorem 1.3. *Assume that the equation* (1.1) *admits a nonuniform exponential dichotomy, and that there exist $c > 0$ and $q > 0$ such that*

$$\|f(t, u) - f(t, v)\| \leq c\|u - v\|(\|u\|^q + \|v\|^q)$$

for every $t \geq 0$ *and* $u, v \in X$. *If*

$$\lambda + a + (a+b)/q < 0 \quad and \quad \lambda + b < \mu, \tag{1.7}$$

then there exists a Lipschitz function $\varphi \colon U \to F$, *where* $U \subset \mathbb{R}_0^+ \times E$ *is an open neighborhood of the line* $\mathbb{R}_0^+ \times \{0\}$, *such that its graph* $\mathcal{W} \subset \mathbb{R} \times X$ *has the following properties:*

1. $(t, 0) \in \mathcal{W}$ *for every* $t \geq 0$;
2. $\mathcal{W}$ *is forward invariant under the semiflow* Ψ_τ *on* $\mathbb{R}_0^+ \times X$ *generated by the autonomous system*

$$t' = 1, \quad v' = A(t)v + f(t, v);$$

3. *there exists* $D > 0$ *such that for every* $(s, u), (s, v) \in \mathcal{W}$ *and* $\tau \geq 0$, *we have*

$$\|\Psi_\tau(s, u) - \Psi_\tau(s, v)\| \leq De^{\lambda \tau + as}\|u - v\|.$$

We refer to Section 4.2 for a detailed formulation. We observe that the Lipschitz invariant manifolds constructed in Theorem 1.3 are in fact as smooth as the vector filed. We refer to Chapters 5 and 6 for details.

Note that the first inequality in (1.7) is satisfied for a given $a < |\lambda|$ provided that q, the order of the perturbation, is sufficiently large. Furthermore, both inequalities in (1.7) are automatically satisfied when a and b are sufficiently small. The "small" exponentials e^{as} and e^{bt} in (1.5), that are not present in the case of a uniform exponential dichotomy, are the main cause of difficulties. On the other hand, it turns out that the smallness of the nonuniformity is a rather common phenomenon from the point of view of ergodic theory: almost all linear variational equations obtained from a measure-preserving flow on a smooth Riemannian manifold admit a nonuniform exponential dichotomy with arbitrarily small nonuniformity (see Theorem 10.6).

Our definition of weak nonuniform exponential dichotomy in (1.5) is inspired in the notion of uniform exponential dichotomy and in the notion of nonuniformly hyperbolic trajectory (see Sections 4.3 and 5.2). Our work is also a contribution to the theory of nonuniformly hyperbolic dynamics. We refer to [1, 3] for detailed expositions of parts of the theory and to the survey [2] for a detailed description of its contemporary status. The theory goes back to the landmark works of Oseledets [65] and Pesin [70, 71, 72]. Since then it became an important part of the general theory of dynamical systems and a principal tool in the study of stochastic behavior. We note that the nonuniform hyperbolicity conditions can be expressed in terms of the Lyapunov exponents. For example, almost all trajectories of a dynamical system preserving a finite invariant measure with nonzero Lyapunov exponents are nonuniformly hyperbolic.

Among the most important properties due to nonuniform hyperbolicity is the existence of stable and unstable manifolds, and their absolute continuity property established by Pesin in [70]. The theory also describes the ergodic properties of dynamical systems with a finite invariant measure absolutely continuous with respect to the volume [71], and expresses the Kolmogorov–Sinai entropy in terms of the Lyapunov exponents by the Pesin entropy formula [71] (see also [55]). In another direction, combining the nonuniform hyperbolicity with the nontrivial recurrence guaranteed by the existence of a finite invariant measure, the fundamental work of Katok [48] revealed a very rich and complicated orbit structure, including an exponential growth rate for the number of periodic points measured by the topological entropy, and an approximation by uniformly hyperbolic horseshoes of the entropy of an invariant measure (see also [50]).

Here we concentrate our attention on the stable manifold theorem. We first briefly describe the relevant references. The proof by Pesin in [70] is an elaboration of the classical work of Perron. His approach was extended by Katok and Strelcyn in [51] for maps with singularities. In [80], Ruelle obtained a proof of the stable manifold theorem based on the study of perturbations of products of matrices in Oseledets' multiplicative ergodic theorem [65]. Another proof is due to Pugh and Shub in [78] with an elaboration of the classical work of Hadamard using graph transform techniques. In [37] Fathi, Herman and Yoccoz provided a detailed exposition of the stable manifold theorem essentially following the approaches of Pesin and Ruelle. We refer to [3] for further details. There exist also versions of the stable manifold theorem for dynamical systems in infinite-dimensional spaces. In [81] Ruelle established a corresponding version in Hilbert spaces, following his approach in [80]. In [58] Mañé considered transformations in Banach spaces under some compactness and invertibility assumptions, including the case of differentiable maps with compact derivative at each point. The results of Mañé were extended by Thieullen in [92] for a class of transformations satisfying a certain asymptotic compactness. We refer the reader to the book [42] for a detailed discussion of the geometric theory of dynamical systems in infinite-dimensional spaces.

We note that in the above works the dynamics is assumed to be of class $C^{1+\varepsilon}$ for some $\varepsilon > 0$. On the other hand, in [77] Pugh constructed a C^1 diffeomorphism in a manifold of dimension 4, that is not of class $C^{1+\varepsilon}$ for any ε, and for which there exists no invariant manifold tangent to a given stable space such that the trajectories along the invariant manifold travel with exponential speed. We refer to [3] for a detailed description of the diffeomorphism. Nevertheless, although this example shows that the hypothesis $\varepsilon > 0$ is crucial in the stable manifold theorem it does not forbid the existence of families of C^1 dynamics which are not of class $C^{1+\varepsilon}$ for any ε but for which there still exist stable manifolds. Indeed, Theorem 5.1 implies the existence of invariant stable manifolds for the nonuniformly hyperbolic trajectories of a large family of maps that, in general, are *at most* of class C^1. A detailed presentation is given in Section 5.3.

There are some differences between our approach and the usual approach in the theory of nonuniformly hyperbolic dynamics. In particular, we start from a linear equation $v' = A(t)v$ instead of a linear variational equation

$v' = A_x(t)v$ with $A_x(t) = d_{\varphi_t x}F$, obtained from a particular solution $\varphi_t x$ of a given autonomous equation $x' = F(x)$. Here φ_t is the flow generated by the autonomous equation. Another feature of our approach is that we deal directly with flows instead of considering time-1 maps as it is sometimes customary in the theory of hyperbolic dynamics. This allows us to give a direct proof, dealing *simultaneously* with all the times along each orbit. On one hand, our approach to the proof of Theorem 1.3 could be considered classical, and consists in using the differential equation (1.6) to express the forward invariance of the manifold W under the dynamics to conclude that φ must satisfy a fixed point problem. However, the extra small exponentials in a nonuniform exponential dichotomy substantially complicate this approach and the implementation requires several new ideas (see Section 4.4 for details). We also obtain in a very direct manner explicit quantitative information on the size of the neighborhoods on which we must choose an initial condition so that the solution satisfies a given bound. In fact, this information is put from the beginning in the space on which we look for the fixed point. Finally, we want to consider *semiflows* and not only flows. In particular, it is thus in general impossible to introduce the same adapted Lyapunov norms as in the case of flows. On the other hand, we still require some appropriate device that can play a similar role in the case of semiflows. This causes several additional difficulties.

1.3 Topological conjugacies

A fundamental problem in the study of the local behavior of a map or a flow is whether the linearization of the system along a given solution approximates well the solution itself in some open neighborhood. In other words, we look for an appropriate local change of variables, called a conjugacy, that can take the system to a linear one. Moreover, as a means to distinguish the dynamics in a neighborhood of the solution further than in the topological category, we would like the change of variables to be as regular as possible. For example, we would like to know whether it is possible to distinguish between different types of nodes. The problem goes back to the pioneering work of Poincaré, which can be interpreted today as looking for an *analytic* change of variables that takes the initial system to a linear one. The work of Sternberg [89, 90] showed that there are algebraic obstructions, expressed in terms of resonances between the eigenvalues of the linear approximation, that prevent the existence of conjugacies with a prescribed high regularity (see also [19, 20, 87, 61] for further related work).

We concentrate here our attention in the case of hyperbolic fixed points. For simplicity of the exposition we consider maps instead of flows. We refer to Chapter 7 for a detailed presentation in the case of flows. Consider the dynamics generated by the map

$$F(v) = Av + f(v) \tag{1.8}$$

in a Banach space X. We assume that A is a linear operator, and that f is a C^1 map with $f(0) = 0$ and $d_0 f = 0$. In this setting, the linearization problem corresponds to ask whether the behavior of the trajectories of (1.8) in some open neighborhood of zero somehow approximates well the behavior of the trajectories of the linear map A. It is well-known that this is the case when A admits an exponential dichotomy: by the Grobman–Hartman theorem, under mild additional assumptions on the perturbation f, locally the two dynamics are *topologically* conjugate, that is, there exists a local homeomorphism h in a neighborhood of 0 such that

$$A \circ h = h \circ F. \tag{1.9}$$

Since $A^n \circ h = h \circ F^n$ for each $n \in \mathbb{N}$, the conjugacy map h takes trajectories of the nonlinear map F into trajectories of the linear operator A, and thus h acts essentially as a dictionary between the two dynamics. In the two-dimensional case the situation is different: in [44] Hartman showed that for a C^2 diffeomorphism it always exists a C^1 conjugacy. On the other hand, in [43] he also gave an example of a diffeomorphism with resonances in $\mathbb{R}^3$ for which there exists no C^1 conjugacy. The original references for the Grobman–Hartman theorem (in the case of uniformly hyperbolic dynamics) are Grobman [38, 39] and Hartman [43, 45]. Using the ideas in Moser's proof in [63] of the structural stability of Anosov diffeomorphisms, the Grobman–Hartman theorem was extended to Banach spaces independently by Palis [66] and Pugh [76]. We note that in the case of continuous time a version of the Grobman–Hartman theorem for nonautonomous differential equations $v' = A(t)v$ was obtained by Palmer in [67] (with the exception of the Hölder continuity of the conjugacy), although only for uniformly hyperbolic dynamics, that is, assuming the existence of a (uniform) exponential dichotomy.

We also want to consider the case of nonautonomous dynamics, where at each time $m \in \mathbb{Z}$ we apply a different map

$$F_m(v) = A_m v + f_m(v).$$

Here each f_m is a Lipschitz map with sufficiently small Lipschitz constant, and with $f_m(0) = 0$. In this case, instead of looking for a single homeomorphism h as in (1.9), we look for a sequence of homeomorphisms h_m such that for each $m \in \mathbb{Z}$ we have the identity

$$A_m \circ h_m = h_{m+1} \circ F_m, \tag{1.10}$$

or equivalently the commutative diagram in Figure 1.1. We emphasize that our work includes as a particular case the classical work for an autonomous uniformly hyperbolic dynamics defined by a map F as in (1.9).

We would like to point out that it is easy to rewrite the identity in (1.10) in a similar manner to that in (1.9) by defining appropriate extensions of the maps from X to $\mathbb{Z} \times X$. Namely, consider the maps $\mathcal{A}$ and $\mathcal{F}$ defined for each $(m, v) \in \mathbb{Z} \times X$ by

$$\mathcal{A}(m,v) = (m+1, A_m v) \quad \text{and} \quad \mathcal{F}(m,v) = (m+1, F_m(v)).$$

One can easily verify that the map $\mathcal{H}$ defined by $\mathcal{H}(m,v) = (m, h_m(v))$, where h_m are the conjugacies in (1.10), satisfies

$$\mathcal{A} \circ \mathcal{H} = \mathcal{H} \circ \mathcal{F}.$$

We note that in the case of continuous time a version of the Grobman–Hartman theorem for nonautonomous differential equations $v' = A(t)v$ was obtained by Palmer in [67] (with the exception of the Hölder continuity of the conjugacy), although only for uniformly hyperbolic dynamics, that is, assuming the existence of a (uniform) exponential dichotomy.

$$\begin{array}{ccccccccc}
\longrightarrow & X & \xrightarrow{F_{m-1}} & X & \xrightarrow{F_m} & X & \xrightarrow{F_{m+1}} & X & \longrightarrow \\
& \downarrow{\scriptstyle h_{m-1}} & & \downarrow{\scriptstyle h_m} & & \downarrow{\scriptstyle h_{m+1}} & & \downarrow{\scriptstyle h_{m+2}} & \\
\longrightarrow & X & \xrightarrow{A_{m-1}} & X & \xrightarrow{A_m} & X & \xrightarrow{A_{m+1}} & X & \longrightarrow
\end{array}$$

Fig. 1.1. Sequence of conjugacies h_m for the problem in (1.10).

We also show that the topological conjugacies h_m are Hölder continuous and have Hölder continuous inverses. We note that in the classical autonomous case of uniform exponential dichotomies (see (1.9)), the Hölder regularity of the conjugacies seems to have been known by some experts for quite some time, although apparently, to the best of our knowledge, no published proof can be found in the literature. In particular, the Hölder property was claimed by van Strien in [96, Proposition 4.6] in 1990, but it was observed in [79] (see also [40]) that there are some problems in the proof that are not yet overcome. This should be compared with the discussion in [40], where the authors indicate that the statement of the Hölder regularity is contained in a preprint of Belitskiĭ [21] (the preprint apparently circulates since 1994 but it remains unpublished; we note that a careful inspection of the arguments in [21] indicates lack of care with some points, being for example unclear what norms are used). We also mention a paper of Tan [91] announcing a forthcoming work with a proof (the draft goes back to 1999 but it was also never published). On the other hand, for the above mentioned example by Hartman in [43] of a diffeomorphism in $\mathbb{R}^3$ for which there exists no C^1 conjugacy, it was shown by Rayskin in [79] that there exists a C^α conjugacy for any $\alpha \in (0,1)$. In fact she obtained this result for a class of C^3 diffeomorphisms in $\mathbb{R}^3$ (namely, those for which A is diagonal, and such that $A + f$ leaves invariant the x and y axes, and the xy and yz planes). It is also conjectured in [79] that this statement (existence of a C^α conjugacy for any $\alpha \in (0,1)$) should be true for any diffeomorphism $A + f$ of $\mathbb{R}^3$ with $d_0 f = 0$. In another direction it was shown by Guysinsky, Hasselblatt, and Rayskin in [40] that for C^∞ diffeomorphisms in $\mathbb{R}^n$ the conjugacy g in the Grobman–Hartman theorem is differentiable at zero with $d_0 g$ being the identity.

1.4 Center manifolds, symmetry and reversibility

Center manifold theorems are powerful tools in the analysis of the behavior of dynamical systems. For example, consider the differential equation (1.6) in a given Banach space. We assume that $f(t, 0) = 0$ for every t. One can ask wether the behavior of the solutions of (1.6) in a neighborhood of zero somehow imitates that of the linear equation (1.1). This is certainly the case when (1.1) admits an exponential dichotomy: by the Grobman–Hartman theorem, locally the two dynamics are topologically conjugate. When the equation (1.1) possesses some elliptic directions one can still establish the existence of center manifolds that are tangent to the vector space generated by these directions. However, the situation is not so simple anymore. Namely, the behavior on the center manifold substantially depends on the nonlinearity f and in general the manifolds need not imitate the behavior on the vector space.

Nevertheless, the understanding of the behavior of solutions of (1.6) plays a crucial role in dynamics, for example in the study of the stability of solutions of a given differential equation. Namely, when the equation (1.1) possesses no unstable directions, all solutions converge exponentially to the center manifold. Therefore, the stability of the system is completely determined by the behavior on the center manifold. Accordingly, one often considers a reduction to the center manifold, and determines the quantitative behavior on it. This has also the advantage of reducing the dimension of the system. We refer the reader to the book [22] for details and references. In particular, using normal forms there is also the possibility of an appropriate classification as well as of giving a description of the allowed bifurcations. Incidentally, since one needs to be able to approximate the center manifolds to sufficiently high order, it is also important to discuss their regularity and to understand how to approximate them up to a given order.

In its classical formulation, the center manifold theorem applies to flows for which the linear equation in (1.1) admits a uniform exponential trichotomy. This means that the exponential estimates for the norms of solutions of the equation (1.1) are assumed to be independent of the initial time for which we consider the solution (see (1.11) and (1.12)). For simplicity, we assume here that the operator $A(t)$ has a block form with respect to some fixed decomposition $E \oplus F_1 \oplus F_2$ of the Banach space, with E, F_1, and F_2 corresponding respectively to the central, stable, and unstable directions (the general case is considered in Chapter 8). Then the solution of (1.1) with $v(s) = v_s$ can be written in the form

$$v(t) = (U(t, s), V_1(t, s), V_2(t, s))v_s,$$

where $U(t, s)$, $V_1(t, s)$, and $V_2(t, s)$ are the evolution operators associated respectively with the three blocks of $A(t)$.

We say that the equation in (1.1) admits a *uniform exponential trichotomy* if there exist constants $\underline{b} > a \geq 0$, $d > c \geq 0$, and $D > 0$ such that for every $s, t \in \mathbb{R}$ with $t \geq s$,

$$\|U(t, s)\| \leq De^{a(t-s)}, \quad \|V_2(t, s)^{-1}\| \leq De^{-b(t-s)}, \qquad (1.11)$$

and for every $s, t \in \mathbb{R}$ with $t \leq s$,

$$\|U(t, s)\| \leq De^{c(s-t)}, \quad \|V_1(t, s)^{-1}\| \leq De^{-d(s-t)}. \qquad (1.12)$$

We can now formulate the classical center manifold theorem. For simplicity we denote by ∂ the partial derivative with respect to u.

Theorem 1.4. *Assume that:*

1. *$v' = A(t)v$ has only global solutions and admits a uniform exponential trichotomy in the Banach space X;*
2. *A and f are of class C^k for some $k \in \mathbb{N}$, with $u \mapsto \partial^k f(t, u)$ Lipschitz, and with $f(t, 0) = 0$ and $\partial f(t, 0) = 0$ for every $t \in \mathbb{R}$;*
3. *$\partial^j f$ is bounded for $j = 1, \ldots, k$.*

If $ka < b - a$ and $kc < d - c$, then there exists a manifold $\mathcal{V}$ of class C^k containing the line $\mathbb{R} \times \{0\}$ and satisfying $T_{(s,0)}\mathcal{V} = \mathbb{R} \times E$ for every $s \in \mathbb{R}$.

It is difficult to give an original reference for the first published version of Theorem 1.4, but the statement should be considered classical. This is also due to the fact that the linear equation in (1.1) is nonautonomous: the modifications which are necessary in the approach for autonomous systems in order to obtain center manifolds for nonautonomous systems are not substantial, but several authors considered only the autonomous case. Theorem 1.4 is also a particular case of Theorem 1.5 that considers the general setting of nonuniform exponential trichotomies.

The study of center manifolds can be traced back to the works of Pliss [73] and Kelley [52]. A very detailed exposition in the case of autonomous equations is given in [93], adapting results in [95]. See also [62, 94] for the case of equations in infinite-dimensional spaces. We refer the reader to [23, 26, 27, 93] for more details and further references.

Our goal is to weaken the condition concerning the existence of a uniform exponential trichotomy, and find the weakest hypotheses under which one can construct center manifolds for the equation (1.6). In particular, we do not require the linear equation (1.1) to possess a uniform exponential behavior (either in the central, stable, or unstable directions). We still use some amount of partial hyperbolicity to establish the existence of the center manifolds, but this hyperbolicity can be spoiled exponentially along each solution as the initial time changes. Namely, we say that the equation in (1.1) admits a *nonuniform exponential trichotomy* if there exist constants $b > a \geq 0$, $d > c \geq 0$, $\theta > 0$, and $D > 0$ such that for every $s, t \in \mathbb{R}$ with $t \geq s$,

$$\|U(t,s)\| \leq De^{a(t-s)+\theta|s|}, \quad \|V_2(t,s)^{-1}\| \leq De^{-b(t-s)+\theta|t|}, \tag{1.13}$$

and for every $s, t \in \mathbb{R}$ with $t \leq s$,

$$\|U(t,s)\| \leq De^{c(s-t)+\theta|s|}, \quad \|V_1(t,s)^{-1}\| \leq De^{-d(s-t)+\theta|t|}. \tag{1.14}$$

The constant θ measures the nonuniformity of the exponential behavior.

We can now formulate a prototype of our center manifold theorem, where we replace uniform exponential trichotomy (see Theorem 1.4) by nonuniform exponential trichotomy.

Theorem 1.5. *Assume that:*

1. *$v' = A(t)v$ has only global solutions and admits a weak nonuniform exponential trichotomy in the Banach space X;*
2. *A and f are of class C^k for some $k \in \mathbb{N}$, with $u \mapsto \partial^k f(t,u)$ Lipschitz, and with $f(t,0) = 0$ and $\partial f(t,0) = 0$ for every $t \in \mathbb{R}$;*
3. *$(t,u) \mapsto e^{(k+2)\theta|t|}\partial^j f(t,u)$ is bounded for $j = 1, \ldots, k$, and the Lipschitz constant of $u \mapsto e^{(k+2)\theta|t|}\partial^k f(t,u)$ is independent of t.*

If $(k+1)(a+\theta) < b$ and $(k+1)(c+\theta) < d$, then there exists a manifold $\mathcal{V}$ of class C^k containing the line $\mathbb{R} \times \{0\}$ and satisfying $T_{(s,0)}\mathcal{V} = \mathbb{R} \times E$ for every $s \in \mathbb{R}$.

Theorem 1.5 is a simple consequence of Theorem 8.2, starting by making the rescaling $(t,u) \mapsto (t,\delta u)$ with δ sufficiently small.

Our approach to the proof of Theorem 10.20 could be considered classical, and consists again in using the differential equation to express the invariance of the center manifold under the dynamics and conclude that it must be the graph of a function satisfying a certain fixed point problem. However, the extra small exponentials in a nonuniform exponential trichotomy (see (1.13) and (1.14)) substantially complicate this approach and the implementation requires several nontrivial changes. In particular, we need to consider two fixed-point problems—one to obtain an a priori estimate for the speed of decay of the central component of the solutions along a given graph, and the other to obtain the graph which is the center manifold. In order to obtain the required estimates in the fixed point problems, we need sharp bounds for the derivatives of the central component of the solutions, and for the derivatives of the vector field along a given graph. For this we use a multivariate version of the Faà di Bruno formula in [30] for the derivatives of a composition. See Section 6.3.2 for details. We also use a result in [34] (see Proposition 6.3), that goes back to a lemma of Henry in [46]. This result allows us to establish the existence and *simultaneously* the regularity of the center manifolds using a single fixed point problem, instead of a fixed point problem for each of the successive higher-order derivatives. Essentially, the result says that the closed unit ball in the space $C^{k,\delta}$ of functions of class C^k between two Banach spaces with Hölder continuous k-th derivative with Hölder exponent δ is closed with respect to the C^0-topology. This allows us to consider contraction maps solely using the supremum norm instead of any norm involving also the derivatives. See [28] for a related approach in the particular case of uniform exponential behavior.

In Chapter 9 we consider the notions of reversible and equivariant equation. We show that the (time) reversibility and equivariance in a given flow descends respectively to the reversibility and equivariance in any center manifold. We emphasize that we consider the general case of nonautonomous equations.

1.5 Lyapunov regularity and stability theory

Let us first consider the finite-dimensional setting by which our work was inspired. We are interested in the study of the persistence of the asymptotic stability of the zero solution of a nonautonomous linear differential equation (1.1) under a perturbation f of the original equation as in (1.6).

We recall that there are examples, going back to Perron, showing that an arbitrarily small perturbation (1.6) of an asymptotically stable nonautonomous linear equation (1.1) may be unstable, and in fact may be exponentially unstable in some directions, even if all Lyapunov exponents of the linear equation (1.1) are negative. It is of course possible to provide additional assumptions of general nature under which the stability persists. This is the case for example with the assumption of uniform asymptotic stability for the linear equation, although this requirement is dramatically restrictive for a nonautonomous system. Incidentally, this assumption is analogous to the restrictive requirement of existence of an exponential dichotomy for the evolution operator of a nonautonomous equation in the case when there exist simultaneously positive and negative Lyapunov exponents. It is thus desirable to look for general assumptions that are substantially weaker than uniform asymptotic stability, under which one can still establish the persistence of stability of the zero solution of (1.6), when the perturbation f is sufficiently small. This is the case of the so-called notion of regularity introduced by Lyapunov in his doctoral thesis [57] (the expression is his own), which unfortunately seems nowadays apparently overlooked in the theory of differential equations (either related to stability or otherwise).

We now briefly recall the classical notion of Lyapunov regularity, or regularity for short, in the finite-dimensional setting. We first introduce the Lyapunov exponent associated to the linear differential equation (1.1) in $\mathbb{R}^n$. We assume that $A(t)$ depends continuously on t, and that all solutions of (1.1) are global. The Lyapunov exponent $\lambda \colon \mathbb{R}^n \to \mathbb{R} \cup \{-\infty\}$ is defined by

$$\lambda(x_0) = \limsup_{t \to +\infty} \frac{1}{t} \log \|x(t)\|, \tag{1.15}$$

where $x(t)$ denotes the solution of (1.1) with $x(0) = x_0$. To introduce the notion of regularity we also need to consider the adjoint equation

$$y' = -A(t)^* y, \tag{1.16}$$

where $A(t)^*$ denotes the transpose of $A(t)$. The associated Lyapunov exponent $\mu \colon \mathbb{R}^n \to \mathbb{R} \cup \{-\infty\}$ is defined by

$$\mu(y_0) = \limsup_{t \to +\infty} \frac{1}{t} \log \|y(t)\|,$$

where $y(t)$ denotes the solution of (1.16) with $y(0) = y_0$. It follows from the abstract theory of Lyapunov exponents (see Section 10.1) that the function λ can take at most n values on $\mathbb{R}^n \setminus \{0\}$, say $-\infty \le \lambda_1 < \cdots < \lambda_p$ for some integer $p \le n$. Furthermore, for each $i = 1, \ldots, p$ the set

$$E_i = \{x \in \mathbb{R}^n : \lambda(x) \le \lambda_i\}$$

is a linear space. We consider the values $\lambda'_1 \le \cdots \le \lambda'_n$ of the Lyapunov exponent λ on $\mathbb{R}^n \setminus \{0\}$ counted with multiplicities, obtained by repeating each value λ_i a number of times equal to $\dim E_i - \dim E_{i-1}$ (with $E_0 = \{0\}$). In a similar manner we can consider the values $\mu'_1 \ge \cdots \ge \mu'_n$ of the Lyapunov exponent μ on $\mathbb{R}^n \setminus \{0\}$ counted with multiplicities. We say that the linear equation (1.1) is *Lyapunov regular* if

$$\lambda'_i + \mu'_i = 0 \text{ for } i = 1, \ldots, n.$$

It is well known that if all values of the Lyapunov exponent are negative then the zero solution of (1.1) is asymptotically stable. However, there may still exist arbitrarily small perturbations $f(t, x)$ with $f(t, 0) = 0$ such that the zero solution of (1.6) is not asymptotically stable. An explicit example in $\mathbb{R}^2$ is the equation $(u', v') = A(t)(u, v)$, with the diagonal matrix

$$A(t) = \begin{pmatrix} -15 - 14(\sin \log t + \cos \log t) & 0 \\ 0 & -15 + 14(\sin \log t + \cos \log t) \end{pmatrix},$$

and the perturbation $f(t, (u, v)) = (0, u^4)$. In this example, one can show that the Lyapunov exponent λ in (1.15) is constant and equal to -1, but there exists a solution $(u(t), v(t))$ of the perturbed system (1.6), that is, of the equation $(u', v') = A(t)(u, v) + (0, u^4)$, with

$$\limsup_{t \to +\infty} \frac{1}{t} \log \|(u(t), v(t))\| > 0$$

(we refer to [1] for full details about the example). In other words, assuming that all values of the Lyapunov exponent λ are negative is not sufficient to guarantee that the asymptotic stability of the linear equation (1.1) persists under sufficiently small perturbations. On the other hand, if (1.1) is *Lyapunov regular*, then for any sufficiently small perturbation $f(t, x)$ with $f(t, 0) = 0$ for every $t \geq 0$, the zero solution of the perturbed equation (1.6) is asymptotically stable (see Theorem 12.5).

It should be noted that while Lyapunov regularity requires much from the structure of the original linear equation, it is substantially weaker than the requirement of uniform asymptotic stability (note that a priori Lyapunov regularity also requires much from the structure of the associated adjoint equation, although there are alternative characterizations of regularity that do not use the adjoint equation; see Section 10.4). More precisely, consider the evolution operator $U(t, s)$ associated to (1.1), satisfying $x(t) = U(t, s)x(s)$ for each $t \geq s$, where $x(t)$ is a solution of (1.1). When the linear system (1.1) is Lyapunov regular and all values of the Lyapunov exponent λ are negative one can show that for every $\beta > 0$ there exist positive constants c and α such that

$$\|U(t, s)\| \leq ce^{-\alpha(t-s)+\beta s} \text{ for every } t \geq s.$$

However, in general one cannot take $\beta = 0$, and thus the system need not be uniformly asymptotically stable. In particular,

$$\|x(t)\| \leq ce^{\beta s}e^{-\alpha(t-s)}\|x(s)\|,$$

where the constant $ce^{\beta s}$ deteriorates exponentially along the orbit of a solution. This means that the "size" of the neighborhood at time s where the exponential stability of the zero solution is guaranteed may decay with exponential rate, although small when compared to the Lyapunov exponents by choosing β sufficiently small.

While the notion of Lyapunov regularity makes considerable demands on the linear system, it turns out that within the context of ergodic theory it is typical under fairly general assumptions. Here we formulate only one of the major results in this direction, which in fact is one of the fundamental pieces at the basis of the so-called smooth ergodic theory or Pesin theory (see [1]). Recall that a finite measure ν in $\mathbb{R}^n$ is *invariant* under the flow $\{\varphi_t\}_{t \in \mathbb{R}}$ if

$$\nu(\varphi_t(A)) = \nu(A) \text{ for every measurable set } A \subset \mathbb{R}^n \text{ and } t \in \mathbb{R}.$$

The following statement is a particular version of the celebrated multiplicative ergodic theorem of Oseledets in [65]. It is a simple consequence of the general theory, as described for example in [1].

Theorem 1.6 (see [65]). *Consider a differential equation $x' = F(x)$ in $\mathbb{R}^n$ with F of class C^1, and assume that it generates a flow $\{\varphi_t\}_{t \in \mathbb{R}}$ which preserves a finite measure ν with compact support in $\mathbb{R}^n$. Then for ν-almost every $x \in \mathbb{R}^n$ the linear variational equation*

$$y' = A_x(t)y \text{ with } A_x(t) = d_{\varphi_t x}F \tag{1.17}$$

is Lyapunov regular.

We refer to [1] for a detailed exposition of the multiplicative ergodic theorem. We remark that since the general solution of the equation (1.17) is given by $y(t) = (d_x\varphi_t)y_0$, with $y_0 \in \mathbb{R}^n$, the Lyapunov exponent λ in (1.15) associated to (1.17) coincides with the "usual" Lyapunov exponent associated to each solution $\varphi_t(x)$ of $x' = F(x)$ along a direction y_0, that is,

$$\chi(x, y_0) = \limsup_{t \to +\infty} \frac{1}{t} \log\|(d_x\varphi_t)y_0\|.$$

We can apply Theorem 1.6 for example to any Hamiltonian equation and the associated invariant Liouville–Lebesgue measure. More generally, any flow defined by a differentiable vector field with zero divergence preserves Lebesgue measure. This happens in particular with the geodesic flow on the unit tangent bundle of a smooth manifold.

Theorem 1.6 and its related versions should be considered strong motivations to study Lyapunov regular systems, in view of the ubiquity of these systems at least in the measurable category. Furthermore, and this is another motivation for our study, there exist several related results in the infinite-dimensional setting. Namely, it turns out that the notion of Lyapunov regularity in a finite-dimensional space has several important geometric consequences, related to the existence of exponential growth rates of norms, angles, and volumes (for details see Chapters 10 and 11). Ruelle [81] was the first to obtain related "geometric" results in Hilbert spaces. Later on Mañé [58] considered transformations in Banach spaces under some compactness assumptions (including the case of differentiable maps with compact derivative at each point). The results of Mañé were extended by Thieullen in [92] for a class of transformations satisfying a certain asymptotic compactness. In view of the regularity theory in finite-dimensional spaces one should ask, and this is another motivation for our study, whether the above "geometric" results in the infinite-dimensional setting have behind them an analogous (infinite-dimensional) regularity theory, which additionally reduces to the classical theory when applied to the finite-dimensional setting. We shall show that this is indeed the case (see Chapter 11). Note that the answer to this question largely depends on finding an appropriate generalization of the notion of Lyapunov regularity for infinite-dimensional spaces.

Part I

Exponential dichotomies

2

Exponential dichotomies and basic properties

The classical notion of exponential dichotomy, which we call *uniform exponential dichotomy*, is a considerable restriction for the dynamics and it is important to look for more general types of hyperbolic behavior. These generalized notions can be much more typical than the notion of uniform exponential dichotomy. This is precisely the case of the notion of nonuniform exponential dichotomy, that we introduce in Section 2.1. In particular, we show that essentially *any* (nonautonomous) linear differential equation admits such a dichotomy (see Section 2.3). We note that each uniform exponential dichotomy is also a nonuniform exponential dichotomy.

2.1 Nonuniform exponential dichotomies

We introduce in this section the concept of *nonuniform* exponential dichotomy for a linear differential equation. We follow closely [13] although now in the infinite-dimensional setting of Banach spaces.

Let X be a Banach space and let $A\colon J \to B(X)$ be a continuous function on some interval $J \subset \mathbb{R}$, where $B(X)$ is the set of bounded linear operators on X. Consider the initial value problem

$$v' = A(t)v, \quad v(s) = v_s, \tag{2.1}$$

with $s \in J$ and $v_s \in X$. We always assume in the book that

$$\text{each solution of (2.1) is defined for every } t \in J. \tag{2.2}$$

We write the unique solution of the initial value problem in (2.1) in the form $v(t) = T(t,s)v(s)$, where $T(t,s)$ is the associated evolution operator. We clearly have

$$T(t,s)T(s,r) = T(t,r) \quad \text{and} \quad T(t,t) = \mathrm{Id}$$

for every t, s, $r \in J$. In particular $T(t,s)$ is invertible and $T(t,s)^{-1} = T(s,t)$ for every t, $s \in J$.

Definition 2.1. *Consider an interval $J \subset \mathbb{R}$. We say that the linear equation $v' = A(t)v$ admits a* nonuniform exponential dichotomy *in J if there exists a function $P\colon J \to B(X)$ such that $P(t)$ is a projection for each $t \in J$, with*

$$P(t)T(t,s) = T(t,s)P(s) \quad \text{for every } t, s \in J \text{ with } t \geq s, \qquad (2.3)$$

and there exist constants

$$\bar{a} < 0 \leq \underline{b}, \quad a, b \geq 0, \quad \text{and} \quad D_1, D_2 \geq 1 \qquad (2.4)$$

such that for every $t, s \in J$ with $t \geq s$,

$$\|T(t,s)P(s)\| \leq D_1 e^{\bar{a}(t-s)+a|s|}, \quad \|T(t,s)^{-1}Q(t)\| \leq D_2 e^{-\underline{b}(t-s)+b|t|}, \quad (2.5)$$

where $Q(t) = \mathrm{Id} - P(t)$ is the complementary projection for each $t \in J$.

We emphasize that the notion of nonuniform exponential dichotomy occurs naturally in the theory of differential equations (see Section 2.3 and Chapter 10). More generally, one can introduce the notion of dichotomy for a given evolution operator $T(t, s)$, without considering any associated linear equation, and thus, in particular, without any a priori assumption on the regularity of the map $(t, s) \mapsto T(t, s)$. The definition of nonuniform exponential dichotomy mimics the classical notion of (uniform) exponential dichotomy. Namely, we recall that the linear equation $v' = A(t)v$ admits a *(uniform) exponential dichotomy* if there exists a function P as in Definition 2.1 and constants D, $\beta > 0$ such that for $t, s \in J$ with $t \geq s$,

$$\|T(t,s)P(s)\| \leq D e^{-\beta(t-s)}, \quad \|T(t,s)^{-1}Q(t)\| \leq D e^{-\beta(t-s)}.$$

Observe that the constant $\underline{b}$ may be zero, which is not the case in the notion of (uniform) exponential dichotomy.

We also consider a strong form of nonuniform exponential dichotomy. Consider constants

$$\underline{a} \leq \bar{a} < 0 \leq \underline{b} \leq \bar{b} \quad \text{and} \quad a, b \geq 0. \qquad (2.6)$$

Definition 2.2. *Consider an interval $J \subset \mathbb{R}$. We say that the linear equation $v' = A(t)v$ admits a* strong nonuniform exponential dichotomy *in J if there exists a function $P\colon J \to B(X)$ such that $P(t)$ is a projection for each $t \in J$ and (2.3) holds, and there exist constants as in (2.6) and $D_1, D_2 \geq 1$ such that for every $t, s \in J$ with $t \geq s$,*

$$\|T(t,s)P(s)\| \leq D_1 e^{\bar{a}(t-s)+a|s|}, \quad \|T(t,s)^{-1}P(t)\| \leq D_1 e^{-\underline{a}(t-s)+a|t|},$$
$$\|T(t,s)Q(s)\| \leq D_2 e^{\bar{b}(t-s)+b|s|}, \quad \|T(t,s)^{-1}Q(t)\| \leq D_2 e^{-\underline{b}(t-s)+b|t|}. \qquad (2.7)$$

Clearly, any strong nonuniform exponential dichotomy is also a nonuniform exponential dichotomy. On the other hand, even when X is finite-dimensional, a linear equation admitting a nonuniform exponential dichotomy may not admit a strong nonuniform exponential dichotomy. We give a simple example. Consider the equation in $\mathbb{R}^2$ given by

$$x' = -x, \quad y' = ty. \tag{2.8}$$

One can easily verify that (2.8) admits a nonuniform exponential dichotomy in $\mathbb{R}^+$ with $\bar{a} = -1$, $\underline{b} = 0$, $a = b = 0$, and $D_1 = D_2 = 1$. Furthermore, the second inequality in (2.7) holds with $\underline{a} = -1$. But we would have to take $\bar{b} = +\infty$ so that the third inequality in (2.7) could hold. Thus, equation (2.8) does not admit a strong nonuniform exponential dichotomy in $\mathbb{R}^+$.

The first four constants in (2.6) play the role of Lyapunov exponents for the solutions of the linear system in (2.1): they correspond, respectively, to the largest and smallest contraction, and to the smallest and largest expansion in the linear system. The constants a and b in (2.6) measure the nonuniformity of the dichotomy, and are also closely related to the Lyapunov exponents. We can indeed obtain estimates for the actual values of the six numbers in (2.6) in terms of the Lyapunov exponents (see Theorem 10.6). The inequality $\bar{a} < 0$ in (2.6) means that there exists at least one negative Lyapunov exponent; one can of course introduce an entirely analogous version of dichotomy when $\bar{a} \leq 0 < \underline{b}$. While this certainly causes an asymmetry in the notion of nonuniform dichotomy, it turns out that for the stable (respectively unstable) manifold theory we do not require the condition $\underline{b} > 0$ (respectively $\bar{a} < 0$). We refer to Chapters 4, 5, and 6 for details.

We now present an example of nonuniform exponential dichotomy that is not uniform. Let $\omega > a > 0$ be real parameters and consider the equation in $\mathbb{R}^2$ given by

$$u' = (-\omega - at \sin t)u, \quad v' = (\omega + at \sin t)v. \tag{2.9}$$

Proposition 2.3. *The linear equation (2.9) admits a nonuniform exponential dichotomy in $\mathbb{R}$ that is not a uniform exponential dichotomy.*

Proof. It is easy to verify that $u(t) = U(t,s)u(s)$ and $v(t) = V(t,s)v(s)$, where

$$U(t,s) = e^{-\omega t + \omega s + at \cos t - as \cos s - a \sin t + a \sin s},$$
$$V(t,s) = e^{\omega t - \omega s - at \cos t + as \cos s + a \sin t - a \sin s}.$$

The evolution operator $T(t,s)$ associated to (2.9) is given by

$$T(t,s)(u,v) = (U(t,s)u, V(t,s)v).$$

We consider the projection $P(t) \colon \mathbb{R}^2 \to \mathbb{R}^2$ defined by $P(t)(u,v) = u$. Clearly, (2.3) holds. It remains to show that there exists $D \leq e^{2a}$ such that

$$U(t,s) \leq De^{(-\omega + a)(t-s) + 2a|s|} \text{ for } t \geq s, \tag{2.10}$$

and

$$V(s,t) \le De^{-(\omega+a)(t-s)+2a|t|} \text{ for } t \ge s. \tag{2.11}$$

We first note that

$$U(t,s) = e^{(-\omega+a)(t-s)+at(\cos t-1)-as(\cos s-1)+a(\sin s-\sin t)}. \tag{2.12}$$

For $t, s \ge 0$ it follows from (2.12) that $U(t,s) \le e^{2a}e^{(-\omega+a)(t-s)+2as}$. Furthermore, if $t = 2k\pi$, $s = (2l-1)\pi$ with $k, l \in \mathbb{N}$ then

$$U(t,s) = e^{(-\omega+a)(t-s)+2as}. \tag{2.13}$$

For $t \ge 0$ and $s \le 0$ it follows from (2.12) that $U(t,s) \le e^{2a}e^{(-\omega+a)(t-s)}$. Finally, for $s \le t \le 0$ it follows from (2.12) that

$$U(t,s) \le e^{2a}e^{(-\omega+a)(t-s)+2a|t|} \le e^{2a}e^{(-\omega+a)(t-s)+2a|s|}.$$

This establishes (2.10). The bound for $V(t,s)$ in (2.11) can be obtained in a similar manner. Furthermore, if $t = -2k\pi$, $s = -(2l-1)\pi$ with $k, l \in \mathbb{N}$, then

$$V(s,t) = e^{-(\omega+a)(t-s)+2a|t|}. \tag{2.14}$$

By (2.10) and (2.11), the linear equation (2.9) admits a nonuniform exponential dichotomy. It follows from (2.13) and (2.14) that the exponentials $e^{2a|s|}$ and $e^{2a|t|}$ in (2.10) and (2.11) cannot be removed by making D or $\omega - a$ sufficiently large. This shows that the exponential dichotomy is not uniform. $\square$

2.2 Stable and unstable subspaces

Assume that the linear equation $v' = A(t)v$ admits a nonuniform exponential dichotomy in J. We consider the linear subspaces

$$E(t) = P(t)X \quad \text{and} \quad F(t) = Q(t)X \tag{2.15}$$

for each $t \in J$. We call $E(t)$ and $F(t)$ respectively the *stable* and *unstable* *subspaces* at time t (although strictly speaking $F(t)$ should be called unstable space only when $\underline{b} > 0$). Clearly

$$X = E(t) \oplus F(t) \text{ for every } t \in J,$$

and the dimensions $\dim E(t)$ and $\dim F(t)$ are independent of t.

The unique solution of (2.1) can be written in the form

$$v(t) = (U(t,s)\xi, V(t,s)\eta) \in E(t) \times F(t) \text{ for } t, s \in J \text{ with } t \ge s, \tag{2.16}$$

with $v_s = (\xi, \eta) \in E(s) \times F(s)$, where

$$U(t,s) := T(t,s)P(s) = T(t,s)P(s)^2 = P(t)T(t,s)P(s),$$
$$V(t,s) := T(t,s)Q(s) = T(t,s)Q(s)^2 = Q(t)T(t,s)Q(s).$$

Using (2.3), one can easily verify that

$$U(t,s)E(s) = E(t) \quad \text{and} \quad V(t,s)F(s) = F(t)$$

for every t, $s \in J$. In particular, when the stable and unstable subspaces are *independent* of t, that is, $E(t) = E$ and $F(t) = F$ for every t, the operator $T(t,s)$ has a block form with respect to the direct sum $E \oplus F$, namely

$$T(t,s) = \begin{pmatrix} U(t,s) & 0 \\ 0 & V(t,s) \end{pmatrix}.$$

Furthermore, the linear operators

$$U(t,s)\colon E(s) \to E(t) \quad \text{and} \quad V(t,s)\colon F(s) \to F(t)$$

are invertible. Without danger of confusion we denote the corresponding inverses by $U(t,s)^{-1}$ and $V(t,s)^{-1}$. Clearly,

$$U(t,s)^{-1} = U(s,t) \quad \text{and} \quad V(t,s)^{-1} = V(s,t)$$

for every t, $s \in J$. Note that we can replace the inequalities (2.7) in the notion of strong nonuniform exponential dichotomy by

$$\|U(t,s)\| \le D_1 e^{\overline{a}(t-s)+a|s|}, \quad \|U(t,s)^{-1}\| \le D_1 e^{-\underline{a}(t-s)+a|t|},$$
$$\|V(t,s)\| \le D_2 e^{\overline{b}(t-s)+b|s|}, \quad \|V(t,s)^{-1}\| \le D_2 e^{-\underline{b}(t-s)+b|t|}. \tag{2.17}$$

Similarly, we can replace the inequalities (2.5) in the notion of nonuniform exponential dichotomy by the first and the last inequalities in (2.17).

Setting $t = s$ in (2.5) we obtain

$$\|P(t)\| \le D_1 e^{a|t|} \quad \text{and} \quad \|Q(t)\| \le D_2 e^{b|t|}$$

for every $t \in J$. We now define

$$\alpha(t) = \inf\{\|x - y\| : x \in E(t), y \in F(t), \|x\| = \|y\| = 1\}. \tag{2.18}$$

Proposition 2.4. *For each $t \in J$ we have*

$$\frac{1}{\|P(t)\|} \le \alpha(t) \le \frac{2}{\|P(t)\|} \quad \text{and} \quad \frac{1}{\|Q(t)\|} \le \alpha(t) \le \frac{2}{\|Q(t)\|}.$$

Proof. We only establish the inequalities with $P(t)$. The inequalities with $Q(t)$ follow from the symmetry in the definition of $\alpha(t)$. For the first inequality, note that $P(t)(x - y) = x$ for each x and y as in (2.18). Hence,

$$1 = \|P(t)(x - y)\| \leq \|P(t)\| \cdot \|x - y\|,$$

which gives the lower bound for $\alpha(t)$. For the second inequality, observe that for $v, w \in X$ with $\bar{v} = P(t)v \neq 0$ and $\bar{w} = Q(t)w \neq 0$ we have

$$\left\| \frac{\bar{v}}{\|\bar{v}\|} - \frac{\bar{w}}{\|\bar{w}\|} \right\| = \frac{\|(\bar{v} - \bar{w})\|\bar{w}\| + \bar{w}(\|\bar{w}\| - \|\bar{v}\|)\|}{\|\bar{v}\| \cdot \|\bar{w}\|} \leq \frac{2\|\bar{v} - \bar{w}\|}{\|\bar{v}\|}.$$

Note that $P(t)(\bar{v} - \bar{w}) = \bar{v}$. Given $\varepsilon > 0$ we can select $v, w \in X$ such that for $z = \bar{v} - \bar{w}$ we have

$$\frac{\|z\|}{\|P(t)z\|} \leq \frac{1}{\|P(t)\|} + \varepsilon.$$

Therefore,

$$\left\| \frac{\bar{v}}{\|\bar{v}\|} - \frac{\bar{w}}{\|\bar{w}\|} \right\| \leq \frac{2\|z\|}{\|P(t)z\|} \leq \frac{2}{\|P(t)\|} + 2\varepsilon.$$

Since ε is arbitrary we obtain the upper bound for $\alpha(t)$. $\square$

In particular, $\alpha(t) \geq e^{-\kappa|t|}/D$ where $\kappa = \min\{a, b\}$ and $D = \min\{D_1, D_2\}$. One can easily verify that in the case of Hilbert spaces

$$\alpha(t) = 2\sin(\theta(t)/2),$$

where $\theta(t)$ is the angle between the subspaces $E(t)$ and $F(t)$. Thus,

$$\theta(t) \geq 2\sin(\theta(t)/2) \geq e^{-\kappa|t|}/D,$$

that is, $\theta(t)$ cannot decrease more than exponentially. In the general case of Banach spaces, the number $\alpha(t)$ in (2.18) can also be thought of as an "angle" between the subspaces $E(t)$ and $F(t)$. We refer to [47] for a related discussion.

2.3 Existence of dichotomies and ergodic theory

We discuss here briefly the existence of nonuniform exponential dichotomies and the relation of the notion with ergodic theory. A detailed description will be given in latter chapters. Incidentally, we note that the proofs of Theorems 2.5 and 2.6 depend on results in latter chapters. However, the theorems are used nowhere in the book, and thus there is no danger of circular reasoning.

The following statement shows that nonuniform dichotomies are very common at least in finite-dimensional spaces. Set $X = \mathbb{R}^n$. We assume that there exists a decomposition $\mathbb{R}^n = E \oplus F$ (independent of t), with respect to which $A(t)$ is a $n \times n$ matrix with the block form

$$A(t) = \begin{pmatrix} B(t) & 0 \\ 0 & C(t) \end{pmatrix} \quad \text{for } t \geq 0. \tag{2.19}$$

Given $v_0 \in \mathbb{R}^n$, we define the *Lyapunov exponent* of v_0 by

$$\lambda(v_0) = \limsup_{t \to +\infty} \frac{1}{t} \log \|v(t)\|,$$

where $v(t)$ is the solution of (2.1) with $v(0) = v_0$. We make the convention that $\log 0 = -\infty$.

Theorem 2.5 ([13]). *Assume that $A(t)$ is a $n \times n$ matrix with the block form in (2.19) for $t \geq 0$. If $\lambda(v_0) < 0$ for every $v_0 \in E$, and $\lambda(v_0) \geq 0$ for every $v_0 \in F \setminus \{0\}$, then the linear equation $v' = A(t)v$ admits a strong nonuniform exponential dichotomy in $\mathbb{R}^+$.*

Theorem 2.5 is an immediate consequence of Theorem 10.6, that also addresses the question of estimating the numbers in (2.6). Namely, these numbers will be related with the Lyapunov exponents. We refer to Section 12.4 for a related discussion in the infinite-dimensional setting of Hilbert spaces.

In the construction of invariant manifolds in the following chapters, we need the numbers a and b in (2.7) or (2.17) to be *sufficiently small* when compared to the "Lyapunov exponents", that is, to the constants $\underline{a}$, $\overline{a}$, $\underline{b}$, and $\overline{b}$ in (2.6). We note that, however, we never need them to be zero (in which case we would have a uniform dichotomy). This smallness assumption means that the nonuniformity of the dichotomy is sufficiently small when compared to these constants. It turns out that at least from the point of view of ergodic theory, the constants a and b can be made arbitrarily small almost always. To formulate a rigorous statement, we recall that a flow $\Psi_t \colon M \to M$ is said to preserve a measure μ on M if $\mu(\Psi_t A) = \mu(A)$ for every measurable set $A \subset M$ and every $t \in \mathbb{R}$.

Theorem 2.6. *If F is a vector field of class C^1 on a compact smooth Riemannian manifold M whose flow Ψ_t preserves a finite measure μ on M, then for μ-almost every $x \in M$ the evolution operator $T(t, s) = d_x\Psi_t(d_x\Psi_s)^{-1}$ defined by the linear variational equation*

$$v' = A_x(t)v, \quad \text{with } A_x(t) = d_{\Psi_t x}F$$

admits a strong nonuniform exponential dichotomy in $\mathbb{R}$ with arbitrarily small a and b.

Proof. The statement can be obtained as an immediate consequence of Theorem 10.22. For this it is sufficient to observe that the map $x \mapsto \|d_x\Psi_t\|$ is continuous and positive on the compact manifold M, and thus the condition (10.62) holds automatically. $\square$

We refer to Section 10.6 for more general statements, whose proofs require however the multiplicative ergodic theorem.

3

Robustness of nonuniform exponential dichotomies

We give in this chapter conditions for the robustness of *nonuniform* exponential dichotomies in Banach spaces, in the sense that the existence of an exponential dichotomy for a linear equation $v' = A(t)v$ persists under sufficiently small linear perturbations. We also establish the continuous dependence with the perturbation of the constants in the notion of dichotomy and the "angles" between the stable and unstable subspaces. The proofs exhibit (implicitly) the exponential dichotomies of the perturbed equations in terms of fixed points of certain contractions. We emphasize that we do not need the notion of admissibility (with respect to bounded nonlinear perturbations). We also establish related results in the case of strong nonuniform exponential dichotomies. All the results are obtained in Banach spaces. The presentation follows closely [18].

3.1 Robustness in semi-infinite intervals

We discuss in this section the robustness of nonuniform exponential dichotomies in semi-infinite intervals. We refer to Section 3.3 for the case of dichotomies in $\mathbb{R}$.

3.1.1 Formulation of the results

We continue to consider the equation in (2.1), and we assume that (2.2) holds. Also, we continue to denote by $T(t, s)$ the associated evolution operator. When equation (2.1) has a nonuniform exponential dichotomy, we say that the dichotomy is *robust* in a given class of (sufficiently small) perturbations if for B in this class the equation

$$v' = [A(t) + B(t)]v \tag{3.1}$$

still has a nonuniform exponential dichotomy.

With the notation in (2.4) we set

$$c = \min\{-\bar{a}, \underline{b}\}, \quad \vartheta = \max\{a, b\}, \quad \text{and} \quad D = \max\{D_1, D_2\}. \tag{3.2}$$

We also write

$$\tilde{c} = c\sqrt{1 - 2\delta D/c} \quad \text{and} \quad \tilde{D} = \frac{D}{1 - \delta D/(\tilde{c} + c)}. \tag{3.3}$$

The following is the main robustness result. We consider dichotomies in an interval $J = [\varrho, +\infty)$ with $\varrho \le 0$.

Theorem 3.1 ([18]). *Let $A, B: J \to B(X)$ be continuous functions such that:*

1. *equation (2.1) admits a nonuniform exponential dichotomy in the interval J with $\vartheta < c$;*
2. *$\|B(t)\| \le \delta e^{-2\vartheta|t|}$ for every $t \in J$.*

If δ is sufficiently small, then equation (3.1) admits a nonuniform exponential dichotomy in J, with the constants c, ϑ, and D replaced respectively by $\tilde{c}$, 2ϑ, and $4D\tilde{D}$.

The proof of Theorem 3.1 is given at the end of Section 3.1.2. Note that setting $\theta = 2\delta D/c$ the constants in (3.3) satisfy

$$\tilde{c} = c\left(1 - \frac{1}{2}\theta - \frac{1}{8}\theta^2 - \cdots\right) = c - \delta D - \frac{\delta^2 D^2}{2c} - \cdots$$

and

$$\tilde{D} = D\left(1 + \frac{1}{4}\theta + \frac{1}{8}\theta^2 + \cdots\right) = D + \frac{\delta D^2}{2c} + \frac{\delta^2 D^3}{2c^2} + \cdots.$$

It should be noted that, by considering the constants in (2.4) instead of those in (3.2), in general we could in principle obtain better estimates for the expansion and contraction rates of the nonuniform exponential dichotomy for the perturbed equation (3.1). However, particularly due to the nonuniformity, this would require heavier computations which would hide the main principles of our approach. Instead, we prefer to present a clear approach aimed at showing the "qualitative nature" of the robustness of nonuniform exponential dichotomies, without a lengthy discussion about optimal constants.

We also establish a weaker statement with slightly weaker hypotheses. Namely, in the following theorem we obtain norm bounds along the stable and unstable directions for the perturbed equation (3.1). However, we give no information about the norms of the projections for the perturbed equation. This requires slightly stronger hypotheses and is included in the statement of Theorem 3.1. We shall denote by $\hat{T}(t, s)$ the evolution operator associated to equation (3.1).

Theorem 3.2 ([18]). *Let $A, B\colon J \to B(X)$ be continuous functions such that equation (2.1) admits a nonuniform exponential dichotomy in J with $\vartheta < c$, and assume that $\|B(t)\| \le \delta e^{-\vartheta|t|}$ for every $t \in J$. If*

$$\theta = 2\delta D/c < 1, \tag{3.4}$$

then there exist projections $\hat{P}(t)\colon X \to X$ for each $t \in J$ such that

$$\hat{T}(t,s)\hat{P}(s) = \hat{P}(t)\hat{T}(t,s), \quad t \ge s, \tag{3.5}$$

and

$$\|\hat{T}(t,s)|\operatorname{Im}\hat{P}(s)\| \le \tilde{D}e^{-\tilde{c}(t-s)+\vartheta|s|}, \quad t \ge s, \tag{3.6}$$

$$\|\hat{T}(t,s)|\operatorname{Im}\hat{Q}(s)\| \le \tilde{D}e^{-\tilde{c}(s-t)+\vartheta|s|}, \quad s \ge t, \tag{3.7}$$

where $\hat{Q}(t) = \operatorname{Id} - \hat{P}(t)$ is the complementary projection of $\hat{P}(t)$.

We will start by proving this weaker statement in the following section. We note that to obtain Theorem 3.1 from Theorem 3.2 it remains to obtain sharp bounds for the norms of the projections $\hat{P}(t)$ and $\hat{Q}(t)$.

Some of the arguments are inspired by work of Popescu in [75] in the case of uniform exponential dichotomies. Incidentally, we note that he uses the notion that is sometimes called admissibility, while we do not need this notion, of course independently of its interest in other situations. This notion goes back to Perron and refers to the characterization of the existence of an exponential dichotomy in terms of the existence and uniqueness of bounded solutions of the equation $v' = A(t)v + f(t)$ for $f(t)$ in a certain class of bounded nonlinear perturbations. This property is called the *admissibility* of the pair of spaces in which we respectively take the perturbation and look for the solutions. One can also consider the admissibility of other pairs of spaces.

We note that in the case of uniform exponential dichotomies the study of robustness has a long history. In particular, the robustness was discussed by Massera and Schäffer [59] (building on earlier work of Perron [69]; see also [60]), Coppel [31], and in the case of Banach spaces by Dalec'kiĭ and Kreĭn [33], with different approaches and successive generalizations. The continuous dependence of the projections for the exponential dichotomies of the perturbed equations was obtained by Palmer [68]. For more recent works we refer to [25, 64, 75, 74] and the references therein (since we are dealing with nonuniform exponential dichotomies we refrain to be more detailed on the literature). In particular, Chow and Leiva [25] and Pliss and Sell [74] considered the context of linear skew-product semiflows and gave examples of applications in the infinite-dimensional setting, including to parabolic partial differential equations and functional differential equations. We emphasize that all these works consider only the case of *uniform* exponential dichotomies.

3.1.2 Proofs

We shall divide the proof of Theorem 3.2 into several steps. We first prove some auxiliary results.

Step 1. Construction of bounded solutions

Set
$$G = \{(t, s) \in J \times J : t \geq s\},$$
and consider the space
$$\mathcal{C} = \{U : G \to B(X) : U \text{ is continuous and } \|U\| < \infty\} \tag{3.8}$$
with the norm
$$\|U\| = \sup\{\|U(t, s)\|e^{-\vartheta|s|} : (t, s) \in G\}. \tag{3.9}$$

Lemma 3.3. *The equation $Z' = (A(t) + B(t))Z$ has a unique solution $U \in \mathcal{C}$ such that for each $(t, s) \in G$,*

$$
U(t, s) = T(t, s)P(s) + \int_s^t T(t, \tau)P(\tau)B(\tau)U(\tau, s)\, d\tau
$$
$$
- \int_t^\infty T(t, \tau)Q(\tau)B(\tau)U(\tau, s)\, d\tau. \tag{3.10}
$$

Proof of the lemma. Assume that some function $U \in \mathcal{C}$ satisfies (3.10). Then $t \mapsto U(t, s)$ is differentiable (since $t \mapsto T(t, s)$ is differentiable), and taking derivatives with respect to t in (3.10) a simple computation shows that $t \mapsto U(t, s)\xi$, $t \geq s$ is a solution of equation (3.1) for each $\xi \in X$. Thus, we must show that the operator L defined by

$$
(LU)(t, s) = T(t, s)P(s) + \int_s^t T(t, \tau)P(\tau)B(\tau)U(\tau, s)\, d\tau
$$
$$
- \int_t^\infty T(t, \tau)Q(\tau)B(\tau)U(\tau, s)\, d\tau \tag{3.11}
$$

has a unique fixed point in the space $\mathcal{C}$. We have

$$
\|(LU)(t, s)\| \leq \|T(t, s)P(s)\|
$$
$$
+ \int_s^t \|T(t, \tau)P(\tau)\| \cdot \|B(\tau)\| \cdot \|U(\tau, s)\|\, d\tau
$$
$$
+ \int_t^\infty \|T(t, \tau)Q(\tau)\| \cdot \|B(\tau)\| \cdot \|U(\tau, s)\|\, d\tau \tag{3.12}
$$
$$
\leq De^{-c(t-s)+\vartheta|s|} + D\delta e^{\vartheta|s|}\|U\| \int_s^t e^{-c(t-\tau)}\, d\tau
$$
$$
+ D\delta e^{\vartheta|s|}\|U\| \int_t^\infty e^{-c(\tau-t)}\, d\tau.
$$

Since $c > 0$, in view of (3.4) this implies that

$$\|LU\| \leq D + \theta\|U\| < \infty.$$

Therefore, we have a well-defined operator $L\colon \mathcal{C} \to \mathcal{C}$. Using the identity (3.11) for $U_1, U_2 \in \mathcal{C}$, and proceeding in a similar manner to that in (3.12) we obtain

$$\|LU_1 - LU_2\| \leq \theta\|U_1 - U_2\|.$$

It follows from (3.4) that L is a contraction, and thus there exists a unique $U \in \mathcal{C}$ such that $LU = U$. This completes the proof of the lemma. $\qquad\square$

Lemma 3.4. *For any $t \geq \tau \geq s$ in J we have*

$$U(t, \tau)U(\tau, s) = U(t, s).$$

Proof of the lemma. Write

$$X(t, s) = T(t, s)P(s)B(s) \quad \text{and} \quad Y(t, s) = T(t, s)Q(s)B(s).$$

Since $P(t)$ and $Q(t)$ are complementary projections, it follows from (2.3) together with (3.10) that $U(t, \tau)U(\tau, s)$ is equal to

$$T(t, \tau)P(\tau)T(\tau, s)P(s)$$
$$+ T(t, \tau)P(\tau)\left(\int_s^\tau X(\tau, u)U(u, s)\, du - \int_\tau^\infty Y(\tau, u)U(u, s)\, du \right)$$
$$+ \left(\int_\tau^t X(t, u)U(u, \tau)\, du - \int_t^\infty Y(t, u)U(u, \tau)\, du \right)U(\tau, s),$$

and thus,

$$U(t, \tau)U(\tau, s) = T(t, s)P(s) + \int_s^\tau X(t, u)U(u, s)\, du$$
$$+ \int_\tau^t X(t, u)U(u, \tau)U(\tau, s)\, du - \int_t^\infty Y(t, u)U(u, \tau)U(\tau, s)\, du.$$

Using again (3.10) this yields

$$U(t, \tau)U(\tau, s) - U(t, s) = \int_s^\tau X(t, u)U(u, s)\, du$$
$$+ \int_\tau^t X(t, u)U(u, \tau)U(\tau, s)\, du$$
$$- \int_t^\infty Y(t, u)U(u, \tau)U(\tau, s)\, du$$
$$- \int_s^t X(t, u)U(u, s)\, du + \int_t^\infty Y(t, u)U(u, s)\, du$$
$$= \int_\tau^t X(t, u)[U(u, \tau)U(\tau, s) - U(u, s)]\, du$$
$$- \int_t^\infty Y(t, u)[U(u, \tau)U(\tau, s) - U(u, s)]\, du.$$

Setting
$$Z(u) = U(u, \tau)U(\tau, s) - U(u, s), \qquad (3.13)$$

we can rewrite the above identity in the form

$$Z(t) = \int_\tau^t X(t, u)Z(u)\, du - \int_t^\infty Y(t, u)Z(u)\, du. \qquad (3.14)$$

For each fixed $\tau \geq s$ in J, we consider the operator N defined by

$$(NW)(t) = \int_\tau^t X(t, u)W(u)\, du - \int_t^\infty Y(t, u)W(u)\, du, \qquad (3.15)$$

in the Banach space

$$\mathcal{E} = \{W \colon [\tau, +\infty) \to B(X) : W \text{ is continuous and } \|W\| < \infty\} \qquad (3.16)$$

with the supremum norm $\|W\| = \sup\{\|W(u)\| : u \in [\tau, +\infty)\}$. By (3.15),

$$\|(NW)(t)\| \leq D \int_\tau^t e^{-c(t-u)+\vartheta|u|} \|B(u)\| \cdot \|W(u)\|\, du$$
$$+ D \int_t^\infty e^{-c(u-t)+\vartheta|u|} \|B(u)\| \cdot \|W(u)\|\, du \leq \theta\|W\|,$$

and thus $N(\mathcal{E}) \subset \mathcal{E}$. Furthermore, proceeding in a similar manner we find that for $W_1, W_2 \in \mathcal{E}$,

$$\|NW_1 - NW_2\| \leq \theta\|W_1 - W_2\|.$$

By hypothesis (3.4), N is a contraction, and hence there is a unique function $W \in \mathcal{E}$ satisfying (3.14). On the other hand, $0 \in \mathcal{E}$ also satisfies (3.14) and thus we must have $W = 0$. By Lemma 3.3, the function Z in (3.13) is in $\mathcal{E}$, and since it also satisfies (3.14), we conclude that for any $t \geq \tau \geq s$ in J,

$$Z(t) = U(t, \tau)U(\tau, s) - U(t, s) = 0.$$

Therefore, $U(t, \tau)U(\tau, s) = U(t, s)$. $\qquad\qquad\qquad\qquad\qquad\square$

Step 2. Projections and invariance of the evolution operator

Recall that $\hat{T}(t, s)$ denotes the evolution operator associated to equation (3.1). For each $t \in J$ we define the linear operators

$$\hat{P}(t) = \hat{T}(t, 0)U(0, 0)\hat{T}(0, t) \quad \text{and} \quad \hat{Q}(t) = \mathrm{Id} - \hat{P}(t). \qquad (3.17)$$

We want to show that the evolution operator admits a nonuniform exponential dichotomy with projections $\hat{P}(t)$. We start by showing that the linear operators $\hat{P}(t)$ are indeed projections, leaving invariant $\hat{T}(t, s)$.

Lemma 3.5. *The operator $\hat{P}(t)$ is a projection for each $t \in J$, and (3.5) holds.*

Proof of the lemma. Set $R = U(0,0)$. By Lemma 3.4, we have $R^2 = R$. Since $\hat{T}(t,t) = \text{Id}$,

$$\hat{P}(t)\hat{P}(t) = \hat{T}(t,0)R\hat{T}(0,t)\hat{T}(t,0)R\hat{T}(0,t)$$
$$= \hat{T}(t,0)R^2\hat{T}(0,t) = \hat{P}(t),$$

and $\hat{P}(t)$ is a projection. Moreover, for $t \geq s$,

$$\hat{P}(t)\hat{T}(t,s) = \hat{T}(t,0)R\hat{T}(0,t)\hat{T}(t,s)$$
$$= \hat{T}(t,s)\hat{T}(s,0)R\hat{T}(0,s) = \hat{T}(t,s)\hat{P}(s).$$

This completes the proof.   □

Step 3. Characterization of bounded solutions

Lemma 3.6. *Given $s \in J$, if $y \colon [s,+\infty) \to X$ is a bounded solution of equation (3.1) with $y(s) = \xi$, then*

$$y(t) = T(t,s)P(s)\xi + \int_s^t T(t,\tau)P(\tau)B(\tau)y(\tau)\,d\tau$$
$$- \int_t^\infty T(t,\tau)Q(\tau)B(\tau)y(\tau)\,d\tau.$$

Proof of the lemma. By the variation of constants formula, for $t \geq s$ in J,

$$P(t)y(t) = T(t,s)P(s)\xi + \int_s^t T(t,\tau)P(\tau)B(\tau)y(\tau)\,d\tau \qquad (3.18)$$

and

$$Q(t)y(t) = T(t,s)Q(s)\xi + \int_s^t T(t,\tau)Q(\tau)B(\tau)y(\tau)\,d\tau. \qquad (3.19)$$

Equivalently, the last formula can be written in the form

$$Q(s)\xi = T(s,t)Q(t)y(t) - \int_s^t T(s,\tau)Q(\tau)B(\tau)y(\tau)\,d\tau. \qquad (3.20)$$

Since $y(t)$ is bounded, we have

$$\|T(s,t)Q(t)y(t)\| \leq CDe^{-c(t-s)+\vartheta|t|},$$

where $C = \sup\{\|y(t)\| : t \geq s \text{ in } J\} < \infty$. Furthermore,

$$\int_s^\infty \|T(s,\tau)Q(\tau)\| \cdot \|B(\tau)\| \cdot \|y(\tau)\|\,d\tau \leq D\delta C \int_s^\infty e^{-c(\tau-s)}\,d\tau = \frac{D\delta C}{c}.$$

Taking limits in (3.20) when $t \to +\infty$, since $a > \vartheta$ we obtain

$$Q(s)\xi = -\int_s^\infty T(s,\tau)Q(\tau)B(\tau)y(\tau)\,d\tau.$$

It follows from (3.19) that

$$Q(t)y(t) = -\int_s^\infty T(t,\tau)Q(\tau)B(\tau)y(\tau)\,d\tau + \int_s^t T(t,\tau)Q(\tau)B(\tau)y(\tau)\,d\tau$$
$$= -\int_t^\infty T(t,\tau)Q(\tau)B(\tau)y(\tau)\,d\tau.$$

The desired statement follows readily from adding this identity to (3.18).

$\square$

Lemma 3.7. *The function* $[s,+\infty) \cap J \ni t \mapsto \hat{P}(t)\hat{T}(t,s)$ *is bounded, and for any* $t \geq s$ *in* J *we have*

$$\hat{P}(t)\hat{T}(t,s) = T(t,s)P(s)\hat{P}(s) + \int_s^t T(t,\tau)P(\tau)B(\tau)\hat{P}(\tau)\hat{T}(\tau,s)\,d\tau$$
$$-\int_t^\infty T(t,\tau)Q(\tau)B(\tau)\hat{P}(\tau)\hat{T}(\tau,s)\,d\tau.$$

Proof of the lemma. By Lemma 3.3, the function $t \mapsto U(t,0)\xi$, $t \geq 0$ is a solution of equation (3.1) with initial condition at time zero equal to $U(0,0)\xi$. Therefore,

$$U(t,0) = \hat{T}(t,0)U(0,0).$$

By Lemma 3.5 (see (3.5)),

$$\hat{P}(t)\hat{T}(t,s) = \hat{T}(t,s)\hat{P}(s)$$
$$= \hat{T}(t,s)\hat{T}(s,0)U(0,0)\hat{T}(0,s) \qquad (3.21)$$
$$= \hat{T}(t,0)U(0,0)\hat{T}(0,s) = U(t,0)\hat{T}(0,s).$$

Again by Lemma 3.3, for each $\xi \in X$ the function

$$y(t) = \hat{P}(t)\hat{T}(t,s)\xi = U(t,0)\hat{T}(0,s)\xi$$

is a solution of (3.1). Furthermore, by the definition of the space $\mathcal{C}$ in (3.8)–(3.9) this solution is bounded for $t \geq s$, and by (3.21),

$$y(s) = U(s,0)\hat{T}(0,s)\xi = \hat{P}(s)\hat{T}(s,s)\xi = \hat{P}(s)\xi.$$

The desired identity follows now readily from Lemma 3.6.

$\square$

Step 4. Auxiliary bounds

Lemma 3.8. *Given $s \in \mathbb{R}$ and $\varsigma \in (s, +\infty]$, let $x \colon [s, \varsigma) \to [0, +\infty)$ be a continuous function satisfying*

$$x(t) \le De^{-c(t-s)+\vartheta|s|}\gamma + \delta D \int_s^t e^{-c(t-\tau)}x(\tau)\,d\tau + \delta D \int_t^\varsigma e^{-c(\tau-t)}x(\tau)\,d\tau \tag{3.22}$$

for every $t \in [s, \varsigma)$, and assumed to be bounded when $\varsigma = +\infty$. Then

$$x(t) \le \widetilde{D}\gamma e^{-\widetilde{c}(t-s)+\vartheta|s|}, \quad t \in [s, \varsigma).$$

Proof of the lemma. We will show that $x(t) \le \Phi(t)$, where $\Phi(t)$ is any bounded continuous function satisfying the integral equation

$$\Phi(t) = De^{-c(t-s)+\vartheta|s|}\gamma + \delta D \int_s^t e^{-c(t-\tau)}\Phi(\tau)\,d\tau + \delta D \int_t^\varsigma e^{-c(\tau-t)}\Phi(\tau)\,d\tau \tag{3.23}$$

for $t \ge s$. Clearly, $\Phi(t)$ satisfies the differential equation

$$z'' - c^2(1-\theta)z = 0. \tag{3.24}$$

Notice that $-\widetilde{c} = -c\sqrt{1-\theta}$ is the negative root of the corresponding characteristic equation. In order that Φ is a bounded function when $\varsigma = +\infty$, we must have $\Phi(t) = \Phi(s)e^{-\widetilde{c}(t-s)}$ (when $\varsigma < +\infty$ we simply take $\Phi(t)$ to be of this form). Furthermore, by (3.23), substituting $\Phi(t)$ and setting $t = s$,

$$\Phi(s) = De^{\vartheta|s|}\gamma + \delta D\Phi(s) \int_s^\varsigma e^{-(c+\widetilde{c})(\tau-s)}\,d\tau = De^{\vartheta|s|}\gamma + \Phi(s)\frac{\delta D}{c+\widetilde{c}}.$$

Since $c + \widetilde{c} > 0$, this yields

$$\Phi(s) = \frac{D}{1 - \delta D/(\widetilde{c}+c)}e^{\vartheta|s|}\gamma,$$

and thus

$$\Phi(t) = \widetilde{D}\gamma e^{-\widetilde{c}(t-s)+\vartheta|s|}.$$

We now set $z(t) = x(t) - \Phi(t)$ for $t \ge s$. It follows from (3.22) and (3.23) that

$$z(t) \le \delta D \int_s^t e^{-c(t-\tau)}z(\tau)\,d\tau + \delta D \int_t^\varsigma e^{-c(\tau-t)}z(\tau)\,d\tau.$$

Set also $z = \sup_{t \ge s} z(t)$. Since the functions x and Φ are bounded, z is finite, and taking the supremum in the above inequality we obtain

$$z \le \delta Dz \sup_{t \ge s} \int_s^t e^{-c(t-\tau)}\,d\tau + \delta Dz \sup_{t \ge s} \int_t^\varsigma e^{-c(\tau-t)}\,d\tau.$$

Hence, $z \le \theta z$. It follows from (3.4) that $z \le 0$, and thus $z(t) \le 0$, that is, $x(t) \le \Phi(t)$ for $t \ge s$. $\qquad\square$

Lemma 3.9. *Given* $s \in \mathbb{R}$ *and* $\varrho \in [-\infty, s)$, *let* $y \colon (\varrho, s] \to [0, +\infty)$ *be a continuous function satisfying*

$$y(t) \leq De^{-c(s-t)+\vartheta|s|}\gamma + \delta D \int_{\varrho}^{t} e^{-c(t-\tau)}y(\tau)\,d\tau + \delta D \int_{t}^{s} e^{-c(\tau-t)}y(\tau)\,d\tau$$

$$(3.25)$$

for $t \in (\varrho, s]$, *and assumed to be bounded when* $\varrho = -\infty$. *Then*

$$y(t) \leq \tilde{D}\gamma e^{-\tilde{c}(s-t)+\vartheta|s|}, \quad t \in (\varrho, s].$$

Proof of the lemma. Proceeding in a similar manner to that in the proof of Lemma 3.8 we can show that $y(t) \leq \Psi(t)$, where $\Psi(t)$ is any bounded continuous function satisfying

$$\Psi(t) = De^{-c(s-t)+\vartheta|s|}\gamma + \delta D \int_{\varrho}^{t} e^{-c(t-\tau)}\Psi(\tau)\,d\tau + \delta D \int_{t}^{s} e^{-c(\tau-t)}\Psi(\tau)\,d\tau$$

for $t \leq s$. Note first that $\Psi(t)$ also satisfies the differential equation (3.24). Substituting $\Psi(t) = \Psi(s)e^{-\tilde{c}(s-t)}$ in the above identity and setting $t = s$ we obtain

$$\Psi(s) = De^{\vartheta|s|}\gamma + \delta D\Psi(s) \int_{\varrho}^{s} e^{-(c+\tilde{c})(s-\tau)}\,d\tau \leq De^{\vartheta|s|}\gamma + \Psi(s)\frac{\delta D}{c+\tilde{c}}.$$

Hence,

$$\Psi(s) \leq \frac{D}{1 - \delta D/(c+\tilde{c})}e^{\vartheta|s|}\gamma,$$

and $\Psi(t) \leq \Psi(s)e^{-\tilde{c}(s-t)}$. Proceeding in a similar manner to that in Lemma 3.8 we find that

$$y(t) \leq \Psi(t) \leq \tilde{D}\gamma e^{-\tilde{c}(s-t)+\vartheta|s|}.$$

This completes the proof of the lemma. □

Step 5. Norm bounds for the evolution operator

We now estimate the norms of $\hat{T}(t, s)|\operatorname{Im}\hat{P}(s)$ for $t \geq s$ and $\hat{T}(t, s)|\operatorname{Im}\hat{Q}(s)$ for $t \leq s$. We recall that the constants $\tilde{c}$ and $\tilde{D}$ are given by (3.3).

Lemma 3.10. *The inequality* (3.6) *holds for every* $t \geq s$ *in* J.

Proof of the lemma. Let $\xi \in X$. Setting

$$x(t) = \|\hat{P}(t)\hat{T}(t, s)\xi\|$$

for $t \geq s$, and $\gamma = \|\hat{P}(s)\xi\|$ it follows from Lemma 3.7 and (2.5) that the function x is bounded, and satisfies the inequality in (3.22) with $\varrho = +\infty$. Therefore, by Lemma 3.8,

$$\|\hat{P}(t)\hat{T}(t,s)\xi\| \leq \tilde{D}e^{-\tilde{c}(t-s)+\vartheta|s|}\|\hat{P}(s)\xi\|, \quad t \geq s.$$

By Lemma 3.5 we have

$$\hat{P}(t)\hat{T}(t,s) = \hat{T}(t,s)\hat{P}(s) = \hat{T}(t,s)\hat{P}(s)\hat{P}(s),$$

and hence, setting $\eta = \hat{P}(s)\xi$,

$$\|\hat{T}(t,s)\hat{P}(s)\eta\| \leq \tilde{D}e^{-\tilde{c}(t-s)+\vartheta|s|}\|\eta\|, \quad t \geq s.$$

This establishes the desired inequality. $\square$

Lemma 3.11. *The inequality* (3.7) *holds for every* $t \leq s$ *in* J.

Proof of the lemma. We first derive an equation for $\hat{Q}(t)\hat{T}(t,s)$. By the variation of constants formula, that is,

$$\hat{T}(t,s) = T(t,s) + \int_s^t T(t,\tau)B(\tau)\hat{T}(\tau,s)\,d\tau,$$

the function $y(t) = \hat{T}(t,0)\hat{Q}(0)$ satisfies

$$y(t) = T(t,0)\hat{Q}(0) + \int_0^t T(t,\tau)B(\tau)y(\tau)\,d\tau. \tag{3.26}$$

On the other hand, using (3.10) with $t = s = 0$,

$$\hat{P}(0) = U(0,0) = P(0) - \int_0^\infty Q(0)T(0,\tau)B(\tau)U(\tau,0)\,d\tau. \tag{3.27}$$

Since $P(0)$ and $Q(0)$ are complementary projections, by (3.27) we have

$$P(0)\hat{P}(0) = P(0). \tag{3.28}$$

Therefore,

$$Q(0)\hat{Q}(0) = (\mathrm{Id} - P(0))(\mathrm{Id} - \hat{P}(0)) = \mathrm{Id} - \hat{P}(0) = \hat{Q}(0). \tag{3.29}$$

It follows from (3.26) that

$$y(s) = T(s,0)\hat{Q}(0) + \int_0^s T(s,\tau)B(\tau)y(\tau)\,d\tau$$
$$= T(s,0)Q(0)\hat{Q}(0) + \int_0^s T(s,\tau)B(\tau)y(\tau)\,d\tau. \tag{3.30}$$

Multiplying (3.30) on the left by $T(t,s)Q(s)$ and using again (3.29), we obtain

$$T(t,s)Q(s)y(s) = T(t,0)Q(0)\hat{Q}(0) + \int_0^s T(t,\tau)Q(\tau)B(\tau)y(\tau)\,d\tau$$
$$= T(t,0)\hat{Q}(0) + \int_0^s T(t,\tau)Q(\tau)B(\tau)y(\tau)\,d\tau. \tag{3.31}$$

Combining (3.26) and (3.31) yields

$$
\begin{aligned}
y(t) &= T(t,s)Q(s)y(s) - \int_0^s T(t,\tau)Q(\tau)B(\tau)y(\tau)\,d\tau \\
&\quad + \int_0^t T(t,\tau)B(\tau)y(\tau)\,d\tau \\
&= T(t,s)Q(s)y(s) + \int_0^t T(t,\tau)P(\tau)B(\tau)y(\tau)\,d\tau \\
&\quad - \int_t^s T(t,\tau)Q(\tau)B(\tau)y(\tau)\,d\tau.
\end{aligned}
\tag{3.32}
$$

On the other hand, it follows readily from Lemma 3.5 that

$$
\hat{Q}(t)\hat{T}(t,s) = \hat{T}(t,s)\hat{Q}(s).
\tag{3.33}
$$

Since $y(\tau) = \hat{T}(\tau,0)\hat{Q}(0)$, we obtain

$$
y(\tau)\hat{T}(0,s) = \hat{Q}(\tau)\hat{T}(\tau,s).
$$

Thus, multiplying (3.32) on the right by $\hat{T}(0,s)$ we find that for $t \le s$,

$$
\begin{aligned}
\hat{Q}(t)\hat{T}(t,s) &= T(t,s)Q(s)\hat{Q}(s) + \int_0^t T(t,\tau)P(\tau)B(\tau)\hat{Q}(\tau)\hat{T}(\tau,s)\,d\tau \\
&\quad - \int_t^s T(t,\tau)Q(\tau)B(\tau)\hat{Q}(\tau)\hat{T}(\tau,s)\,d\tau.
\end{aligned}
\tag{3.34}
$$

Let now $\xi \in X$, and set

$$
y(t) = \|\hat{T}(t,s)\hat{Q}(s)\xi\|
$$

for $t \le s$ in J, and $\gamma = \|\hat{Q}(s)\xi\|$. It follows readily from (3.34) and (3.33) that the function y satisfies the inequality (3.25). Using Lemma 3.9 and proceeding in a similar manner to that in the proof of Lemma 3.10 we readily obtain the desired inequality.   □

We can now establish the robustness results.

Proof of Theorem 3.2. We have shown above that there exist projections $\hat{P}(t)$ (see (3.17)) leaving invariant the evolution operator $\hat{T}(t,s)$ (see Lemma 3.5). The corresponding norms bounds for $\hat{T}(t,s)|\operatorname{Im}\hat{P}(t)$ and $\hat{T}(t,s)|\operatorname{Im}\hat{Q}(t)$ are given respectively by Lemmas 3.10 and 3.11. This completes the proof of the theorem.   □

Proof of Theorem 3.1. Applying Theorem 3.2 we obtain projections $\hat{P}(t)$ satisfying (3.5) as well as the norm bounds in (3.6) and (3.7). We also obtain norm bounds for the projections. We note that this is the only place in the proof where the assumption $\|B(t)\| \le \delta e^{-\vartheta|t|}$ in Theorem 3.2 must be replaced by the new assumption $\|B(t)\| \le \delta e^{-2\vartheta|t|}$.

Lemma 3.12. *Provided that δ is sufficiently small, for any $t \in J$ we have*

$$\|\hat{P}(t)\| \leq 4De^{\vartheta|t|} \quad and \quad \|\hat{Q}(t)\| \leq 4De^{\vartheta|t|}. \tag{3.35}$$

Proof of the lemma. By Lemma 3.7 with $t = s$, since $P(t)$ and $Q(t)$ are complementary projections,

$$Q(t)\hat{P}(t) = -\int_t^\infty T(t,\tau)Q(\tau)B(\tau)\hat{P}(\tau)\hat{T}(\tau,t)\,d\tau. \tag{3.36}$$

By Lemma 3.10 and Lemma 3.5 (see (3.5)) we have that for $\tau \geq t$ in J,

$$\|\hat{P}(\tau)\hat{T}(\tau,t)\| \leq \tilde{D}e^{-\tilde{c}(\tau-t)+\vartheta|t|}\|\hat{P}(t)\|. \tag{3.37}$$

By (3.36), using the second inequality in (2.5) we obtain

$$\|Q(t)\hat{P}(t)\| \leq \delta D\tilde{D}\|\hat{P}(t)\| \int_t^\infty e^{-c(\tau-t)+\vartheta|\tau|}e^{-2\vartheta|\tau|}e^{-\tilde{c}(\tau-t)+\vartheta|t|}\,d\tau$$
$$\leq \delta D\tilde{D}\|\hat{P}(t)\| \int_t^\infty e^{-(c+\tilde{c}-\vartheta)(\tau-t)}\,d\tau = \frac{D\tilde{D}\delta}{c+\tilde{c}-\vartheta}\|\hat{P}(t)\|, \tag{3.38}$$

since $c > \vartheta$. Similarly, it follows from (3.34) with $t = s$ that

$$P(t)\hat{Q}(t) = \int_0^t T(t,\tau)P(\tau)B(\tau)\hat{Q}(\tau)\hat{T}(\tau,t)\,d\tau. \tag{3.39}$$

By Lemma 3.11 and (3.33) we have that for $\tau \leq t$ in J,

$$\|\hat{Q}(\tau)\hat{T}(\tau,t)\| \leq \tilde{D}e^{-\tilde{c}(t-\tau)+\vartheta|t|}\|\hat{Q}(t)\|. \tag{3.40}$$

By (3.39), using the first inequality in (2.5) we obtain

$$\|P(t)\hat{Q}(t)\| \leq \delta D\tilde{D}\|\hat{Q}(t)\| \int_0^t e^{-c(t-\tau)+\vartheta|\tau|}e^{-2\vartheta|\tau|}e^{-\tilde{c}(t-\tau)+\vartheta|t|}\,d\tau$$
$$\leq \delta D\tilde{D}\|\hat{Q}(t)\| \int_0^t e^{-(c+\tilde{c}-\vartheta)(t-\tau)}\,d\tau = \frac{D\tilde{D}\delta}{c+\tilde{c}-\vartheta}\|\hat{Q}(t)\|. \tag{3.41}$$

Observe now that

$$\hat{P}(t) - P(t) = \hat{P}(t) - P(t)\hat{P}(t) - P(t) + P(t)\hat{P}(t)$$
$$= (\mathrm{Id} - P(t))\hat{P}(t) - P(t)(\mathrm{Id} - \hat{P}(t)) \tag{3.42}$$
$$= Q(t)\hat{P}(t) - P(t)\hat{Q}(t).$$

It follows from (3.38) and (3.41) that

$$\|\hat{P}(t) - P(t)\| \leq \frac{\delta D\tilde{D}}{c+\tilde{c}-\vartheta}(\|\hat{P}(t)\| + \|\hat{Q}(t)\|). \tag{3.43}$$

On the other hand, by (2.5) with $t = s$, we have

$$\|P(t)\| \leq De^{\vartheta|t|} \quad \text{and} \quad \|Q(t)\| \leq De^{\vartheta|t|}.$$

It follows from (3.43) that

$$\|\hat{P}(t)\| \leq \|\hat{P}(t) - P(t)\| + \|P(t)\|$$

$$\leq \frac{\delta D\tilde{D}}{c + \tilde{c} - \vartheta}(\|\hat{P}(t)\| + \|\hat{Q}(t)\|) + De^{\vartheta|t|},$$

and since $\|\hat{Q}(t) - Q(t)\| = \|\hat{P}(t) - P(t)\|$ we also have

$$\|\hat{Q}(t)\| \leq \|\hat{P}(t) - P(t)\| + \|Q(t)\|$$

$$\leq \frac{\delta D\tilde{D}}{c + \tilde{c} - \vartheta}(\|\hat{P}(t)\| + \|\hat{Q}(t)\|) + De^{\vartheta|t|}.$$

Therefore,

$$\|\hat{P}(t)\| + \|\hat{Q}(t)\| \leq \frac{2\delta D\tilde{D}}{c + \tilde{c} - \vartheta}(\|\hat{P}(t)\| + \|\hat{Q}(t)\|) + 2De^{\vartheta|t|},$$

and

$$\left(1 - \frac{2\delta D\tilde{D}}{c + \tilde{c} - \vartheta}\right)(\|\hat{P}(t)\| + \|\hat{Q}(t)\|) \leq 2De^{\vartheta|t|}.$$

Taking δ sufficiently small so that $2\delta D\tilde{D}/(c + \tilde{c} - \vartheta) \leq 1/2$ we obtain

$$\|\hat{P}(t)\| + \|\hat{Q}(t)\| \leq 4De^{\vartheta|t|}.$$

This yields the desired inequalities.   □

Combining (3.37) with (3.35) we find that for $\tau \geq t$ in J,

$$\|\hat{P}(\tau)\hat{T}(\tau, t)\| \leq \tilde{D}e^{-\tilde{c}(\tau - t) + \vartheta|t|}\|\hat{P}(t)\| \leq 4D\tilde{D}e^{-\tilde{c}(\tau - t) + 2\vartheta|t|}.$$

Similarly, combining (3.40) with (3.35) we find that for $\tau \leq t$ in J,

$$\|\hat{Q}(\tau)\hat{T}(\tau, t)\| \leq \tilde{D}e^{-\tilde{c}(t - \tau) + \vartheta|t|}\|\hat{Q}(t)\| \leq 4D\tilde{D}e^{-\tilde{c}(t - \tau) + 2\vartheta|t|}.$$

This completes the proof of the theorem.   □

3.2 Stable and unstable subspaces

We now consider the stable and unstable subspaces $E(t)$ and $F(t)$ in (2.15), and we study how they vary with the perturbation $B(t)$. We recall that the dimensions $\dim E(t)$ and $\dim F(t)$ are independent of t. Under the hypotheses of

Theorem 3.1, equation (3.1) admits a nonuniform exponential dichotomy, and using the associated projections $\hat{P}(t)$ (see (3.17)) we define linear subspaces

$$\hat{E}(t) = \hat{P}(t)X \quad \text{and} \quad \hat{F}(t) = \hat{Q}(t)X \qquad (3.44)$$

for each $t \in J$. These are respectively the *stable* and *unstable subspaces* at time t associated to the exponential dichotomy of equation (3.1). The dimensions $\dim \hat{E}(t)$ and $\dim \hat{F}(t)$ are also independent of t.

Proposition 3.13. *Under the hypotheses of Theorem 3.1, we have* $\dim \hat{E}(t) = \dim E(t)$ *and* $\dim \hat{F}(t) = \dim F(t)$ *for each* $t \in J$.

Proof. In view of the above discussion, it is sufficient to consider $t = 0$. Fix $\tau \in J$ and set

$$Z(t) = U(t, \tau)(P(\tau) - \text{Id}), \quad t \geq \tau.$$

Using (3.10) it is simple to verify that Z satisfies (3.14). By Lemma 3.3, we have $Z \in \mathcal{E}$ (see (3.16)) and proceeding as in the proof of Lemma 3.4 we find that $Z = 0$, that is,

$$Z(t) = U(t, \tau)(P(\tau) - \text{Id}) = U(t, \tau)P(\tau) - U(t, \tau) = 0.$$

In particular, setting $t = \tau = 0$,

$$\hat{P}(0)P(0) = \hat{P}(0). \qquad (3.45)$$

We now consider the linear operators

$$S = \text{Id} - P(0) + \hat{P}(0) \quad \text{and} \quad T = \text{Id} + P(0) - \hat{P}(0).$$

It follows easily from (3.28) and (3.45) that $ST = \text{Id}$. Therefore, S is invertible and $S^{-1} = T$. Furthermore, a simple computation shows that

$$SP(0)S^{-1} = SP(0)T = \hat{P}(0),$$

and $P(0)$ and $\hat{P}(0)$ are similar. The same happens with $Q(0)$ and $\hat{Q}(0)$. In particular,

$$\dim E(0) = \dim \hat{E}(0) \quad \text{and} \quad \dim F(0) = \dim \hat{F}(0).$$

This implies the desired statement.   □

We now describe how the spaces $\hat{E}(t)$ and $\hat{F}(t)$ vary with the perturbation $B(t)$, or more precisely with the parameter δ in Theorem 3.1. In view of (3.44) this is equivalent to describe how the projections $\hat{P}(t)$ vary with the perturbation.

Proposition 3.14. *Under the hypotheses of Theorem 3.1, for any* $t \geq s$ *in* J *we have*

$$\|P(t) - \hat{P}(t)\| \leq \delta \frac{8D^2\tilde{D}}{c + \tilde{c}} e^{\vartheta|t|}. \qquad (3.46)$$

In particular, for each fixed $t \in J$ *we have* $\hat{P}(t) \to P(t)$ *as* $\delta \to 0$.

Proof. The inequality (3.46) follows immediately from (3.43) and (3.35). □

We now define "angles" $\alpha(t)$ by (2.18), and

$$\hat{\alpha}(t) = \inf\{\|x - y\| : \inf\{\|x - y\| : x \in \hat{E}(t), y \in \hat{F}(t), \|x\| = \|y\| = 1\}.$$

By Proposition 2.4 we have

$$\frac{1}{\|P(t)\|} \le \alpha(t) \le \frac{2}{\|P(t)\|} \quad \text{and} \quad \frac{1}{\|\hat{P}(t)\|} \le \hat{\alpha}(t) \le \frac{2}{\|\hat{P}(t)\|}. \tag{3.47}$$

It follows from Proposition 3.14 that

$$\left| \|P(t)\| - \|\hat{P}(t)\| \right| \le \|P(t) - \hat{P}(t)\| \le \delta \frac{8D^2 \widetilde{D}}{c + \widetilde{c}} e^{\vartheta |t|}.$$

Hence, by (3.47), for each fixed $t \in J$,

$$\lim_{\delta \to 0} |\alpha(t) - \hat{\alpha}(t)| \le \frac{1}{\|P(t)\|} \le \alpha(t).$$

3.3 Robustness in the line

To establish the robustness of nonuniform exponential dichotomies in $\mathbb{R}$, we first need to consider separately the cases of exponential dichotomies in the intervals $J = [\varrho, +\infty)$ with $\varrho \le 0$, and $J = (-\infty, \varsigma]$ with $\varsigma \ge 0$. The first type of interval was the object of Theorem 3.1. We now consider the second interval simply by reversing time in the proof of this theorem. We continue to consider the constants $\widetilde{c}$ and $\widetilde{D}$ in (3.3).

Theorem 3.15. *The statement in Theorem 3.1 holds for the interval* $J = (-\infty, \varsigma]$ *with* $\varsigma \ge 0$.

Proof. The proof is analogous to the proof of Theorem 3.1, and hence we will only indicate the main differences. Set

$$H = \{(t, s) \in J \times J : t \le s\},$$

and consider the Banach space

$$\mathcal{D} = \{V : H \to B(X) : V \text{ is continuous and } \|V\| < \infty\} \tag{3.48}$$

with the norm

$$\|V\| = \sup\{\|V(t, s)\| e^{-\vartheta |s|} : (t, s) \in H\}.$$

Similar arguments to those in the proofs of Lemmas 3.3 and 3.4 establish the following statement.

Lemma 3.16. *The equation* $Z' = (A(t)+B(t))Z$ *has a unique solution* $V \in \mathcal{D}$ *such that for each* $(t, s) \in H$,

$$V(t,s) = T(t,s)Q(s) - \int_t^s T(t,\tau)Q(\tau)B(\tau)V(\tau,s)\,d\tau$$
$$+ \int_{-\infty}^t T(t,\tau)P(\tau)B(\tau)V(\tau,s)\,d\tau. \tag{3.49}$$

Furthermore, $V(s,\tau)V(\tau,t) = V(s,t)$ *for any* $t \geq \tau \geq s$ *in* H.

Let now $\hat{T}(t,s)$ be the evolution operator associated to equation (3.1). For each $t \in J$ we consider the linear operators

$$\hat{Q}(t) = \hat{T}(t,0)V(0,0)\hat{T}(0,t) \quad \text{and} \quad \hat{P}(t) = \text{Id} - \hat{Q}(t).$$

Lemma 3.17. *The operator* $\hat{P}(t)$ *is a projection, and*

$$\hat{P}(t)\hat{T}(t,s) = \hat{T}(t,s)\hat{P}(s), \quad t \geq s.$$

The proof of the lemma is analogous to the one of Lemma 3.5. To obtain the norm bounds for $\hat{T}(t,s)\hat{P}(s)$ when $t \geq s$ and $\hat{T}(t,s)\hat{Q}(t)$ when $t \leq s$ we start with the following statement.

Lemma 3.18. *Given* $s \in H$, *if* $y\colon (-\infty, s] \to X$ *is a bounded solution of equation* (3.1) *with* $y(s) = \xi$, *then*

$$y(t) = T(t,s)Q(s)\xi - \int_s^t T(t,\tau)Q(\tau)B(\tau)y(\tau)\,d\tau$$
$$+ \int_{-\infty}^t T(t,\tau)P(\tau)B(\tau)y(\tau)\,d\tau.$$

Proof of the lemma. By the variation of constants formula, for $t \leq s$ in H,

$$P(s)\xi = T(s,t)P(t)y(t) + \int_t^s T(s,\tau)P(\tau)B(\tau)y(\tau)\,d\tau \tag{3.50}$$

and

$$Q(s)\xi = T(s,t)Q(t)y(t) + \int_t^s T(s,\tau)Q(\tau)B(\tau)y(\tau)\,d\tau. \tag{3.51}$$

Since $y(t)$ is bounded, we have

$$\|T(s,t)P(t)y(t)\| \leq CDe^{-c(s-t)+\vartheta|t|},$$

where $C = \sup\{\|y(t)\| : t \leq s \text{ in } H\} < \infty$. Furthermore,

$$\int_{-\infty}^s \|T(s,\tau)P(\tau)\| \cdot \|B(\tau)\| \cdot \|y(\tau)\|\,d\tau \leq D\delta C \int_{-\infty}^s e^{-c(s-\tau)}\,d\tau = \frac{D\delta C}{c}.$$

Taking limits in (3.50) when $t \to -\infty$, since $a > \vartheta$ we obtain

$$P(s)\xi = \int_{-\infty}^{s} T(s,\tau)P(\tau)B(\tau)y(\tau)\,d\tau.$$

By (3.51) we conclude that

$$P(t)y(t) = \int_{-\infty}^{s} T(t,\tau)P(\tau)B(\tau)y(\tau)\,d\tau - \int_{t}^{s} T(t,\tau)P(\tau)B(\tau)y(\tau)\,d\tau$$

$$= \int_{-\infty}^{t} T(t,\tau)Q(\tau)B(\tau)y(\tau)\,d\tau.$$

The desired statement follows from this identity and (3.51).   □

Proceeding as in the proof of Lemma 3.7 we obtain the following.

Lemma 3.19. *The function* $(-\infty, s] \cap J \ni t \mapsto \hat{Q}(t)\hat{T}(t, s)$ *is bounded, and for any $t \le s$ in H we have,*

$$\hat{Q}(t)\hat{T}(t,s) = T(t,s)Q(s)\hat{Q}(s) - \int_{t}^{s} T(t,\tau)Q(\tau)B(\tau)\hat{T}(\tau,s)\hat{Q}(s)\,d\tau$$

$$+ \int_{-\infty}^{t} T(t,\tau)P(\tau)B(\tau)\hat{T}(\tau,s)\hat{Q}(s)\,d\tau.$$

Lemma 3.20. *We have*

$$\|\hat{T}(t,s)|\operatorname{Im}\hat{P}(s)\| \le \tilde{D}e^{-\tilde{c}(t-s)+\vartheta|s|}, \quad t \ge s \text{ in } H,$$

$$\|\hat{T}(t,s)|\operatorname{Im}\hat{Q}(s)\| \le \tilde{D}e^{-\tilde{c}(s-t)+\vartheta|s|}, \quad t \le s \text{ in } H.$$

Proof of the lemma. The proof of the second statement is analogous to the proof of Lemma 3.10. Namely, it follows from Lemma 3.19 that for each $\xi \in X$ the function $y \colon (-\infty, s] \to [0, +\infty)$ given by

$$y(t) = \hat{Q}(t)\hat{T}(t,s)\xi$$

is a bounded solution of (3.1). Thus the desired inequality follows readily from Lemma 3.9 with $\varrho = -\infty$.

The proof of the first statement is analogous to the proof of Lemma 3.11. Namely, using similar arguments we can show that for $t \ge s$,

$$\hat{P}(t)\hat{T}(t,s) = T(t,s)P(s)\hat{P}(s) + \int_{0}^{t} T(t,\tau)Q(\tau)B(\tau)\hat{P}(\tau)\hat{T}(\tau,s)\,d\tau$$

$$+ \int_{s}^{t} T(t,\tau)P(\tau)B(\tau)\hat{P}(\tau)\hat{T}(\tau,s)\,d\tau.$$

The desired statement follows now easily from Lemma 3.8.   □

Finally, for any sufficiently small δ, proceeding as in the proof of Lemma 3.12 we obtain the norm bounds

$$\|\hat{P}(t)\| \le 4De^{\vartheta|t|} \quad \text{and} \quad \|\hat{Q}(t)\| \le 4De^{\vartheta|t|}$$

for any $t \in H$. Combined with Lemma 3.20 we find that

$$\|\hat{T}(t,s)\hat{P}(s)\| \le \tilde{D}e^{-\tilde{c}(t-s)+\vartheta|s|}\|\hat{P}(s)\| \le 4D\tilde{D}e^{-\tilde{c}(t-s)+2\vartheta|s|}$$

for $t \ge s$ in H, and that

$$\|\hat{T}(t,s)\hat{Q}(s)\| \le \tilde{D}e^{-\tilde{c}(s-t)+\vartheta|s|}\|\hat{Q}(s)\| \le 4D\tilde{D}e^{-\tilde{c}(s-t)+2\vartheta|s|}.$$

for $t \le s$ in H. This completes the proof of the theorem. $\square$

We now consider the case of dichotomies in the interval $\mathbb{R}$.

Theorem 3.21 ([18]). *The statement in Theorem 3.1 holds for $J = \mathbb{R}$.*

Proof. Consider the sets

$$G = \{(t,s) \in \mathbb{R} \times \mathbb{R} : t \ge s\} \quad \text{and} \quad H = \{(t,s) \in \mathbb{R} \times \mathbb{R} : t \le s\}.$$

We also consider the spaces $\mathcal{C}$ and $\mathcal{D}$ in (3.8) and (3.48) associated respectively with G and H. We note that by repeating arguments in the proofs of Theorems 3.2 and 3.15 using these sets we find that the operators

$$\hat{P}_+(t) := \hat{T}(t,0)U(0,0)\hat{T}(0,t), \quad \hat{Q}_-(t) := \hat{T}(t,0)V(0,0)\hat{T}(0,t)$$

are projections for each $t \in \mathbb{R}$, such that for every $t, s \in \mathbb{R}$,

$$\hat{P}_+(t)\hat{T}(t,s) = \hat{T}(t,s)\hat{P}_+(s), \quad \hat{Q}_-(t)\hat{T}(t,s) = \hat{T}(t,s)\hat{Q}_-(s),$$

and

$$\|\hat{T}(t,s)|\operatorname{Im}\hat{P}_+(s)\| \le \tilde{D}e^{-\tilde{c}(t-s)+\vartheta|s|}, \quad t \ge s, \tag{3.52}$$
$$\|\hat{T}(t,s)|\operatorname{Im}\hat{Q}_-(s)\| \le \tilde{D}e^{-\tilde{c}(s-t)+\vartheta|s|}, \quad t \le s. \tag{3.53}$$

Indeed, notice that Lemmas 3.8 and 3.9 hold respectively for functions of the form $x \colon [s,+\infty) \to [0,+\infty)$ and $y \colon (-\infty,s] \to [0,+\infty)$, for any $s \in \mathbb{R}$. This allows us to establish respectively (3.52) (see the proof of Lemma 3.10) and (3.53) (see the proof of Lemma 3.20). Using the identities

$$P(0)\hat{P}_+(0) = P(0), \quad \hat{P}_+(0)P(0) = \hat{P}_+(0) \tag{3.54}$$

(see (3.28) and (3.45)), and the corresponding

$$Q(0)\hat{Q}_-(0) = Q(0), \quad \hat{Q}_-(0)Q(0) = \hat{Q}_-(0), \tag{3.55}$$

we can establish the following statement.

Lemma 3.22. *If δ is sufficiently small, then the operator $S = \hat{P}_+(0) + \hat{Q}_-(0)$ is invertible.*

Proof of the lemma. Setting $\hat{P}_-(0) = \mathrm{Id} - \hat{Q}_-(0)$, it follows readily from (3.55) that $P(0)\hat{P}_-(0) = \hat{P}_-(0)$. Using also (3.54) we obtain

$$\hat{P}_+(0) + \hat{Q}_-(0) - \mathrm{Id} = \hat{P}_+(0) - P(0) + P(0) - \hat{P}_-(0)$$
$$= \hat{P}_+(0) - P(0)\hat{P}_+(0) + P(0) - P(0)\hat{P}_-(0)$$
$$= Q(0)\hat{P}_+(0) + P(0)\hat{Q}_-(0).$$

By Lemma 3.16,

$$P(0)\hat{Q}_-(0) = P(0)V(0,0) = -\int_{-\infty}^{0} T(0,\tau)P(\tau)B(\tau)V(\tau,0)\,d\tau,$$

and by Lemma 3.3,

$$Q(0)\hat{P}_+(0) = Q(0)U(0,0) = -\int_{0}^{\infty} T(0,\tau)Q(\tau)B(\tau)U(\tau,0)\,d\tau.$$

To estimate the two integrals, we need to obtain the bounds for $U(t,0)$ when $t \geq 0$ and for $V(t,0)$ when $t \leq 0$. Using (3.10) and (2.5) we obtain

$$\|U(t,0)\| \leq \|T(t,0)P(0)\| + \int_{0}^{t} \|T(t,\tau)P(\tau)\| \cdot \|B(\tau)\| \cdot \|U(\tau,0)\|\,d\tau$$
$$+ \int_{t}^{\infty} \|T(t,\tau)Q(\tau)\| \cdot \|B(\tau)\| \cdot \|U(\tau,0)\|\,d\tau$$
$$\leq De^{-ct} + D\delta \int_{0}^{t} e^{-c(t-\tau)}\|U(\tau,0)\|\,d\tau$$
$$+ D\delta \int_{t}^{\infty} e^{-c(\tau-t)}\|U(\tau,0)\|\,d\tau.$$

Setting $x(t) = \|U(t,0)\|$ and $\gamma = 1$, it follows from Lemma 3.8 that

$$\|U(t,0)\| \leq \tilde{D}e^{-\tilde{c}t}, \quad t \geq 0. \tag{3.56}$$

To estimate $V(t,0)$ for $t \geq 0$, we note that using (3.49) and again (2.5),

$$\|V(t,0)\| \leq \|T(t,0)Q(0)\| + \int_{t}^{0} \|T(t,\tau)Q(\tau)\| \cdot \|B(\tau)\| \cdot \|V(\tau,0)\|\,d\tau$$
$$+ \int_{-\infty}^{t} \|T(t,\tau)P(\tau)\| \cdot \|B(\tau)\| \cdot \|V(\tau,0)\|\,d\tau$$
$$\leq De^{ct} + D\delta \int_{t}^{0} e^{-c(\tau-t)}\|V(\tau,0)\|\,d\tau$$
$$+ D\delta \int_{-\infty}^{t} e^{-c(t-\tau)}\|V(\tau,0)\|\,d\tau.$$

Setting $x(t) = \|V(t,0)\|$ and $\gamma = 1$, it follows from Lemma 3.9 (for functions in the interval $(-\infty, s]$) that

$$\|V(t,0)\| \le \widetilde{D}e^{\widetilde{a}t}, \quad t \le 0. \tag{3.57}$$

It follows from (3.56) and (3.57) that

$$\|\hat{P}_+(0) + \hat{Q}_-(0) - \mathrm{Id}\| \le \int_0^\infty \|T(0,\tau)Q(\tau)\| \cdot \|B(\tau)\| \cdot \|U(\tau,0)\|\, d\tau$$

$$+ \int_{-\infty}^0 \|T(0,\tau)P(\tau)\| \cdot \|B(\tau)\| \cdot \|V(\tau,0)\|\, d\tau$$

$$\le \delta D\widetilde{D} \int_0^\infty e^{-(c+\widetilde{c})\tau}\, d\tau + \delta D\widetilde{D} \int_{-\infty}^0 e^{(c+\widetilde{c})\tau}\, d\tau$$

$$\le \frac{2\delta D\widetilde{D}}{c+\widetilde{c}}.$$

Hence, taking δ sufficiently small, we can make $\|\hat{P}_+(0) + \hat{Q}_-(0) - \mathrm{Id}\|$ as small as desired, and thus $S = \hat{P}_+(0) + \hat{Q}_-(0)$ becomes invertible. $\qquad\square$

For each $t \in \mathbb{R}$ we set

$$\widetilde{P}(t) = \hat{T}(t,0)SP(0)S^{-1}\hat{T}(0,t).$$

We have

$$\widetilde{P}(t)^2 = \hat{T}(t,0)\widetilde{P}(0)^2\hat{T}(0,t) = \hat{T}(t,0)SP(0)^2S^{-1}\hat{T}(0,t) = \widetilde{P}(t),$$

and $\widetilde{P}(t)$ is a projection for each t. Furthermore,

$$\hat{T}(t,s)\widetilde{P}(s) = \hat{T}(t,0)SP(0)S^{-1}\hat{T}(0,s) = \widetilde{P}(t)\hat{T}(t,s). \tag{3.58}$$

Thus, to show that equation (3.1) admits a nonuniform exponential dichotomy in $\mathbb{R}$ with projections $\widetilde{P}(t)$, it remains to obtain norm bounds for $\hat{T}(t,s)\widetilde{P}(s)$ when $t \ge s$, and $\hat{T}(t,s)\widetilde{Q}(s)$ when $t \le s$. These will be a consequence of (3.52)–(3.53). Observe first that by (3.54)–(3.55),

$$SP(0) = \hat{P}_+(0)P(0) + \hat{Q}_-(0)P(0) = \hat{P}_+(0),$$

$$SQ(0) = \hat{P}_+(0)Q(0) + \hat{Q}_-(0)Q(0) = \hat{Q}_-(0).$$

Therefore, setting

$$S(t) = \hat{T}(t,0)S\hat{T}(0,t) = \hat{P}_+(t) + \hat{Q}_-(t),$$

we obtain

$$\widetilde{P}(t)S(t) = \hat{T}(t,0)SP(0)S^{-1}S\hat{T}(0,t)$$

$$= \hat{T}(0,t)SP(0)\hat{T}(0,t) = \hat{P}_+(t),$$

and thus also $\widetilde{Q}(t)S(t) = \hat{Q}_-(t)$, where $\widetilde{Q}(t) = \mathrm{Id} - \widetilde{P}(t)$. Therefore,

$$\mathrm{Im}\,\widetilde{P}(t) \supset \mathrm{Im}\,\hat{P}_+(t) \quad \text{and} \quad \mathrm{Im}\,\widetilde{Q}(t) \supset \mathrm{Im}\,\hat{Q}_-(t).$$

Since $S(t)$ is invertible it follows that indeed

$$\mathrm{Im}\,\widetilde{P}(t) = \mathrm{Im}\,\hat{P}_+(t) \quad \text{and} \quad \mathrm{Im}\,\widetilde{Q}(t) = \mathrm{Im}\,\hat{Q}_-(t).$$

By (3.52) we obtain that for $t \geq s$,

$$\begin{aligned}
\|\hat{T}(t,s)\widetilde{P}(s)\| &\leq \|\hat{T}(t,s)|\,\mathrm{Im}\,\widetilde{P}(s)\| \cdot \|\widetilde{P}(s)\| \\
&= \|\hat{T}(t,s)|\,\mathrm{Im}\,\hat{P}_+(s)\| \cdot \|\widetilde{P}(s)\| \qquad (3.59) \\
&\leq \widetilde{D}e^{-\tilde{c}(t-s)+\vartheta|s|}\|\widetilde{P}(s)\|.
\end{aligned}$$

Similarly, it follows from (3.53) that for $t \leq s$,

$$\begin{aligned}
\|\hat{T}(t,s)\widetilde{Q}(s)\| &\leq \|\hat{T}(t,s)|\,\mathrm{Im}\,\hat{Q}_-(s)\| \cdot \|\widetilde{P}(s)\| \\
&\leq \widetilde{D}e^{-\tilde{c}(s-t)+\vartheta|s|}\|\widetilde{Q}(s)\|.
\end{aligned} \qquad (3.60)$$

Lemma 3.23. *Provided that δ is sufficiently small, for any $t \in \mathbb{R}$ we have*

$$\|\widetilde{P}(t)\| \leq 4De^{\vartheta|t|} \quad \text{and} \quad \|\widetilde{Q}(t)\| \leq 4De^{\vartheta|t|}.$$

Proof of the lemma. It follows from (3.59) that for each $\xi \in X$ the function $y(t) = \hat{T}(t,s)\widetilde{P}(s)\xi$, $t \geq s$ is bounded. Since $y(s) = \widetilde{P}(s)\xi$, it follows from Lemma 3.6 that for $t \geq s$,

$$\begin{aligned}
\widetilde{P}(t)\hat{T}(t,s) = T(t,s)P(s)\widetilde{P}(s) &+ \int_s^t T(t,\tau)P(\tau)B(\tau)\widetilde{P}(\tau)\hat{T}(\tau,s)\,d\tau \\
&- \int_t^\infty T(t,\tau)Q(\tau)B(\tau)\widetilde{P}(\tau)\hat{T}(\tau,s)\,d\tau.
\end{aligned}$$

Setting $t = s$, since $P(t)$ and $Q(t)$ are complementary projections we obtain

$$Q(t)\widetilde{P}(t) = -\int_t^\infty T(t,\tau)Q(\tau)B(\tau)\widetilde{P}(\tau)\hat{T}(\tau,t)\,d\tau.$$

By (3.59) and (3.58) we have that for $\tau \geq t$,

$$\|\widetilde{P}(\tau)\hat{T}(\tau,t)\| \leq \widetilde{D}e^{-\tilde{c}(\tau-t)+\vartheta|t|}\|\widetilde{P}(t)\|.$$

Proceeding as in (3.38) we obtain

$$\|Q(t)\widetilde{P}(t)\| \leq \frac{D\widetilde{D}\delta}{c+\tilde{c}-\vartheta}\|\widetilde{P}(t)\|.$$

Similarly, by (3.60), for each $\xi \in X$ the function $y(t) = \hat{T}(t,s)\widetilde{Q}(s)\xi$, $t \leq s$ is bounded. Since $y(s) = \widetilde{Q}(s)\xi$, it follows from Lemma 3.18 that for $t \leq s$,

$$\widetilde{Q}(t)\hat{T}(t,s) = T(t,s)Q(s)\widetilde{Q}(s) - \int_s^t T(t,\tau)Q(\tau)B(\tau)\widetilde{Q}(\tau)\hat{T}(\tau,s)\,d\tau$$
$$+ \int_{-\infty}^t T(t,\tau)P(\tau)B(\tau)\widetilde{Q}(\tau)\hat{T}(\tau,s)\,d\tau.$$

Setting $t = s$ we obtain

$$P(t)\widetilde{Q}(t) = \int_{-\infty}^t T(t,\tau)P(\tau)B(\tau)\widetilde{Q}(\tau)\hat{T}(\tau,t)\,d\tau. \tag{3.61}$$

By (3.60) and (3.58) we have that for $\tau \leq t$,

$$\|\widetilde{Q}(\tau)\hat{T}(\tau,t)\| \leq \widetilde{D}e^{-\tilde{c}(t-\tau)+\vartheta|t|}\|\widetilde{Q}(t)\|.$$

Therefore, in view of (3.61),

$$\|P(t)\widetilde{Q}(t)\| \leq \frac{D\widetilde{D}\delta}{c+\tilde{c}-\vartheta}\|\widetilde{Q}(t)\|.$$

Observe now that replacing $\hat{P}(t)$ by $\widetilde{P}(t)$ in (3.42) we obtain

$$\widetilde{P}(t) - P(t) = Q(t)\widetilde{P}(t) - P(t)\widetilde{Q}(t).$$

The desired statement can be obtained by repeating arguments in the proof of Lemma 3.12, replacing $\hat{P}(t)$ by $\widetilde{P}(t)$ and $\hat{Q}(t)$ by $\widetilde{Q}(t)$. □

The theorem follows readily from (3.59), (3.60), and Lemma 3.23. □

3.4 The case of strong dichotomies

We now consider the problem of robustness of strong nonuniform exponential dichotomies (see Definition 2.2). With the notation in (2.4) we continue to consider the constants c, ϑ and D in (3.2), and the constants $\tilde{c}$ and $\widetilde{D}$ in (3.3). We also set

$$d = \max\{-\underline{a}, \overline{b}\} \quad \text{and} \quad \tilde{d} = d + 2D\delta.$$

The following is a robustness result for strong dichotomies in the line.

Theorem 3.24. Let $A, B \colon \mathbb{R} \to B(X)$ be continuous functions such that:

1. equation (2.1) admits a strong nonuniform exponential dichotomy in $\mathbb{R}$ with $\vartheta < a$;
2. $\|B(t)\| \leq \delta e^{-2\vartheta|t|}$ for every $t \in \mathbb{R}$.

If δ is sufficiently small, then equation (3.1) admits a strong nonuniform exponential dichotomy in $\mathbb{R}$, with the constants c, d, ϑ and D replaced respectively by $\tilde{c}$, $\tilde{d}$, 3ϑ and $8D^2\tilde{D}$.

Proof. Let $\hat{P}(t)$ be the projections associated to the exponential dichotomy in Theorem 3.21, and set $\hat{Q}(t) = \mathrm{Id} - \hat{P}(t)$. In view of Theorem 3.21, it only remains to show that the last two inequalities in (2.7) hold for the evolution operator $\hat{T}(t,s)$ of equation (3.1). By (3.58) we have

$$\hat{P}(t)\hat{T}(t,s) = \hat{T}(t,s)\hat{P}(s), \quad t,s \in \mathbb{R}.$$

Thus, by the variation of constants formula,

$$\hat{P}(t)\hat{T}(t,s) = T(t,s)\hat{P}(s) + \int_s^t T(t,\tau)B(\tau)\hat{P}(\tau)\hat{T}(\tau,s)\,d\tau,$$

$$\hat{Q}(t)\hat{T}(t,s) = T(t,s)\hat{Q}(s) + \int_s^t T(t,\tau)B(\tau)\hat{Q}(\tau)\hat{T}(\tau,s)\,d\tau.$$

$$(3.62)$$

On the other hand, by Theorem 3.21,

$$\|\hat{P}(t)\| \le 4D\tilde{D}e^{2\vartheta|t|} \quad \text{and} \quad \|\hat{Q}(t)\| \le 4D\tilde{D}e^{2\vartheta|t|}.$$

By (2.7), for $t \le s$ we have

$$\|T(t,s)\| = \|T(t,s)(P(s) + Q(s))\|$$
$$\le \|T(t,s)P(s)\| + \|T(t,s)Q(s)\| \le 2De^{b(s-t)+\vartheta|s|}.$$

$$(3.63)$$

Therefore,

$$\|T(t,s)\hat{P}(s)\| \le 8D^2\tilde{D}e^{b(s-t)+3\vartheta|s|}.$$

$$(3.64)$$

Set now

$$x(t) = \|\hat{T}(t,s)\hat{P}(s)\|.$$

Using (3.62), it follows from (3.63) and (3.64) that for $t \le s$,

$$x(t) \le 8D^2\tilde{D}e^{b(s-t)+3\vartheta|s|} + 2D\delta\int_t^s e^{b(\tau-t)-\vartheta|\tau|}x(\tau)\,d\tau.$$

$$(3.65)$$

Let $\Gamma(\tau) = x(\tau)e^{-b(s-\tau)}$. Then, by (3.65),

$$\Gamma(t) \le 8D^2\tilde{D}e^{3\vartheta|s|} + 2D\delta\int_t^s \Gamma(\tau)\,d\tau.$$

Setting $w(z) = \Gamma(s-z)$ we obtain

$$w(s-t) \le 8D^2\tilde{D}e^{3\vartheta|s|} + 2D\delta\int_0^{s-t} w(u)\,du,$$

and by Gronwall's lemma,

$$w(s-t) \le 8D^2 \widetilde{D} e^{3\vartheta|s|+2D\delta(s-t)}.$$

This yields

$$x(t) \le 8D^2 \widetilde{D} e^{(b+2D\delta)(s-t)+3\vartheta|s|}.$$

We now estimate $\|\hat{T}(t,s)\hat{Q}(t)\|$. In a similar manner, using (2.7) we find that for $t \ge s$,

$$\|T(t,s)\| \le 2D e^{b(t-s)+\vartheta|s|}, \tag{3.66}$$

and thus

$$\|T(t,s)\hat{Q}(s)\| \le 8D^2 \widetilde{D} e^{b(t-s)+3\vartheta|s|}. \tag{3.67}$$

Setting

$$y(t) = \|\hat{T}(t,s)\hat{Q}(s)\|,$$

it follows from (3.66) and (3.67) that for $t \ge s$,

$$y(t) \le 8D^2 \widetilde{D} e^{b(t-s)+3\vartheta|s|} + 2D\delta \int_s^t e^{b(t-\tau)-\vartheta|\tau|} y(\tau)\, d\tau.$$

Proceeding as above we find that

$$y(t) \le 8D^2 \widetilde{D} e^{(b+2D\delta)(t-s)+3\vartheta|s|}.$$

This completes the proof of the theorem. $\square$

Stable manifolds and topological conjugacies

Part II

4

Lipschitz stable manifolds

We want to construct stable and unstable invariant manifolds without assuming the existence of a uniform exponential dichotomy for the linear variational equation. Our main objective is to describe the weakest possible setting under which one can construct the invariant manifolds. We still require some amount of hyperbolicity. Namely, we show that under fairly general assumptions the generalized notion of *nonuniform exponential dichotomy* allows us to establish the existence of stable and unstable invariant manifolds. In this chapter we only consider "Lipschitz manifolds", that is, graphs of Lipschitz functions. We refer to Chapters 5 and 6 for the existence of smooth invariant manifolds (respectively in $\mathbb{R}^n$ and in arbitrary Banach spaces), under slightly stronger assumptions. We follow closely [12], although now considering the general case when the stable and unstable subspaces may depend on the time t. Lipschitz center manifolds were obtained with a similar approach in [8]; we refer to Chapter 8 for the construction of smooth center manifolds.

4.1 Setup and standing assumptions

Let X be a Banach space and let $A\colon \mathbb{R}_0^+ \to B(X)$ be a continuous function, where $B(X)$ is the set of bounded linear operators in X. We want to study nonlinear perturbations of the equation $v' = A(t)v$. Let $f\colon \mathbb{R}_0^+ \times X \to X$ be a continuous function such that

$$f(t,0) = 0 \text{ for every } t \geq 0. \tag{4.1}$$

We assume that there exist constants $c > 0$ and $q > 0$ such that

$$\|f(t,u) - f(t,v)\| \leq c\|u - v\|(\|u\|^q + \|v\|^q) \tag{4.2}$$

for every $t \geq 0$ and $u, v \in X$. One can easily verify that when f is differentiable this is equivalent to the existence of constants $c > 0$ and $q > 0$ such that

$$\left\| \frac{\partial f}{\partial v}(t,v) \right\| \leq c\|v\|^q \text{ for every } t \geq 0 \text{ and } v \in X. \tag{4.3}$$

The condition (4.2) may sometimes be obtained by choosing an appropriate cut-off of the perturbation in a neighborhood of $0 \in X$ and using a Taylor expansion of f, provided that the perturbation is sufficiently regular. We note that since all norms in $\mathbb{R}^2$ are equivalent, assuming that $q > 1$, the q-norm

$$\|(u,v)\|_q = (\|u\|^q + \|v\|^q)^{1/q}$$

is equivalent to the 1-norm $\|u\| + \|v\|$. In this case, one can thus replace the factor $\|u\|^q + \|v\|^q$ by $(\|u\| + \|v\|)^q$ in each inequality in (4.2), up to a multiplicative constant.

Consider now the initial value problem

$$v' = A(t)v + f(t,v), \quad v(s) = v_s, \tag{4.4}$$

with $s \geq 0$ and $v_s \in X$. Note that $v(t) \equiv 0$ is a solution of (4.4). The main objective in this chapter (see also Chapters 5 and 6) is to obtain stable and unstable invariant manifolds for the equation (4.4). Given $s \geq 0$ and $v_s = (\xi, \eta) \in E(s) \times F(s)$ we denote by

$$(x(t), y(t)) = (x(t,s,v_s), y(t,s,v_s)) \in E(t) \times F(t) \tag{4.5}$$

the unique solution of the problem (4.4) or, equivalently, of the problem

$$\begin{aligned}
x(\rho) &= U(\rho,s)\xi + \int_s^\rho U(\rho,r)f(r,x(r),y(r))\,dr, \\
y(\rho) &= V(\rho,s)\eta + \int_s^\rho V(\rho,r)f(r,x(r),y(r))\,dr
\end{aligned} \tag{4.6}$$

for $\rho \geq s$ (see (2.16) for the definition of the operators $U(t,s)$ and $V(t,s)$). For each $\tau \geq 0$, we define

$$\Psi_\tau(s,\xi,\eta) = (s + \tau, x(s + \tau, s, \xi, \eta), y(s + \tau, s, \xi, \eta)). \tag{4.7}$$

This is the semiflow generated by the equation in (4.4).

4.2 Existence of Lipschitz stable manifolds

The stable manifolds will be obtained as graphs of Lipschitz functions. We first describe the class of Lipschitz functions that will be considered. Let

$$\alpha = a(1 + 1/q) + b/q, \tag{4.8}$$

with a and b as in (2.6). We fix $\delta > 0$ and we consider the set of initial conditions

$$X_\alpha = X_\alpha(\delta) = \{(s,\xi) : s \geq 0 \text{ and } \xi \in R_s(\delta e^{-\alpha s})\}, \qquad (4.9)$$

where $R_s(\delta) \subset E(s)$ is the open ball of radius $\delta > 0$ centered at zero. We denote by $\mathcal{X}_\alpha$ the space of continuous functions $\varphi \colon X_\alpha \to X$ such that for each $s \geq 0$, we have

$$\varphi(s, R_s(\delta e^{-\alpha s})) \subset F(s), \quad \varphi(s, 0) = 0,$$

and

$$\|\varphi(s,x) - \varphi(s,y)\| \leq \|x - y\| \text{ for every } x,\, y \in R_s(\delta e^{-\alpha s}). \qquad (4.10)$$

Given a function $\varphi \in \mathcal{X}_\alpha$ we consider the graph of φ,

$$\mathcal{W} = \{(s, \xi, \varphi(s,\xi)) : (s,\xi) \in X_\alpha\} \subset \mathbb{R}_0^+ \times X. \qquad (4.11)$$

We refer to $\mathcal{W}$ as a *Lipschitz manifold*. Note that $(s, 0) \in \mathcal{W}$ (since $\varphi(s,0) = 0$) for every $s \geq 0$. In particular, the set $\mathcal{W}$ contains the line $\mathbb{R}_0^+ \times \{0\}$ and the Lipschitz graph

$$\mathcal{W}_s = \{(s, \xi, \varphi(s,\xi)) : \xi \in R_s(\delta e^{-\alpha s})\}$$

for each fixed $s \geq 0$. See Figure 4.1 for an illustration.

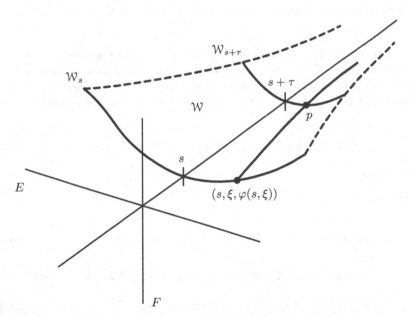

Fig. 4.1. A local stable manifold $\mathcal{W}$ of the origin. In order that $\mathcal{W}$ is invariant under the semiflow Ψ_τ we require that $p = \Psi_\tau(s, \xi, \varphi(s,\xi))$. Here the subspaces $E = E(t)$ and $F = F(t)$ are assumed to be independent of t.

We also consider the constant

$$\beta = \alpha + a = a(2 + 1/q) + b/q, \tag{4.12}$$

and the corresponding sets X_β and $\mathcal{X}_\beta$ defined as before, simply replacing α by β everywhere. We will show that there exists a function $\varphi \in \mathcal{X}_\alpha$ such that for every initial condition $(s, \xi) \in X_\beta \subset X_\alpha$ the corresponding solution of (4.4) is entirely contained in $\mathcal{W}$. This means that for this particular φ the set $\mathcal{W}$ is forward invariant under the semiflow Ψ_τ.

To establish the existence of a stable manifold $\mathcal{W}$ for the origin in equation (4.4), we make the following assumptions:

A1. the function $A \colon \mathbb{R}_0^+ \to B(X)$ is continuous and satisfies (2.2);
A2. the function $f \colon \mathbb{R}_0^+ \times X \to X$ is continuous and satisfies (4.1) and (4.2) for some $c > 0$ and $q > 0$.

We also assume the conditions

$$\bar{a} + \alpha < 0 \quad \text{and} \quad \bar{a} + b < \underline{b}. \tag{4.13}$$

Note that both inequalities in (4.13) are automatically satisfied when a and b are sufficiently small, and that the first is satisfied for a given $a < |\bar{a}|$ provided that q is sufficiently large (that is, provided that the order of the perturbation is sufficiently large).

The following statement concerns the existence of Lipschitz stable manifolds.

Theorem 4.1 ([12]). *Assume that A1–A2 hold. If the equation $v' = A(t)v$ in the Banach space X admits a nonuniform exponential dichotomy in $\mathbb{R}^+$ and the conditions in (4.13) hold, then there exist $\delta > 0$ and a unique function $\varphi \in \mathcal{X}_\alpha$ such that the set $\mathcal{W}$ in (4.11) is forward invariant under the semiflow Ψ_τ, that is,*

$$\text{if } (s, \xi) \in X_\beta \text{ then } \Psi_\tau(s, \xi, \varphi(s, \xi)) \in \mathcal{W} \text{ for every } \tau \geq 0. \tag{4.14}$$

Furthermore:

1. for every $(s, \xi) \in X_\beta$, we have

$$\varphi(s, \xi) = -\int_s^{+\infty} V(\tau, s)^{-1} h(\Psi_{\tau-s}(s, \xi, \varphi(s, \xi))) \, d\tau; \tag{4.15}$$

2. there exists $D > 0$ such that for every $s \geq 0$, $\xi, \bar{\xi} \in R_s(\delta e^{-\beta s})$, and $\tau \geq 0$,

$$\|\Psi_\tau(s, \xi, \varphi(s, \xi)) - \Psi_\tau(s, \bar{\xi}, \varphi(s, \bar{\xi}))\| \leq D e^{\bar{a}\tau + as} \|\xi - \bar{\xi}\|. \tag{4.16}$$

The proof of Theorem 4.1 is given in Section 4.4.

We call the set $\mathcal{W}$ in (4.11) a *(Lipschitz) local stable manifold* or simply a *(Lipschitz) stable manifold* of the origin for the equation (4.4). In particular, setting $\bar{\xi} = 0$ in (4.16) we see that any solution of the initial value problem in (4.4) starting in $\mathcal{W}$, that is, with $v(s) = (\xi, \varphi(s, \xi))$ for some $\xi \in R_s(\delta e^{-\beta s})$, approaches the zero solution with exponential speed $\bar{a}$ (which is independent of ξ). It also follows from Theorem 4.1 that if $\xi \in R_s(\delta e^{-\beta s})$ then

$$y(\rho, \Psi_t(s, \xi, \varphi(s, \xi))) = \varphi(\rho, x(\rho, \Psi_t(s, \xi, \varphi(s, \xi)))) \tag{4.17}$$

for every $\rho \geq s + t$ and $t \geq 0$. The fact that the initial condition ξ must be taken in a neighborhood of exponentially decreasing size $R_s(\delta e^{-\beta s})$, with respect to the initial time s, is a manifestation of the exponential terms e^{as} and e^{bt} in the norm bounds in (2.5) for the operators $U(t, s)$ and $V(t, s)$. Roughly speaking, this means that the size of the neighborhood of initial conditions for which there exists a bounded solution of the differential equation $v' = A(t)v + f(t, v)$ decreases essentially with the same exponential speed with which increases the loss of control of the norm of the operators $U(t, s)$ and $V(t, s)$ for the linear equation $v' = A(t)v$. It should be noted that although these sizes may vary with exponential speed, if the constants a and b are sufficiently small, the speed will be small when compared to the Lyapunov exponents.

4.3 Nonuniformly hyperbolic trajectories

We now explain how Theorem 4.1 can be used to establish the existence of stable manifolds for nonuniformly hyperbolic solutions of a differential equation. Consider a C^1 function $F : \mathbb{R}_0^+ \times X \to X$ and the equation

$$v' = F(t, v). \tag{4.18}$$

Let now $v_0(t)$ be a solution of (4.18). We say that $v_0(t)$ is *nonuniformly hyperbolic* if the linear equation defined by

$$A(t) = \frac{\partial F}{\partial v}(t, v_0(t)) \tag{4.19}$$

admits a nonuniform exponential dichotomy in $\mathbb{R}^+$. We continue to assume that $A(t)$ satisfies (2.2), that is, all solutions of $v' = A(t)v$ are global.

Theorem 4.2. *Assume that F is of class C^1 and let $v_0(t)$ be a nonuniformly hyperbolic solution of (4.18) such that there exist $c > 0$ and $q > 0$ such that for every $t \geq 0$ and $y \in X$,*

$$\left\| \frac{\partial F}{\partial v}(t, y + v_0(t)) - A(t) \right\| \leq c\|y\|^q. \tag{4.20}$$

If the conditions (4.13) hold, then there exist $\delta > 0$ and a unique function $\varphi \in \mathfrak{X}_\beta$ such that the set

$$\mathcal{W} = \{(s, \xi, \varphi(s, \xi)) + (0, v_0(s)) : (s, \xi) \in X_\alpha\} \tag{4.21}$$

satisfies the following properties:

1. $\mathcal{W}$ *is forward invariant under solutions of* (4.18), *that is, if*

$$(s, v_s) \in \{(s, \xi, \varphi(s, \xi)) + (0, v_0(s)) : (s, \xi) \in X_\beta\},$$

then $(t, v(t)) \in \mathcal{W}$ *for every* $t \geq s$, *where* $v(t) = v(t, v_s)$ *is the unique solution of* (4.18) *for* $t \geq s$ *with* $v(s) = v_s$;

2. *there exists* $D > 0$ *such that for every* $s \geq 0$, (s, v_s), $(s, \bar{v}_s) \in \mathcal{W}$, *and* $t \geq s$ *we have*

$$\|v(t, v_s) - v(t, \bar{v}_s)\| \leq De^{\bar{a}(t-s)+as}\|v_s - \bar{v}_s\|.$$

Proof. We shall reduce the study of the equation (4.18) to that of (4.4). For this we consider the change of variables $(t, y) = (t, v - v_0(t))$. Letting $y(t) = v(t) - v_0(t)$, where $v(t)$ is a solution of (4.18), we obtain

$$
\begin{aligned}
y'(t) &= F(t, v(t)) - F(t, v_0(t)) \\
&= F(t, y(t) + v_0(t)) - F(t, v_0(t)) = A(t)y(t) + G(t, y(t)),
\end{aligned}
$$

where

$$G(t, y) = F(t, y + v_0(t)) - F(t, v_0(t)) - A(t)y. \tag{4.22}$$

By hypothesis $A(t)$ satisfies the assumption A1. Furthermore, it follows from (4.22) that G is continuous and clearly $G(t, 0) = 0$ for every $t \geq 0$. It remains to establish property (4.2). For this we note that

$$\|G(t, y) - G(t, z)\| \leq \sup_{r \in [0,1], i=1,\ldots,n} \left\| \frac{\partial G_i}{\partial y}(t, y + r(z - y)) \right\| \cdot \|y - z\|,$$

where $G = (G_1, \ldots, G_n)$. Since

$$\frac{\partial G}{\partial y}(t, y) = \frac{\partial F}{\partial v}(t, y + v_0(t)) - A(t)$$

for every $t \geq 0$ and $y \in X$, we obtain

$$
\begin{aligned}
\|G(t, y) - G(t, z)\| &\leq c \sup_{r \in [0,1]} \|y + r(z - y)\|^q \|y - z\| \\
&\leq c \max\{\|y\|^q, \|z\|^q\} \|y - z\| \\
&\leq c(\|y\|^q + \|z\|^q)\|y - z\|.
\end{aligned}
$$

Thus, the function G satisfies the assumption A2. We can now apply Theorem 4.1 to obtain the desired statement. $\square$

We call the set $\mathcal{W}$ in (4.21) a *(Lipschitz) local stable manifold* or simply a *(Lipschitz) stable manifold* of the solution $v_0(t)$ (of the equation (4.18)).

4.4 Proof of the existence of stable manifolds

4.4.1 Preliminaries

The approach to the proof of Theorem 4.1 can be considered classical, and consists in looking for $\mathcal{W}$ as the graph of a Lipschitz function φ, while using the differential equation to express the forward invariance of the graph under the dynamics to conclude that φ must satisfy a fixed point problem. Nevertheless, to implement this approach in the present context presents additional difficulties, due to the nonuniformity of the exponential behavior of the evolution operators. In particular, the customary application of Gronwall's lemma need not always provide a control of the stable component of the solution.

In view of the desired forward invariance of $\mathcal{W}$ under solutions of the equation in (4.4) (see (4.14)), any solution with initial condition in $\mathcal{W}$ at time $s \geq 0$ must remain in $\mathcal{W}$ for every $t \geq s$ and thus must be of the form

$$(x(t), \varphi(t, x(t))) \in E(t) \times F(t) \text{ for every } t \geq s.$$

In particular, the equations in (4.6) can be replaced by

$$x(t) = U(t, s)x(s) + \int_s^t U(t, \tau)f(\tau, x(\tau), \varphi(\tau, x(\tau))) \, d\tau, \tag{4.23}$$

$$\varphi(t, x(t)) = V(t, s)\varphi(s, x(s)) + \int_s^t V(t, \tau)f(\tau, x(\tau), \varphi(\tau, x(\tau))) \, d\tau. \tag{4.24}$$

We equip the space $\mathcal{X}_\alpha$ (see Section 4.2 for the definition) with the norm

$$\|\varphi\| = \sup \left\{ \|\varphi(t, x)\|/\|x\| : t \geq 0 \text{ and } x \in R_t(\delta e^{-\alpha t}) \setminus \{0\} \right\}. \tag{4.25}$$

Note that $\|\varphi\| \leq 1$ for every $\varphi \in \mathcal{X}_\alpha$. Furthermore, given $t \geq 0$ and $x \neq 0$, we have

$$\|\varphi(t, x)\| \leq \delta e^{-\alpha t}\|\varphi(t, x)\|/\|x\| \leq \delta\|\varphi\| \leq \delta$$

for every $\varphi \in \mathcal{X}_\alpha$. This readily implies that $\mathcal{X}_\alpha$ is a complete metric space with the distance induced by $\|\cdot\|$. For technical reasons, we also consider the space $\mathcal{X}_\alpha^*$ of continuous functions $\varphi \colon X_\alpha^* \to X$, where

$$X_\alpha^* = \{(s, \xi) : s \geq 0 \text{ and } \xi \in E(s)\}, \tag{4.26}$$

such that $\varphi|X_\alpha \in \mathcal{X}_\alpha$ and

$$\varphi(s, \xi) = \varphi(s, \delta e^{-\alpha s}\xi/\|\xi\|) \text{ whenever } s \geq 0 \text{ and } \xi \notin R_s(\delta e^{-\alpha s}).$$

There is a one-to-one correspondence between functions in $\mathcal{X}_\alpha$ and functions in $\mathcal{X}_\alpha^*$: one can easily verify that each function $\varphi \in \mathcal{X}_\alpha$ extends uniquely to a Lipschitz function $\bar{\varphi}$ on $\overline{Z_\alpha}$ with

$$\|\bar{\varphi}(s, x) - \bar{\varphi}(s, y)\| \leq \|x - y\| \text{ for every } x, y \in \overline{R_s(\delta e^{-\alpha s})}. \tag{4.27}$$

In particular $\mathcal{X}_\alpha^*$ is also a Banach space with the norm $\mathcal{X}_\alpha^* \ni \varphi \mapsto \|\varphi|X_\alpha\|$.

Lemma 4.3. *For each $\varphi \in \mathfrak{X}_\alpha^*$ and $s \geq 0$ we have*

$$\|\varphi(s, x) - \varphi(s, y)\| \leq 2\|x - y\| \text{ for every } x, y \in E(s).$$

Proof. In view of (4.27) we may assume that $x \notin \overline{R_s(\delta e^{-\alpha s})}$. We first consider the case when also $y \notin \overline{R_s(\delta e^{-\alpha s})}$. Setting $c = \delta e^{-\alpha s}$, we obtain

$$\|\varphi(s, x) - \varphi(s, y)\| = \left\| \varphi\left(s, c\frac{x}{\|x\|}\right) - \varphi\left(s, c\frac{y}{\|y\|}\right) \right\| \leq c \left\| \frac{x}{\|x\|} - \frac{y}{\|y\|} \right\|.$$

Since

$$\left\| \frac{x}{\|x\|} - \frac{y}{\|y\|} \right\| = \frac{\|(x - y)\|y\| + y(\|y\| - \|x\|)\|}{\|x\| \cdot \|y\|} \leq \frac{2\|x - y\|}{\|x\|},$$

we have

$$\|\varphi(s, x) - \varphi(s, y)\| \leq 2\|x - y\|.$$

Let now $y \in \overline{R_s(\delta e^{-\alpha s})}$ and take $\kappa \in [0, 1)$ such that $z = \kappa x + (1 - \kappa)y$ has norm $\|z\| = c$. Then

$$\begin{aligned}
\|\varphi(s, x) - \varphi(s, y)\| &\leq \|\varphi(s, x) - \varphi(s, z)\| + \|\varphi(s, z) - \varphi(s, y)\| \\
&\leq \|x - z\| + 2\|z - y\| \\
&= \|x - y\| + \|z - y\| \leq 2\|x - y\|.
\end{aligned}$$

This completes the proof. $\square$

We note that in the case of Hilbert spaces one can easily verify that given $\varphi \in \mathfrak{X}_\alpha^*$ and $s \geq 0$ we have

$$\|\varphi(s, x) - \varphi(s, y)\| \leq \|x - y\| \text{ for every } x, y \in E(s).$$

4.4.2 Solution on the stable direction

The proof of Theorem 4.1 is obtained in several steps. We first establish the existence of a unique function $x(t) = x_\varphi(t)$ satisfying (4.23) for each given $\varphi \in \mathfrak{X}_\alpha^*$. By (4.13) we have

$$T_1 := q\bar{a} + a < 0. \tag{4.28}$$

Lemma 4.4. *There exists $R > 0$ such that for every $\delta > 0$ sufficiently small:*

1. *for each $\varphi \in \mathfrak{X}_\alpha^*$, given $(s, \xi) \in X_\alpha$ there exists a unique continuous function $x = x_\varphi: [s, +\infty) \to X$ with $x_\varphi(s) = \xi$, satisfying $x_\varphi(t) \in E(t)$ and (4.23) for every $t \geq s$;*
2. *the function x_φ satisfies*

$$\|x_\varphi(t)\| \leq R e^{\bar{a}(t-s)+as}\|\xi\| \text{ for every } t \geq s. \tag{4.29}$$

Proof. Given $\delta > 0$ and $s \geq 0$, we consider the space

$$\mathcal{B} = \{x \colon [s, +\infty) \to X \text{ continuous:} \ x(t) \in E(t) \text{ for } t \geq s \text{ and } \|x\|' \leq \delta e^{-\alpha s}\},$$

with the norm

$$\|x\|' = \frac{1}{2D_1} \sup\{\|x(t)\| e^{-\overline{a}(t-s)-as} : t \geq s\},$$

where $D_1 \geq 1$ is the constant in (2.5). One can easily verify that with the distance induced by this norm $\mathcal{B}$ is a complete metric space. Given $\varphi \in X_\alpha^*$ and $s \geq 0$, we can define the operator

$$(Jx)(t) = \int_s^t U(t, \tau) f(\tau, x(\tau), \varphi(\tau, x(\tau))) \, d\tau$$

for each $x \in \mathcal{B}$. Clearly, $(Jx)(t) \in E(t)$ for every $t \geq s$. Given $x, y \in \mathcal{B}$ and $\tau \geq s$, it follows from Lemma 4.3 that

$$\|(x(\tau), \varphi(\tau, x(\tau)))\| = \|(x(\tau), \varphi(\tau, x(\tau)) - \varphi(\tau, 0))\| \leq 3\|x(\tau)\|, \qquad (4.30)$$

and

$$\|(x(\tau), \varphi(\tau, x(\tau))) - (y(\tau), \varphi(\tau, y(\tau)))\| \leq 3\|x(\tau) - y(\tau)\|. \qquad (4.31)$$

Therefore, by (4.2),

$$\begin{aligned}
&\|f(\tau, x(\tau), \varphi(\tau, x(\tau))) - f(\tau, y(\tau), \varphi(\tau, y(\tau)))\| \\
&\leq c3^{q+1}\|x(\tau) - y(\tau)\|(\|x(\tau)\|^q + \|y(\tau)\|^q) \qquad (4.32) \\
&\leq 2D_1^{1+q} c\, 6^{q+1} \delta^q e^{\overline{a}(q+1)(\tau-s)-bs}\|x - y\|'.
\end{aligned}$$

By the first inequality in (2.17) we obtain

$$\begin{aligned}
&\|(Jx)(t) - (Jy)(t)\| \\
&\leq \int_s^t \|U(t, \tau)\| \cdot \|f(\tau, x(\tau), \varphi(\tau, x(\tau))) - f(\tau, y(\tau), \varphi(\tau, y(\tau)))\| \, d\tau \\
&\leq 2D_1^{1+q} c\, 6^{q+1} \delta^q \|x - y\|' \int_s^t D_1 e^{\overline{a}(t-\tau)+a\tau} e^{\overline{a}(q+1)(\tau-s)-bs} \, d\tau \\
&\leq 2D_1^{2+q} c\, 6^{q+1} \delta^q \|x - y\|' e^{\overline{a}(t-s)+as-bs} \int_s^\infty e^{(q\overline{a}+a)(\tau-s)} \, d\tau \\
&\leq 2D_1^{2+q} c\, 6^{q+1} \delta^q \|x - y\|' e^{\overline{a}(t-s)+as} \int_s^\infty e^{T_1(\tau-s)} \, d\tau,
\end{aligned}$$

with $T_1 < 0$ as in (4.28). Therefore

$$\|Jx - Jy\|' \leq \theta \|x - y\|', \qquad (4.33)$$

where
$$\theta = c6^{q+1}D_1^{1+q}\delta^q/|T_1|.$$

We now choose $\delta > 0$ sufficiently small, independently of s, so that $\theta < 1/2$. Given $\xi \in R_s(\delta e^{-\alpha s})$ we consider the operator $\bar{J}$ on the space $\mathcal{B}$ defined by

$$(\bar{J}x)(t) = z(t) + (Jx)(t),$$

where $z(t) = U(t,s)\xi \in E(t)$. For $y = 0$ we obtain $Jy = 0$ (note that since $\varphi \in X_\alpha^*$ we have $\varphi(t,0) = 0$ for every $t \geq 0$), and thus, by (4.33), $\|Jx\|' \leq \theta\|x\|'$. By the first inequality in (2.17) we obtain $\|z\|' \leq \|\xi\|/2$ and hence,

$$\|\bar{J}x\|' \leq \|z\|' + \|Jx\|' \leq \frac{1}{2}\|\xi\| + \theta\|x\|'$$
$$\leq \frac{1}{2}\delta e^{-\alpha s} + \frac{1}{2}\delta e^{-\alpha s} = \delta e^{-\alpha s}. \tag{4.34}$$

Therefore, $\bar{J}: \mathcal{B} \to \mathcal{B}$ is a well-defined operator. In view of (4.33),

$$\|\bar{J}x - \bar{J}y\|' = \|Jx - Jy\|' \leq \theta\|x - y\|',$$

and $\bar{J}$ is a contraction. Therefore, there exists a unique function $x = x_\varphi \in \mathcal{B}$ such that $\bar{J}x = x$. It follows from (4.34) that

$$\|x\|' \leq \frac{1}{2}\|\xi\| + \theta\|x\|', \quad \text{and hence,} \quad \|x\|' \leq \frac{\|\xi\|}{2(1-\theta)}.$$

Therefore, for every $t \geq s$,

$$\|x(t)\| \leq 2D_1 e^{\bar{a}(t-s)+as}\frac{\|\xi\|}{2(1-\theta)}.$$

We obtain the desired result with $R = D_1/(1-\theta)$. □

4.4.3 Behavior under perturbations of the data

We now establish some auxiliary results that describe the asymptotic behavior of the function x_φ given by Lemma 4.4, as time approaches infinity, when we change the initial condition ξ or the function φ. Given $\delta > 0$ sufficiently small, $\varphi \in X_\alpha^*$, $s \geq 0$, and initial conditions $\xi, \bar{\xi} \in R_s(\delta e^{-\alpha s})$, we denote by x_φ and $\bar{x}_\varphi$ respectively the unique continuous functions given by Lemma 4.4 such that $x_\varphi(s) = \xi$ and $\bar{x}_\varphi(s) = \bar{\xi}$.

Lemma 4.5. *There exists $K_1 > 0$ such that for every $\delta > 0$ sufficiently small, $\varphi \in X_\alpha^*$, $s \geq 0$, and $\xi, \bar{\xi} \in R_s(\delta e^{-\alpha s})$ we have*

$$\|x_\varphi(t) - \bar{x}_\varphi(t)\| \leq K_1 e^{\bar{a}(t-s)+as}\|\xi - \bar{\xi}\| \text{ for every } t \geq s. \tag{4.35}$$

Proof. Proceeding in a similar manner to that in (4.30), (4.31), and (4.32), for every $\tau \geq s$ we obtain

$$
\begin{aligned}
&\|f(\tau, x_\varphi(\tau), \varphi(\tau, x_\varphi(\tau))) - f(\tau, \bar{x}_\varphi(\tau), \varphi(\tau, \bar{x}_\varphi(\tau)))\| \\
&\leq c3^{q+1}\|x_\varphi(\tau) - \bar{x}_\varphi(\tau)\|(\|x_\varphi(\tau)\|^q + \|\bar{x}_\varphi(\tau)\|^q).
\end{aligned}
\tag{4.36}
$$

By Lemma 4.4 (see (4.29)), we thus have

$$
\begin{aligned}
&\|f(\tau, x_\varphi(\tau), \varphi(\tau, x_\varphi(\tau))) - f(\tau, \bar{x}_\varphi(\tau), \varphi(\tau, \bar{x}_\varphi(\tau)))\| \\
&\leq \eta e^{q\bar{a}(\tau-s)-as-bs}\|x_\varphi(\tau) - \bar{x}_\varphi(\tau)\|,
\end{aligned}
$$

where

$$
\eta = 2c3^{q+1}R^q\delta^q \leq 2c3^{q+1}R^q,
\tag{4.37}
$$

provided that $\delta \leq 1$. Note that the last constant is independent of δ. Setting $\rho(t) = \|x_\varphi(t) - \bar{x}_\varphi(t)\|$ and using the first inequality in (2.17), it follows from (4.23) that

$$
\begin{aligned}
\rho(t) &\leq \|U(t,s)\| \cdot \|\xi - \bar{\xi}\| + \int_s^t \|U(t,\tau)\|\eta e^{q\bar{a}(\tau-s)-as-bs}\rho(\tau)\,d\tau \\
&\leq D_1 e^{\bar{a}(t-s)+as}\|\xi - \bar{\xi}\| + D_1\eta \int_s^t e^{\bar{a}(t-\tau)+T_1(\tau-s)}\rho(\tau)\,d\tau,
\end{aligned}
\tag{4.38}
$$

with $T_1 < 0$ as in (4.28). We can now use Gronwall's lemma for the function $e^{-\bar{a}(t-s)}\rho(t)$ and (4.37) to obtain (assuming that $\delta \leq 1$)

$$
\begin{aligned}
\rho(t) &\leq D_1 e^{as+D_1\eta e^{T_1 s}/|T_1|}e^{\bar{a}(t-s)}\|\xi - \bar{\xi}\| \\
&\leq K_1 e^{\bar{a}(t-s)+as}\|\xi - \bar{\xi}\|,
\end{aligned}
$$

with $K_1 = D_1 \exp[2c3^{q+1}D_1 R^q/|T_1|]$. This completes the proof.   □

Given $\delta > 0$ sufficiently small, φ, $\psi \in \mathfrak{X}_\alpha^*$, and $(s, \xi) \in X_\alpha$, we denote by x_φ and x_ψ the continuous functions given by Lemma 4.4 such that $x_\varphi(s) = x_\psi(s) = \xi$.

Lemma 4.6. *There exists $K_2 > 0$ such that for every $\delta > 0$ sufficiently small, φ, $\psi \in \mathfrak{X}_\alpha^*$, and $(s, \xi) \in X_\alpha$ we have*

$$
\|x_\varphi(t) - x_\psi(t)\| \leq K_2 e^{\bar{a}(t-s)+(a-b)s}\|\xi\| \cdot \|\varphi - \psi\| \text{ for every } t \geq s.
\tag{4.39}
$$

Proof. Proceeding in a similar manner to that in (4.32), we obtain

$$
\begin{aligned}
&\|f(\tau, x_\varphi(\tau), \varphi(\tau, x_\varphi(\tau))) - f(\tau, x_\psi(\tau), \psi(\tau, x_\psi(\tau)))\| \\
&\leq c3^q\|(x_\varphi(\tau) - x_\psi(\tau), \varphi(\tau, x_\varphi(\tau)) - \psi(\tau, x_\psi(\tau)))\| \\
&\quad \times (\|x_\varphi(\tau)\|^q + \|x_\psi(\tau)\|^q).
\end{aligned}
\tag{4.40}
$$

Furthermore, by Lemma 4.3,

$$\|\varphi(\tau, x_\varphi(\tau)) - \psi(\tau, x_\psi(\tau))\|$$
$$\leq \|\varphi(\tau, x_\varphi(\tau)) - \psi(\tau, x_\varphi(\tau))\| + \|\psi(\tau, x_\varphi(\tau)) - \psi(\tau, x_\psi(\tau))\| \qquad (4.41)$$
$$\leq \|x_\varphi(\tau)\| \cdot \|\varphi - \psi\| + 2\|x_\varphi(\tau) - x_\psi(\tau)\|.$$

By (4.29) in Lemma 4.4 we conclude that

$$\|f(\tau, x_\varphi(\tau), \varphi(\tau, x_\varphi(\tau))) - f(\tau, x_\psi(\tau), \psi(\tau, x_\psi(\tau)))\|$$
$$\leq 2c3^q R^q \delta^q e^{q\bar{a}(\tau-s)-as-bs}(\|x_\varphi(\tau)\| \cdot \|\varphi - \psi\| + 3\|x_\varphi(\tau) - x_\psi(\tau)\|). \qquad (4.42)$$

We now proceed in a similar manner to that in (4.38) in order to apply Gronwall's lemma. Set

$$\bar{\rho}(t) = \|x_\varphi(t) - x_\psi(t)\| \quad \text{and} \quad \bar{\eta} = 2c3^q R^q \delta^q. \qquad (4.43)$$

Note that $\bar{\eta} \leq 2c3^q R^q$ provided that $\delta \leq 1$. Note also that the last constant is independent of δ. Using the first inequality in (2.17), (4.29) in Lemma 4.4, and (4.42), it follows from (4.23) that

$$\bar{\rho}(t) \leq \bar{\eta} \int_s^t \|U(t,\tau)\| e^{q\bar{a}(\tau-s)-as-bs} \|x_\varphi(\tau)\| \cdot \|\varphi - \psi\| \, d\tau$$
$$+ 3\bar{\eta} \int_s^t \|U(t,\tau)\| e^{q\bar{a}(\tau-s)-as-bs} \|x_\varphi(\tau) - x_\psi(\tau)\| \, d\tau$$
$$\leq \bar{\eta} D_1 R \|\xi\| \cdot \|\varphi - \psi\| \int_s^t e^{\bar{a}(t-s)+a\tau+q\bar{a}(\tau-s)-bs} \, d\tau$$
$$+ 3\bar{\eta} D_1 \int_s^t e^{\bar{a}(t-\tau)+a\tau} e^{q\bar{a}(\tau-s)-as} \bar{\rho}(\tau) \, d\tau.$$

We conclude that

$$e^{-\bar{a}(t-s)} \bar{\rho}(t) \leq \bar{\eta} D_1 R e^{(a-b)s} \int_s^\infty e^{T_1(\tau-s)} \, d\tau \|\xi\| \cdot \|\varphi - \psi\|$$
$$+ 3\bar{\eta} D_1 \int_s^t e^{T_1(\tau-s)} e^{-\bar{a}(\tau-s)} \bar{\rho}(\tau) \, d\tau,$$

with $T_1 < 0$ as in (4.28). We can now use Gronwall's lemma for the function $e^{-\bar{a}(t-s)} \bar{\rho}(t)$ to obtain (assuming that $\delta \leq 1$)

$$\bar{\rho}(t) \leq \frac{\bar{\eta} D_1 R}{|T_1|} e^{3\bar{\eta} D_1/|T_1|} e^{\bar{a}(t-s)+(a-b)s} \|\xi\| \cdot \|\varphi - \psi\|$$
$$\leq K_2 e^{\bar{a}(t-s)+(a-b)s} \|\xi\| \cdot \|\varphi - \psi\|,$$

where

$$K_2 = \frac{2c3^q D_1 R^{q+1}}{|T_1|} \exp\left(\frac{2c3^{q+1} D_1 R^q}{|T_1|}\right).$$

This completes the proof of the lemma. $\qquad\qquad\qquad\qquad\qquad\qquad\qquad \square$

4.4.4 Reduction to an equivalent problem

In order to show the existence of a function $\varphi \in X_\alpha^*$ satisfying (4.24) when $x = x_\varphi$, we first reduce this problem to another one. We recall that x_φ is the function given by Lemma 4.4 with $x_\varphi(s) = \xi$. Let

$$T_2 := \overline{a} - \underline{b} + b < 0. \tag{4.44}$$

Lemma 4.7. *Given $\delta > 0$ sufficiently small and $\varphi \in X_\alpha^*$, the following properties hold:*

1. *if*

$$\varphi(t, x_\varphi(t)) = V(t, s)\varphi(s, \xi) + \int_s^t V(t, \tau)f(\tau, x_\varphi(\tau), \varphi(\tau, x_\varphi(\tau)))\, d\tau \tag{4.45}$$

for every $(s, \xi) \in X_\alpha$ and $t \geq s$, then

$$\varphi(s, \xi) = -\int_s^\infty V(\tau, s)^{-1}f(\tau, x_\varphi(\tau), \varphi(\tau, x_\varphi(\tau)))\, d\tau \tag{4.46}$$

for every $(s, \xi) \in X_\alpha$ (including the requirement that the integral is well-defined);

2. *if (4.46) holds for every $(s, \xi) \in X_\alpha = X_\alpha(\delta)$, then (4.45) holds for every $(s, \xi) \in X_\beta = X_\beta(\delta/R)$ and $t \geq s$.*

Proof. We first show that the integral in (4.46) is well-defined for each $(s, \xi) \in X_\alpha$. By (4.29) in Lemma 4.4, (4.30), and (4.2), we have

$$\|f(\tau, x_\varphi(\tau), \varphi(\tau, x_\varphi(\tau)))\| \leq c\|(x_\varphi(\tau), \varphi(\tau, x_\varphi(\tau)))\|^{q+1} \leq c3^{q+1}\|x_\varphi(\tau)\|^{q+1}$$
$$\leq c3^{q+1}R^{q+1}e^{(q+1)\overline{a}(\tau-s)+a(q+1)s}\|\xi\|^{q+1}.$$

It follows from the last inequality in (2.17) that

$$\int_s^\infty \|V(\tau, s)^{-1}f(\tau, x_\varphi(\tau), \varphi(\tau, x_\varphi(\tau)))\|\, d\tau$$

$$\leq D_2 c3^{q+1}R^{q+1}\delta^{q+1}e^{-a(1+1/q)s-b(1+1/q)s}\int_s^\infty e^{-\underline{b}(\tau-s)+b\tau+(q+1)\overline{a}(\tau-s)}\, d\tau$$

$$= D_2 c3^{q+1}R^{q+1}\delta^{q+1}e^{bs-a(1+1/q)s-b(1+1/q)s}\int_s^\infty e^{(T_2+q\overline{a})(\tau-s)}\, d\tau.$$

Since $\overline{a} < 0$, we have $T_2 + q\overline{a} < 0$, and thus the last integral is finite. Therefore, the integral in (4.46) is well-defined.

We now assume that (4.45) holds for every $(s, \xi) \in X_\alpha$ and $t \geq s$, and we rewrite the identity in the equivalent form

$$\varphi(s,\xi) = V(t,s)^{-1}\varphi(t,x_\varphi(t)) - \int_s^t V(\tau,s)^{-1}f(\tau,x_\varphi(\tau),\varphi(\tau,x_\varphi(\tau)))\,d\tau.$$

$$(4.47)$$

By Lemma 4.3 and (4.29) in Lemma 4.4, we have

$$\|V(t,s)^{-1}\varphi(t,x_\varphi(t))\| \le 2D_2 e^{-\underline{b}(t-s)+bt}\|x_\varphi(t)\|$$
$$\le 2D_2 e^{-\underline{b}(t-s)+bt}R e^{\overline{a}(t-s)}\delta e^{-as/q-bs/q}$$
$$= 2D_2 R\delta e^{T_2(t-s)+[b(1-1/q)-a/q]s}.$$

Thus, letting $t \to \infty$ in (4.47), we obtain (4.46) for every $(s,\xi) \in X_\alpha$ and $t \ge s$. This establishes the first property.

We now assume that (4.46) holds for every $(s,\xi) \in X_\alpha$ and $t \ge s$. Since $V(t,s)V(\tau,s)^{-1} = V(t,\tau)$ we readily obtain

$$V(t,s)\varphi(s,\xi) + \int_s^t V(t,\tau)f(\tau,x_\varphi(\tau),\varphi(\tau,x_\varphi(\tau)))\,d\tau$$
$$= -\int_t^\infty V(\tau,t)^{-1}f(\tau,x_\varphi(\tau),\varphi(\tau,x_\varphi(\tau)))\,d\tau.$$

$$(4.48)$$

We now show that the right-hand side of (4.48) is equal to $\varphi(t,x_\varphi(t))$. We first define a semiflow F_τ for each $\tau \ge 0$ and $(s,\xi) \in X_\alpha$ by

$$F_\tau(s,\xi) = (s+\tau, x_\varphi(s+\tau,s,\xi)).$$

Note that in view of (4.46),

$$\varphi(s,\xi) = -\int_s^\infty V(\tau,s)^{-1}f(F_{\tau-s}(s,\xi),\varphi(F_{\tau-s}(s,\xi)))\,d\tau. \qquad (4.49)$$

Given $\tau \ge t \ge s$, we have

$$F_{\tau-t}(t,x_\varphi(t)) = F_{\tau-t}(F_{t-s}(s,\xi)) = F_{\tau-s}(s,\xi) = (\tau,x_\varphi(\tau)).$$

Furthermore, when $(s,\xi) \in X_\beta(\delta/R)$ it follows from (4.29) in Lemma 4.4 that

$$\|x_\varphi(t)\| \le R e^{\overline{a}(t-s)+as}\|\xi\| \le \delta e^{(\overline{a}+\alpha)(t-s)-\alpha t} \le \delta e^{-\alpha t}, \qquad (4.50)$$

and thus $(t,x_\varphi(t)) \in X_\alpha(\delta)$ for every $t \ge s$. This shows that eventually making δ smaller if necessary, we can replace (s,ξ) by $(t,x_\varphi(t))$ in (4.49). This yields

$$\varphi(t,x_\varphi(t)) = -\int_t^\infty V(\tau,t)^{-1}f(F_{\tau-t}(t,x_\varphi(t)),\varphi(F_{\tau-t}(t,x_\varphi(t))))\,d\tau$$
$$= -\int_t^\infty V(\tau,t)^{-1}f(\tau,x_\varphi(\tau),\varphi(\tau,x_\varphi(\tau)))\,d\tau.$$

$$(4.51)$$

Combining (4.48) and (4.51), we conclude that (4.45) holds for every $(s,\xi) \in X_\beta$ and $t \ge s$. This completes the proof of the lemma. $\qquad\square$

4.4.5 Construction of the stable manifolds

We now put together all the information given by the former lemmas to establish the existence of a function $\varphi \in X_\alpha^*$ satisfying (4.24) when $x = x_\varphi$ (with the function x_φ given by Lemma 4.4).

Lemma 4.8. *Given $\delta > 0$ sufficiently small, there exists a unique function $\varphi \in X_\alpha^*$ such that (4.46) holds for every $(s, \xi) \in X_\alpha$.*

Proof. We look for a fixed point of the operator Φ defined for each $\varphi \in X_\alpha^*$ by

$$(\Phi\varphi)(s, \xi) = -\int_s^\infty V(\tau, s)^{-1} f(\tau, x_\varphi(\tau), \varphi(\tau, x_\varphi(\tau))) \, d\tau \qquad (4.52)$$

when $(s, \xi) \in X_\alpha$, where x_φ is the unique continuous function given by Lemma 4.4 such that $x_\varphi(s) = \xi$, and by

$$(\Phi\varphi)(s, \xi) = (\Phi\varphi)(s, \delta e^{-\alpha s} \xi / \|\xi\|)$$

otherwise. We recall that X_α^* is a complete metric space for the distance induced by the norm $\|\cdot\|$ in (4.25) (or more precisely by the norm $\varphi \mapsto \|\varphi|X_\alpha\|$). Note that when $\xi = 0$ we have $x_\varphi(t) = 0$ for every $\varphi \in X_\alpha^*$ and $t \geq s$. Thus, in view of (4.1), $(\Phi\varphi)(t, 0) = 0$.

Given $\varphi \in X_\alpha^*$, $s \geq 0$, and $\xi, \bar{\xi} \in R_s(\delta e^{-\alpha s})$, let now x_φ and $\bar{x}_\varphi$ be the unique continuous functions given by Lemma 4.4 such that respectively $x_\varphi(s) = \xi$ and $\bar{x}_\varphi(s) = \bar{\xi}$. Using (4.29) in Lemma 4.4, (4.35) in Lemma 4.5, and (4.36),

$$\|f(\tau, x_\varphi(\tau), \varphi(\tau, x_\varphi(\tau))) - f(\tau, \bar{x}_\varphi(\tau), \varphi(\tau, \bar{x}_\varphi(\tau)))\|$$
$$\leq c3^{q+1} K_1 R^q e^{(q+1)\bar{a}(\tau-s)+a(q+1)s} \|\xi - \bar{\xi}\| (\|\xi\|^q + \|\bar{\xi}\|^q).$$

Since $\xi, \bar{\xi} \in R_s(\delta e^{-\alpha s})$ we obtain

$$\|f(\tau, x_\varphi(\tau), \varphi(\tau, x_\varphi(\tau))) - f(\tau, \bar{x}_\varphi(\tau), \varphi(\tau, \bar{x}_\varphi(\tau)))\|$$
$$\leq \eta K_1 e^{(q+1)\bar{a}(\tau-s)-bs} \|\xi - \bar{\xi}\|,$$

with η as in (4.37). Using the last inequality in (2.17) we conclude that

$$\|(\Phi\varphi)(s, \xi) - (\Phi\varphi)(s, \bar{\xi})\|$$
$$\leq \int_s^\infty \|V(\tau, s)^{-1}\| \cdot \|f(\tau, x_\varphi(\tau), \varphi(\tau, x_\varphi(\tau))) - f(\tau, \bar{x}_\varphi(\tau), \varphi(\tau, \bar{x}_\varphi(\tau)))\| \, d\tau$$
$$\leq D_2 \eta K_1 \|\xi - \bar{\xi}\| \int_s^\infty e^{-\underline{b}(\tau-s)+b\tau} e^{(q+1)\bar{a}(\tau-s)-bs} \, d\tau$$
$$= D_2 \eta K_1 \|\xi - \bar{\xi}\| \int_s^\infty e^{(T_2+q\bar{a})(\tau-s)} \, d\tau,$$

with $T_2 < 0$ as in (4.44). Choosing $\delta > 0$ sufficiently small we have

$$\sigma = \frac{D_2 \eta K_1}{|T_2 + q\bar{a}|} = \frac{2D_2 c 3^{q+1} R^q K_1 \delta^q}{|T_2 + q\bar{a}|} < 1.$$

In particular,

$$\|(\Phi\varphi)(t,\xi) - (\Phi\varphi)(t,\bar{\xi})\| \le \|\xi - \bar{\xi}\|$$

for every $\xi, \bar{\xi} \in R_s(\delta e^{-\alpha s})$. This shows that $\Phi(\mathfrak{X}_\alpha^*) \subset \mathfrak{X}_\alpha^*$, and hence, the operator $\Phi \colon \mathfrak{X}_\alpha^* \to \mathfrak{X}_\alpha^*$ is well-defined.

We now show that $\Phi \colon \mathfrak{X}_\alpha^* \to \mathfrak{X}_\alpha^*$ is a contraction. Given $\varphi, \psi \in \mathfrak{X}_\alpha^*$, and $(s,\xi) \in X_\alpha$, let x_φ and x_ψ be the unique continuous functions given by Lemma 4.4 such that $x_\varphi(s) = x_\psi(s) = \xi$. By (4.40) and (4.41),

$$\begin{aligned} &\|f(\tau, x_\varphi(\tau), \varphi(\tau, x_\varphi(\tau))) - f(\tau, x_\psi(\tau), \psi(\tau, x_\psi(\tau)))\| \\ &\le c3^q \|(x_\varphi(\tau) - x_\psi(\tau), \varphi(\tau, x_\varphi(\tau)) - \psi(\tau, x_\psi(\tau)))\|(\|x_\varphi(\tau)\|^q + \|x_\psi(\tau)\|^q) \\ &\le c3^q (\|x_\varphi(\tau)\| \cdot \|\varphi - \psi\| + 3\|x_\varphi(\tau) - x_\psi(\tau)\|)(\|x_\varphi(\tau)\|^q + \|x_\psi(\tau)\|^q). \end{aligned}$$

It follows from (4.39) in Lemma 4.6 that

$$\begin{aligned} &\|f(\tau, x_\varphi(\tau), \varphi(\tau, x_\varphi(\tau))) - f(\tau, x_\psi(\tau), \psi(\tau, x_\psi(\tau)))\| \\ &\le \bar{\eta} e^{q\bar{a}(\tau-s) - as - bs}(\|x_\varphi(\tau)\| \cdot \|\varphi - \psi\| + 3\|x_\varphi(\tau) - x_\psi(\tau)\|) \\ &\le \bar{\eta} e^{(q+1)\bar{a}(\tau-s) - bs}(R + 3K_2 e^{-bs})\|\xi\| \cdot \|\varphi - \psi\|, \end{aligned}$$

with $\bar{\eta}$ as in (4.43). Setting $G = R + 3K_2$, we conclude that

$$\begin{aligned} &\|(\Phi\varphi)(s,\xi) - (\Phi\psi)(s,\xi)\| \\ &\le \int_s^\infty \|V(\tau,s)^{-1}\| \cdot \|f(\tau, x_\varphi(\tau), \varphi(\tau, x_\varphi(\tau))) - f(\tau, x_\psi(\tau), \psi(\tau, x_\psi(\tau)))\| \, d\tau \\ &\le D_2 \bar{\eta} G \|\xi\| \cdot \|\varphi - \psi\| \int_s^\infty e^{-\underline{b}(\tau-s) + b\tau} e^{(q+1)\bar{a}(\tau-s) - bs} \, d\tau \\ &= D_2 \bar{\eta} G \|\xi\| \cdot \|\varphi - \psi\| \int_s^\infty e^{(T_2 + q\bar{a})(\tau-s)} \, d\tau \\ &\le \frac{D_2 \bar{\eta} G}{|T_2 + q\bar{a}|} \|\xi\| \cdot \|\varphi - \psi\|. \end{aligned}$$

Provided that $\delta > 0$ is sufficiently small, we have

$$\bar{\theta} = \frac{D_2 \bar{\eta} G}{|T_2 + q\bar{a}|} = \frac{2D_2 c 3^q R^q \delta^q (R + 3K_2)}{|T_2 + q\bar{a}|} < 1.$$

Therefore

$$\|\Phi\varphi_1 - \Phi\varphi_2\| \le \bar{\theta}\|\varphi_1 - \varphi_2\|,$$

and $\Phi \colon \mathfrak{X}_\alpha^* \to \mathfrak{X}_\alpha^*$ is a contraction in the complete metric space $\mathfrak{X}_\alpha^*$. Hence, there exists a unique function $\varphi \in \mathfrak{X}_\alpha^*$ satisfying $\Phi\varphi = \varphi$. In particular, in view of (4.52), the identity (4.46) holds for every $(s,\xi) \in X_\alpha$. This completes the proof of the lemma. $\square$

We can now establish Theorem 4.1.

Proof of Theorem 4.1. As explained in the beginning of Section 4.4.1, in view of the required forward invariance property in (4.14), to show the existence of a (Lipschitz) stable manifold $\mathcal{W}$ is equivalent to find a function φ satisfying (4.23) and (4.24) in some appropriate domain. If follows from Lemma 4.4 that for each fixed $\varphi \in \mathcal{X}_\alpha^*$ there exists a unique function $x = x_\varphi$ satisfying (4.23) and thus it remains to solve (4.24) setting $x = x_\varphi$ or, equivalently, to solve (4.45) in Lemma 4.7. This lemma indicates that this can be reduced to solve the equation in (4.46), that is, to find $\varphi \in \mathcal{X}_\alpha^*$ such that (4.46) holds for every $(s, \xi) \in X_\alpha$. More precisely, it follows from the second property in Lemma 4.7 that if (4.46) holds for every $(s, \xi) \in X_\alpha$, then (4.45) holds for every $(s, \xi) \in X_\beta(\delta/R)$ and $t \geq s$. Finally, Lemma 4.8 shows that there exists a unique function $\varphi \in \mathcal{X}_\alpha^*$ such that (4.46) holds for every $(s, \xi) \in X_\alpha$.

Furthermore, by (4.50), provided that δ is sufficiently small and $(s, \xi) \in X_\beta$ we have $(t, x_\varphi(t)) \in X_\alpha$ for every $t \geq s$. This ensures that we can replace the function φ in (4.23)–(4.24) by the restriction $\varphi|X_\alpha$. In other words, there exists a unique function $\varphi \in \mathcal{X}_\alpha^*$ such that the corresponding set $\mathcal{W}$ in (4.11) obtained from the function $\varphi|X_\alpha$ is forward invariant under the semiflow Ψ_τ for initial conditions with $(s, \xi) \in X_\beta$, and we obtain the invariance property in (4.14).

We now establish the remaining properties in the theorem. The first property is an immediate consequence of the above discussion (or of the first property in Lemma 4.7). To prove the second property, we denote by x_φ and $\bar{x}_\varphi$ the unique continuous functions given by Lemma 4.4 such that respectively $x_\varphi(s) = \xi$ and $\bar{x}_\varphi(s) = \bar{\xi}$. It follows from Lemma 4.5 that

$$
\begin{aligned}
&\|\Psi_\tau(s, \xi, \varphi(s, \xi)) - \Psi_\tau(s, \bar{\xi}, \varphi(s, \bar{\xi}))\| \\
&= \|(t, x_\varphi(t), \varphi(t, x_\varphi(t))) - (t, \bar{x}_\varphi(t), \varphi(t, \bar{x}_\varphi(t)))\| \\
&\leq 2\|x_\varphi(t) - \bar{x}_\varphi(t)\| \leq 2K_1 e^{\bar{a}\tau + as}\|\xi - \bar{\xi}\|
\end{aligned}
\tag{4.53}
$$

for every $\tau = t - s \geq 0$. Again, since $\xi, \bar{\xi} \in R_s(\delta e^{-\beta s})$, in view of (4.50) we can replace the function φ in (4.53) by its restriction to X_α. This completes the proof of the theorem. □

4.5 Existence of Lipschitz unstable manifolds

We now consider the case of unstable manifolds. The theory is entirely analogous and the proofs can be obtained by reversing time in the former notions and arguments. As such, we formulate the corresponding result concerning the existence of unstable manifolds without proof.

We first briefly describe the corresponding setup. Consider a continuous function $A: \mathbb{R}_0^- \to B(X)$, with $\mathbb{R}_0^- = (-\infty, 0]$, such that all solutions of the equation $v' = A(t)v$ are global in the past. This happens, for example, if

$$\limsup_{t \to -\infty} \frac{1}{|t|} \log^+ \|A(t)\| = 0.$$

In an analogous manner to that for positive time, we assume here that there exists a function $P \colon \mathbb{R}_0^- \to B(X)$ such that $P(t)$ is a projection for each $t \geq 0$ and (2.3) holds for $t \leq s < 0$, and that there exist constants $\overline{b} < 0 \leq \underline{a}$, $a, b \geq 0$, and $D_1, D_2 \geq 1$ such that for every $t \leq s \leq 0$,

$$\|T(t,s)^{-1}P(t)\| \leq D_1 e^{-\underline{a}|t-s|+a|t|} \quad \text{and} \quad \|T(t,s)Q(s)\| \leq D_2 e^{\overline{b}|t-s|+b|s|}. \tag{4.54}$$

We also consider a continuous function $f \colon \mathbb{R}_0^- \times X \to X$ such that $f(t,0) = 0$ for every $t \leq 0$, and there exist $c > 0$ and $q > 0$ such that (4.2) holds for every $t \leq 0$ and $u, v \in X$. We consider the semiflow Ψ_τ (now with $\tau \leq 0$) generated by the equation (4.4) or equivalently by the system in (4.6) for $\rho \leq s$ in the corresponding maximal interval of definition.

Again we look for an unstable manifold as the graph of a Lipschitz function. For this we define the new constants

$$\alpha' = b(1 + 1/q) + a/q \quad \text{and} \quad \beta' = b(2 + 1/q) + a/q,$$

and given $\delta > 0$, we consider a space $\mathfrak{X}_{\alpha'}^u$ of continuous functions obtained as in Section 4.2, replacing positive time by negative time: consider the set

$$X_{\alpha'}^u = \{(s, \xi) : s \leq 0 \text{ and } \xi \in R_s(\delta e^{-\alpha'|s|})\}, \tag{4.55}$$

and let $\mathfrak{X}_{\alpha'}^u$ be the space of continuous functions $\psi \colon X_{\alpha'}^u \to X$ such that for each $s \leq 0$, we have

$$\psi(s, R_s(\delta e^{-\alpha'|s|})) \subset F(s), \quad \psi(s, 0) = 0,$$

and

$$\|\psi(s, x) - \psi(s, y)\| \leq \|x - y\| \text{ for every } x, y \in R_s(\delta e^{-\alpha'|s|}).$$

We also consider the set $X_{\beta'}^u$ obtained as in (4.55) with α' replaced by β'.

We can now formulate the result on the existence of Lipschitz unstable manifolds.

Theorem 4.9. *For the equation $v' = A(t)v$ in the Banach space X, if (4.54) holds and the conditions $\overline{b} + \alpha' < 0$ and $\overline{b} + a < \underline{a}$ hold, then there exist $\delta > 0$ and a unique function $\psi \in \mathfrak{X}_{\alpha'}^u$ such that the set*

$$\mathcal{W}^u = \{(s, \psi(s, \xi), \xi) : (s, \xi) \in X_{\alpha'}^u\} \subset \mathbb{R}_0^- \times X \tag{4.56}$$

is invariant under the semiflow Ψ_τ, that is,

$$\text{if } (s, \xi) \in X_{\beta'}^u \text{ then } \Psi_\tau(s, \psi(s, \xi), \xi) \in \mathcal{W}^u \text{ for every } \tau \leq 0.$$

Furthermore:

1. *for every $(s, \xi) \in X_{\beta'}^u$, we have*

$$\psi(s, \xi) = \int_{-\infty}^{s} U(\tau, s)^{-1} f(\Psi_{\tau-s}(s, \psi(s, \xi), \xi)) \, d\tau;$$

2. *there exists $D > 0$ such that for every $s \leq 0$, ξ, $\bar{\xi} \in R_s(\delta e^{-\beta'|s|})$, and $\tau \leq 0$,*

$$\|\Psi_\tau(s, \psi(s, \xi), \xi) - \Psi_\tau(s, \psi(s, \bar{\xi}), \bar{\xi})\| \leq De^{\overline{b}|\tau| + b|s|} \|\xi - \bar{\xi}\|.$$

We call the set $\mathcal{W}^u$ in (4.56) a *(Lipschitz) local unstable manifold* or simply a *(Lipschitz) unstable manifold* of the origin for the equation (4.4).

5

Smooth stable manifolds in $\mathbb{R}^n$

In this chapter we start the study of the regularity of the Lipschitz manifolds constructed in Chapter 4. We only consider stable manifolds. As in Section 4.5, the theory for unstable manifolds is analogous, and the proofs can be readily obtained by reversing the time. We only consider in this chapter the case of finite-dimensional spaces. This is due to the method of proof of the smoothness of the invariant manifolds, which uses in a decisive manner the compactness of the closed unit ball in $\mathbb{R}^n$ (in the proof of Lemma 5.11). The proof is based on the construction of an invariant family of cones, in a similar manner to that in the classical hyperbolic theory, although now using an appropriate family of Lyapunov norms. The family of cones allows us to obtain an invariant distribution which coincides with the tangent bundle of the invariant manifold. This also allows us to discuss the continuity of the distribution, and thus the continuity of the tangent spaces, that corresponds to the smoothness of the invariant manifold. We note that we deal directly with the semiflows instead of first considering time-1 maps as it is sometimes customary in hyperbolic dynamics. The infinite-dimensional case is treated in Chapter 6 with an entirely different approach, although at the expense of requiring more regularity for the vector field. The material in this chapter is taken from [6] (for Sections 5.1–5.4) and [5] (for Sections 5.5–5.6), although now considering the general case when the stable and unstable subspaces may depend on the time t.

5.1 C^1 stable manifolds

We consider the same setup and notations as in Section 4.1, although we shall require slightly more restrictive assumptions on the constants in (2.6).

Set $\vartheta = \max\{a, b\}$. We consider the conditions

$$q\overline{a} + 4\vartheta < \min\{\overline{a} - \overline{b}, (2 - q)\vartheta\} \quad \text{and} \quad \overline{a} + \vartheta < \underline{b}, \tag{5.1}$$

which clearly imply the conditions in (4.13). The second inequality in (5.1) is slightly stronger than the second inequality in (4.13). On the other hand, the first inequality in (5.1) is of different type. It can be written as:

$$q\bar{a} + 4\vartheta < \bar{a} - \bar{b} \quad \text{and} \quad \bar{a} + \vartheta + 2\vartheta/q < 0. \tag{5.2}$$

The former implies the first inequality in (4.13) while the latter requires a certain "spectral gap" (we note that the second condition in (5.2) is not used in the proof of the C^1 regularity of the stable manifolds, but only to know a priori, via Theorem 4.1, the existence of a Lipschitz stable manifold). We note that the first inequality in (5.1) is satisfied for a given ϑ provided that q is sufficiently large (that is, provided that the order of the perturbation f is sufficiently large), while the second is always satisfied when ϑ is sufficiently small.

In this chapter we always take the space X in Sections 4.1 and 4.2 to be $\mathbb{R}^n$. We want to show that the Lipschitz stable manifold $\mathcal{W}$ given by Theorem 4.1 is a smooth manifold of class C^1. For technical reasons we need to slightly reduce the size of the neighborhood $R_s(\delta e^{-\beta s})$, with β as in (4.12). Namely, we fix the new exponent

$$\gamma = \beta + 3\vartheta/q \geq \beta \tag{5.3}$$

and for each $\varrho > 0$, we consider the subset $\mathcal{V} \subset \mathcal{W}$ given by

$$\mathcal{V} = \{(s, \xi, \varphi(s, \xi)) : (s, \xi) \in X_\gamma \text{ with } s > \varrho\} \subset \mathbb{R}^+ \times \mathbb{R}^n, \tag{5.4}$$

with X_γ defined by (4.9). We also replace the conditions A1 and A2 in Section 4.2 by new conditions, now for the finite-dimensional space $X = \mathbb{R}^n$. Let $M_n(\mathbb{R})$ be the set of $n \times n$ matrices with real entries. We assume that:

B1. the function $A \colon \mathbb{R}_0^+ \to M_n(\mathbb{R})$ is of class C^1 and satisfies (2.2);
B2. the function $f \colon \mathbb{R}_0^+ \times \mathbb{R}^n \to \mathbb{R}^n$ is of class C^1 and satisfies (4.1) and (4.2) for some $c > 0$ and $q > 1$.

Note that we replaced the requirement $q > 0$ in condition A1 by the stronger requirement $q > 1$. As observed in Section 4.1, due to the differentiability of f, the condition (4.2) is equivalent to the existence of constants $c > 0$ and $q > 1$ such that (4.3) holds.

The following result establishes the C^1 regularity of $\mathcal{V}$.

Theorem 5.1 ([6]). *Assume that B1–B2 hold. If the equation $v' = A(t)v$ admits a strong nonuniform exponential dichotomy in $\mathbb{R}^+$ and the conditions in (5.1) hold, then for each $\varrho > 0$ there exists $\delta > 0$ such that for the unique function $\varphi \in \mathfrak{X}_\alpha$ given by Theorem 4.1, the set $\mathcal{V}$ in (5.4) is a smooth manifold of class C^1, containing the line $(\varrho, +\infty) \times \{0\}$ and satisfying $T_{(s,0)}\mathcal{V} = \mathbb{R} \times E(s)$ for every $s > \varrho$.*

The proof of Theorem 5.1 is given in Section 5.4. We observe that only the first and the last inequalities in (2.17) are used in the proof of Theorem 4.1. The proof of Theorem 5.1 uses, in addition, the third inequality in (2.17).

We call the set $\mathcal{V}$ in (5.4) a *local stable manifold* or simply a *stable manifold* of the origin for the equation (4.4). We note that in general we are not able to take $\varrho = 0$ in (5.4). The explanation is given at the end of Section 5.4.6, when we can already refer to the appropriate places in the proof. On the other hand, if the functions $A(t)$ and $f(t, x)$ are defined for every $t > -\varepsilon$, for some fixed $\varepsilon > 0$, then we can replace ϱ in (5.4), as well as in the statement of Theorem 5.1, by any number in $(-\varepsilon, 0)$.

A version of Theorem 5.1 in the case of discrete time is established in [9].

5.2 Nonuniformly hyperbolic trajectories

We now explain how Theorem 5.1 can be used to show the existence of smooth stable manifolds for solutions of a given differential equation (possibly nonautonomous) that exhibit nonuniformly hyperbolic behavior.

Consider a C^1 function $F \colon \mathbb{R}_0^+ \times \mathbb{R}^n \to \mathbb{R}^n$. We continue to assume that $A(t)$ in (4.19) satisfies (2.2).

We obtain stable manifolds for nonuniformly hyperbolic solutions of (4.18) (see Section 4.3 for the definition). We use the same notation as in Section 5.1.

Theorem 5.2. *Assume that F is of class C^1 and let $v_0(t)$ be a nonuniformly hyperbolic solution of (4.18) such that:*

1. *the function $t \mapsto A(t)$ is of class C^1;*
2. *there exist $c > 0$ and $q > 1$ such that (4.20) holds for every $t \geq 0$ and $y \in \mathbb{R}^n$.*

If the conditions (5.1) hold, then for each $\varrho > 0$ there exist $\delta > 0$ and a unique function $\varphi \in \mathcal{X}_\gamma$ such that the set

$$\mathcal{V} = \{(s, \xi, \varphi(s, \xi)) + (0, v_0(s)) : (s, \xi) \in X_\gamma \text{ with } s > \varrho\} \qquad (5.5)$$

is a C^1 manifold with the following properties:

1. *$(s, v_0(s)) \in \mathcal{V}$ and $T_{(s, v_0(s))}\mathcal{V} = \mathbb{R} \times E(s)$ for every $s > \varrho$;*
2. *$\mathcal{V}$ is forward invariant under solutions of (4.18), that is, if $s > \varrho$ and*

$$(s, v_s) \in \{(s, \xi, \varphi(s, \xi)) + (0, v_0(s)) : (s, \xi) \in X_{\gamma+a} \text{ with } s > \varrho\},$$

then $(t, v(t)) \in \mathcal{V}$ for every $t \geq s$, where $v(t) = v(t, v_s)$ is the unique solution of (4.18) for $t \geq s$ with $v(s) = v_s$;
3. *there exists $D > 0$ such that for every $s > \varrho$, $(s, v_s), (s, \bar{v}_s) \in \mathcal{V}$, and $t \geq s$ we have*

$$\|v(t, v_s) - v(t, \bar{v}_s)\| \leq D e^{\bar{a}(t-s)+as}\|v_s - \bar{v}_s\|.$$

Proof. The proof follows closely arguments in the proof of Theorem 4.2. We consider again the change of variables $(t, y) = (t, v - v_0(t))$. Letting $y(t) = v(t) - v_0(t)$, where $v(t)$ is a solution of (4.18), we obtain

$$y'(t) = A(t)y(t) + G(t, y(t)),$$

with $G(t, y)$ as in (4.22). This reduces the study of the equation (4.18) to that of (4.4). By hypothesis $A(t)$ satisfies the assumption B1 in Section 5.1. Furthermore, it follows from (4.22) that G is of class C^1 and clearly $G(t, 0) = 0$ for every $t \geq 0$. Proceeding as in the proof of Theorem 4.2 we establish property (4.2). Thus, the function G satisfies the assumption B2 in Section 5.1. We can now apply Theorems 4.1 and 5.1 to obtain the desired statement.

$\square$

We call the set $\mathcal{V}$ in (5.5) a *local stable manifold* or simply a *stable manifold* of the solution $v_0(t)$ (of the equation (4.18)). Note that if F is of class C^2 and there exist $c > 0$ and $q > 1$ such that

$$\left\| \frac{\partial^2 F}{\partial v^2}(t, y + v_0(t)) \right\| \leq c\|y\|^q$$

for every $t \geq 0$ and $y \in \mathbb{R}^n$, then the hypotheses in Theorem 5.2 hold.

5.3 Example of a C^1 flow with stable manifolds

In [77] Pugh gave an explicit example of a C^1 diffeomorphism which is not $C^{1+\varepsilon}$ for any $\varepsilon \in (0, 1)$, for which the statement in the stable manifold theorem fails. Of course that this does not mean that all C^1 diffeomorphisms and flows with nonuniformly hyperbolic trajectories and which lack higher regularity have no stable or unstable invariant manifolds. Indeed, we illustrate with an example that this is not the case.

Proposition 5.3 ([6]). *There exists a C^1 function $F \colon \mathbb{R}_0^+ \times \mathbb{R}^n \to \mathbb{R}^n$ satisfying the hypotheses of Theorem 5.2 which is not $C^{1+\varepsilon}$ for any $\varepsilon \in (0, 1)$.*

Proof. We assume that $F(t, 0) = 0$ for every $t \geq 0$, and we consider the constant solution $v_0 = 0$. We want to exhibit a C^1 function F such that:

1. $H := \frac{\partial F}{\partial v}$ is continuous but not Hölder continuous;
2. the function $t \mapsto H(t, 0)$ is of class C^1;
3. $\|H(t, y) - H(t, 0)\| \leq c\|y\|^q$ for every $t \geq 0$ and $y \in \mathbb{R}^n$, and some constants $c > 0$ and $q > 1$.

For this we consider a continuous function $\rho \colon \mathbb{R}^+ \to [0, 1]$ and a sequence $p_n \in \mathbb{R}^+$ decreasing to zero such that:

1. ρ is of class C^1 outside the points p_n;
2. ρ is Hölder continuous with Hölder exponent at most p_n in some open neighborhood of p_n for each $n \in \mathbb{N}$.

We now define a function H by

$$H(t, y) = H(t, 0) + f(y)\rho(\|y\|)$$

for each $y \neq 0$, where $t \mapsto H(t,0) = A(t)$ and $f \colon \mathbb{R}^n \to \mathbb{R}^n$ are any C^1 functions such that $v' = A(t)v$ has only global solutions and

$$\|f(y)\| \leq c\|y\|^q \text{ for every } y \in \mathbb{R}^n.$$

One can easily verify that $y \mapsto H(t,y)$ is Hölder continuous with exponent at most p_n outside the ball of radius p_n centered at the origin. Thus, H is not Hölder continuous, although it is continuous at $(t,0)$ and thus continuous. Integrating H, while imposing the condition $F(t,0) = 0$ for every $t \geq 0$, we find a function F as desired: namely, F satisfies the hypotheses of Theorem 5.2, but since H is not Hölder continuous, F is not of class $C^{1+\varepsilon}$ for any $\varepsilon \in (0,1)$.
□

In particular, in view of Theorem 5.2, for the vector field F in Proposition 5.3 the nonuniformly hyperbolic solutions of (4.18) possess smooth stable manifolds. A version of Proposition 5.3 in the case of discrete time is established in [9].

5.4 Proof of the C^1 regularity

We establish in this section the C^1 regularity of the Lipschitz manifold $\mathcal{V}$ in (5.4). In view of clarity we separate the proof into several steps.

5.4.1 A priori control of derivatives and auxiliary estimates

Here we establish several estimates that are needed in the proof of Theorem 5.1. We will always assume that B1 and B2 hold. In particular, f is now of class C^1 and the conditions (4.1)–(4.4) hold for some $c > 0$ and $q > 1$. We will also assume that the conditions (5.1) hold. These are standing assumptions that will be used throughout Sections 5.4.1–5.4.6.

We first give a bound for the derivatives of the perturbation. For each $t \geq 0$ we consider the direct sum $\mathbb{R}^n = E(t) \oplus F(t)$, and the components

$$(x,y) = (x(t), y(t)) \in E(t) \times F(t).$$

For simplicity, we will denote by $\partial f / \partial x$ and $\partial f / \partial y$ the partial derivatives with respect to $x(t)$ and $y(t)$ for each given $t \geq 0$, that is, given $v \in E(t)$,

$$\frac{\partial f}{\partial x}(t, x(t), y(t))v = \lim_{h \to 0} \frac{f(t, x(t) + hv, y(t)) - f(t, x(t), y(t))}{h},$$

and given $v \in F(t)$,

$$\frac{\partial f}{\partial y}(t, x(t), y(t))v = \lim_{h \to 0} \frac{f(t, x(t), y(t) + hv) - f(t, x(t), y(t))}{h}.$$

Lemma 5.4. *We have*

$$\max\left\{\left\|\frac{\partial f}{\partial x}\right\|, \left\|\frac{\partial f}{\partial y}\right\|\right\} \le 2c\|(x,y)\|^q.$$

Proof. We first consider the derivative $\partial f/\partial x$. Since f is differentiable, it follows from (4.2) that for every $v \in E(t)$,

$$\left\|\frac{\partial f}{\partial x}v\right\| = \lim_{h \to 0}\frac{\|f(t, x + hv, y) - f(t, x, y)\|}{|h|}$$

$$\le c\lim_{h \to 0}\frac{\|x + hv - x\|(\|(x + hv, y)\|^q + \|(x, y)\|^q)}{|h|} \le 2c\|(x,y)\|^q\|v\|.$$

Therefore, $\|\partial f/\partial x\| \le 2c\|(x,y)\|^q$. Proceeding in a similar manner with the derivative $\partial f/\partial y$ we obtain the desired statement. $\square$

We now obtain norm bounds for the derivatives of the solution with respect to the initial conditions (note that by the hypotheses B1 and B2, the solution of (4.6) is indeed of class C^1 in the initial conditions). Fix $s \ge 0$ and $t \ge 0$. For each $\rho \ge s + t$, we set

$$p_\rho = (\rho, \Psi_t(s, \xi, \varphi(s, \xi))), \quad q_\rho = (\rho, x(p_\rho), y(p_\rho)), \tag{5.6}$$

and

$$S_\rho = \left\|\frac{\partial x}{\partial \xi}|_{p_\rho}\right\| + \left\|\frac{\partial y}{\partial \xi}|_{p_\rho}\right\|, \quad T_\rho = \left\|\frac{\partial x}{\partial \eta}|_{p_\rho}\right\| + \left\|\frac{\partial y}{\partial \eta}|_{p_\rho}\right\|,$$

where x and y are the functions in (4.5). Since Ψ_τ is a semiflow, we have the identities

$$x(p_\rho) = x(\rho, s, \xi, \varphi(s, \xi)) \quad \text{and} \quad y(p_\rho) = y(\rho, s, \xi, \varphi(s, \xi)). \tag{5.7}$$

The following lemma gives several exponential bounds, in particular for S_ρ and T_ρ, which are essential in the proof of Theorem 5.1. As described above, for technical reasons we need to slightly reduce the size of the neighborhood $R_s(\delta e^{-\beta s})$. Namely, we consider the new neighborhood $R_s(\delta e^{-\gamma s}) \subset R_s(\delta e^{-\beta s})$. In view of the first inequality in (5.1), we have

$$q\bar{a} + a < 0 \quad \text{and} \quad q\bar{a} + b < 0. \tag{5.8}$$

Let also

$$d = 2a + 2b + aq + \max\{a, b\},$$

and

$$\theta' = c2^{q+1}D^q\delta^q \quad \text{and} \quad \theta = c2^{q+1}D^q\delta^q(D_1 + D_2), \tag{5.9}$$

with D as in (4.16) and D_1, D_2 as in (2.17).

Lemma 5.5. *Given $\delta > 0$ sufficiently small, for each $(s, \xi) \in X_\gamma$, $t \geq 0$, and $\rho \geq s + t$ we have*

$$\left\| \frac{\partial f}{\partial x}\big|_{q_\rho} \frac{\partial x}{\partial \xi}\big|_{p_\rho} + \frac{\partial f}{\partial y}\big|_{q_\rho} \frac{\partial y}{\partial \xi}\big|_{p_\rho} \right\| \leq \theta' e^{q\bar{a}(\rho - s) - ds} S_\rho, \tag{5.10}$$

$$\left\| \frac{\partial f}{\partial x}\big|_{q_\rho} \frac{\partial x}{\partial \eta}\big|_{p_\rho} + \frac{\partial f}{\partial y}\big|_{q_\rho} \frac{\partial y}{\partial \eta}\big|_{p_\rho} \right\| \leq \theta' e^{q\bar{a}(\rho - s) - ds} T_\rho, \tag{5.11}$$

and

$$S_\rho \leq D_1 e^{(\bar{b} + \theta)(\rho - s - t) + a(s + t)}, \quad T_\rho \leq D_2 e^{(\bar{b} + \theta)(\rho - s - t) + b(s + t)}. \tag{5.12}$$

Proof. Set $x(r) = x(r, s, \xi, \eta)$ and $y(r) = y(r, s, \xi, \eta)$ (see (4.5)). Taking derivatives with respect to ξ and η in (4.6), given $\rho \geq s + t$ we obtain

$$
\begin{aligned}
\frac{\partial x}{\partial \xi}\big|_{p_\rho} &= U(\rho, s + t) + \int_{s+t}^{\rho} U(\rho, r) \left(\frac{\partial f}{\partial x} \frac{\partial x}{\partial \xi}\big|_{p_r} + \frac{\partial f}{\partial y} \frac{\partial y}{\partial \xi}\big|_{p_r} \right) dr, \\
\frac{\partial y}{\partial \xi}\big|_{p_\rho} &= \int_{s+t}^{\rho} V(\rho, r) \left(\frac{\partial f}{\partial x} \frac{\partial x}{\partial \xi}\big|_{p_r} + \frac{\partial f}{\partial y} \frac{\partial y}{\partial \xi}\big|_{p_r} \right) dr, \\
\frac{\partial x}{\partial \eta}\big|_{p_\rho} &= \int_{s+t}^{\rho} U(\rho, r) \left(\frac{\partial f}{\partial x} \frac{\partial x}{\partial \eta}\big|_{p_r} + \frac{\partial f}{\partial y} \frac{\partial y}{\partial \eta}\big|_{p_r} \right) dr, \\
\frac{\partial y}{\partial \eta}\big|_{p_\rho} &= V(\rho, s + t) + \int_{s+t}^{\rho} V(\rho, r) \left(\frac{\partial f}{\partial x} \frac{\partial x}{\partial \eta}\big|_{p_r} + \frac{\partial f}{\partial y} \frac{\partial y}{\partial \eta}\big|_{p_r} \right) dr,
\end{aligned}
\tag{5.13}
$$

with the partial derivatives of f computed at q_r (see (5.6)). Recall that $y(p_r) = \varphi(r, x(p_r))$ for every $r \geq s + t$ (see (4.17)), and thus $\|(x, y)(p_r)\| \leq 2\|x(p_r)\|$. By Lemma 5.4 we conclude that

$$\left\| \frac{\partial f}{\partial x}\big|_{q_r} \frac{\partial x}{\partial \xi}\big|_{p_r} + \frac{\partial f}{\partial y}\big|_{q_r} \frac{\partial y}{\partial \xi}\big|_{p_r} \right\| \leq c 2^{q+1} \|x(p_r)\|^q S_r.$$

Since $\xi \in R_s(\delta e^{-\gamma s}) \subset R_s(\delta e^{-\beta s})$, it follows from (5.7) and Theorem 4.1 (making $\bar{\xi} = 0$ in (4.16)) that

$$\|x(p_r)\|^q = \|x(r, s, \xi, \varphi(s, \xi))\|^q \leq D^q \delta^q e^{q\bar{a}(r - s) - ds}.$$

Thus,

$$\left\| \frac{\partial f}{\partial x}\big|_{q_r} \frac{\partial x}{\partial \xi}\big|_{p_r} + \frac{\partial f}{\partial y}\big|_{q_r} \frac{\partial y}{\partial \xi}\big|_{p_r} \right\| \leq \theta' e^{q\bar{a}(r - s) - ds} S_r,$$

which is the inequality in (5.10). We can obtain (5.11) in a similar manner. If follows from the first identity in (5.13), (2.6), and (5.10) that

$$\left\| \frac{\partial x}{\partial \xi}\big|_{p_\rho} \right\| \leq D_1 e^{\bar{a}(\rho - s - t) + a(s + t)} + \theta' \int_{s+t}^{\rho} D_1 e^{\bar{a}(\rho - r) + ar} e^{q\bar{a}(r - s) - ds} S_r \, dr.$$

In a similar manner, using the second identity in (5.13) and again (5.10), we obtain

$$\left\|\frac{\partial y}{\partial \xi}|_{p_\rho}\right\| \leq \theta' \int_{s+t}^{\rho} D_2 e^{\bar{b}(\rho-r)+br} e^{q\bar{a}(r-s)-ds} S_r\, dr.$$

Therefore,

$$S_\rho \leq D_1 e^{\bar{a}(\rho-s-t)+a(s+t)} + \theta' \int_{s+t}^{\rho} D_1 e^{\bar{a}(\rho-r)+ar} e^{q\bar{a}(r-s)-ds} S_r\, dr$$

$$+ \theta' \int_{s+t}^{\rho} D_2 e^{\bar{b}(\rho-r)+br} e^{q\bar{a}(r-s)-ds} S_r\, dr.$$

Using (5.8) and the fact that $\bar{a} < \bar{b}$ (see (2.6)), we conclude that

$$S_\rho \leq D_1 e^{\bar{a}(\rho-s-t)+a(s+t)}$$

$$+ \theta' \int_{s+t}^{\rho} D_1 e^{\bar{a}(\rho-r)} S_r\, dr + \theta' \int_{s+t}^{\rho} D_2 e^{\bar{b}(\rho-r)} S_r\, dr \qquad (5.14)$$

$$\leq D_1 e^{\bar{b}(\rho-s-t)+a(s+t)} + \theta e^{\bar{b}(\rho-s-t)} \int_{s+t}^{\rho} e^{-\bar{b}(r-s-t)} S_r\, dr.$$

We now write $\tilde{S}_r = e^{-\bar{b}(r-s-t)} S_r$ for each $r \geq s+t$. It follows from (5.14) that

$$\tilde{S}_\rho \leq D_1 e^{a(s+t)} + \theta \int_{s+t}^{\rho} \tilde{S}_r\, dr$$

for every $\rho \geq s+t$. Therefore, using Gronwall's lemma we conclude that

$$\tilde{S}_\rho \leq D_1 e^{a(s+t)} e^{\theta(\rho-s-t)}$$

for every $\rho \geq s+t$. This establishes the third inequality in the lemma.

In a similar manner, using the third and fourth identities in (5.13), together with (2.6) and (5.11) we obtain

$$T_\rho \leq D_2 e^{\bar{b}(\rho-s-t)+b(s+t)} + \theta' \int_{s+t}^{\rho} D_1 e^{\bar{a}(\rho-r)} e^{ar} e^{q\bar{a}(r-s)-ds} T_r\, dr$$

$$+ \theta' \int_{s+t}^{\rho} D_2 e^{\bar{b}(\rho-r)+br} e^{q\bar{a}(r-s)-ds} T_r\, dr.$$

Using (5.8) and the fact that $\bar{a} < \bar{b}$, this yields

$$T_\rho \leq D_2 e^{\bar{b}(\rho-s-t)+b(s+t)} + \theta e^{\bar{b}(\rho-s-t)} \int_{s+t}^{\rho} e^{-\bar{b}(r-s-t)} T_r\, dr.$$

Writing $\tilde{T}_r = e^{-\bar{b}(r-s-t)} T_r$ for each $r \geq s+t$, we conclude that

$$\tilde{T}_\rho \leq D_2 e^{b(s+t)} + \theta \int_{s+t}^{\rho} \tilde{T}_r\, dr$$

for every $\rho \geq s+t$. Thus, it follows from Gronwall's lemma that

$$\tilde{T}_\rho \leq D_2 e^{b(s+t)} e^{\theta(\rho-s-t)}$$

for every $\rho \geq s+t$. This establishes the last inequality in (5.12). $\square$

We note that by the dependence of θ' and θ on δ (see (5.9)), these two constants can be made arbitrarily small by making δ sufficiently small. In particular, the exponent $\bar{b} + \theta$ in (5.12) can be made arbitrarily close to $\bar{b}$.

The following statement considers a function F which occurs in the construction of the invariant families of cones in Section 5.4.3. The value $F(x)$ is essentially the size of the cone at time x.

Lemma 5.6. *Consider the function $F \colon \mathbb{R}_0^+ \to \mathbb{R}$ defined by*

$$F(x) = \frac{e^{a_1 x} + \nu(1 - e^{a_2 x})}{1 - \nu(1 - e^{a_2 x})},$$

where $a_1, a_2 < 0$, $a_1 \geq a_2$, and $\nu \in (0, 1)$. If

$$\nu < \frac{a_1}{2a_2}, \tag{5.15}$$

then $F(x) < 1$ for every $x > 0$.

Proof. Assume first that $a_1 = a_2 = a$. Then

$$F'(x) = \frac{a(1 - 2\nu)e^{ax}}{(1 - \nu(1 - e^{ax}))^2}$$

and it follows from (5.15) that $F'(x) < 0$. Therefore, $F(x) < F(0) = 1$ for every $x > 0$. Assume now that $a_1 > a_2$. In this case,

$$F'(x) = \frac{a_1 e^{a_1 x} - \nu a_1 e^{a_1 x} - \nu a_2 e^{a_2 x} - \nu a_2 e^{(a_1+a_2)x} + \nu a_1 e^{(a_1+a_2)x}}{(1 - \nu(1 - e^{a_2 x}))^2}.$$

Since $a_1 > a_2$ and $a_2 < 0$ we have $-\nu a_2 e^{a_2 x} < -\nu a_2 e^{a_1 x}$ and thus,

$$F'(x) \leq \frac{e^{a_1 x}(a_1(1 - \nu) - \nu a_2 + \nu(a_1 - a_2)e^{a_2 x})}{(1 - \nu(1 - e^{a_2 x}))^2}.$$

Furthermore, again since $a_1 > a_2$ and $x \geq 0$, we have $\nu(a_1 - a_2)e^{a_2 x} \leq \nu(a_1 - a_2)$ and hence,

$$F'(x) \leq \frac{e^{a_1 x}(a_1(1 - \nu) - \nu a_2 + \nu(a_1 - a_2))}{(1 - \nu(1 - e^{a_2 x}))^2} \leq \frac{e^{a_1 x}(a_1 - 2\nu a_2)}{(1 - \nu(1 - e^{a_2 x}))^2}.$$

It follows from (5.15) that $F'(x) < 0$. Therefore, $F(x) < F(0) = 1$ for every $x > 0$. This completes the proof of the lemma. $\square$

5.4.2 Lyapunov norms

Due to the nonuniformity of the norm bounds for the operators $U(t, s)$ and $V(t, s)$ in the notion of nonuniform exponential dichotomy (see (2.17)), we introduce here a new family of Lyapunov norms. However, since we are dealing in general with semiflows instead of flows, it is impossible to introduce the Lyapunov norms which are standard in the case of flows (see [3]). On the other hand, to study the regularity of the invariant manifolds in terms of an invariant family of cones we still must show the invariance of the whole family, which is somewhat delicate for *small* time, that is, before sufficient time has passed so that the contraction given by $\bar{a}$ in (2.17) overcomes the nonuniformity (which depends on the initial time s). This prevents us from using time-1 maps, at least without some appropriate preliminary preparation. We prefer to deal from the beginning with the original semiflow and we overcome the above difficulties by introducing appropriate families of Lyapunov norms, although with a new procedure developed here for the case of semiflows.

We fix $\varrho > 0$ and $s \geq \varrho$, and given $r \geq s$ and $(v, w) \in E(r) \times F(r)$ we define the new norms

$$
\begin{aligned}
\|v\|_r' &= \int_r^{+\infty} \|U(\sigma, r)v\| e^{a'(\sigma - r)} \, d\sigma, \\
\|w\|_r' &= \int_{s-\varrho}^r \|V(r, \sigma)^{-1} w\| e^{-b'(\sigma - r)} \, d\sigma,
\end{aligned}
\tag{5.16}
$$

where

$$
a' = -\bar{a} - \varsigma > 0, \quad b' = \underline{b} + \varsigma > 0
\tag{5.17}
$$

for some $\varsigma > 0$ such that

$$
\varsigma \leq a \quad \text{and} \quad \varsigma \neq \bar{b} - \underline{b} - b.
\tag{5.18}
$$

Clearly, the integrals in (5.16) are finite. We also set

$$
\|(v, w)\|_r' = \|v\|_r' + \|w\|_r' \text{ for each } (v, w) \in E(r) \times F(r).
$$

This choice of norms is certainly not unique; in particular one can easily change them so that $\|(v, w)\|_r'$ is obtained from an inner product for which $E(r)$ and $F(r)$ are orthogonal (recall that we are now in $\mathbb{R}^n$). Nevertheless, the resulting stable distribution, which is the essential element in the present proof of the C^1 regularity of the manifold $\mathcal{V}$, is independent of the choice of norms.

We now consider the relation between the norms $\|\cdot\|$ and $\|\cdot\|'$.

Lemma 5.7. *For every $s \geq \varrho$, $r \geq s$, and $(v, w) \in E(r) \times F(r)$ we have*

$$
C_1 e^{-ar} \|v\| \leq \|v\|_r' \leq \frac{D_1}{\varsigma} e^{ar} \|v\|,
\tag{5.19}
$$

$$
C_2 e^{-br} \|w\| \leq \|w\|_r' \leq \frac{D_2}{\varsigma} e^{br} (e^{\varsigma(r+\varrho-s)} - 1) \|w\|,
\tag{5.20}
$$

where

$$
C_1 = \frac{1 - e^{\underline{a} - \bar{a} - \varsigma - a}}{D_1(\bar{a} - \underline{a} + \varsigma + a)} \quad \text{and} \quad C_2 = \frac{e^{(\underline{b} - \bar{b} + \varsigma + b)\varrho} - 1}{D_2(\underline{b} - \bar{b} + \varsigma + b)}.
\tag{5.21}
$$

Proof. By the definition of the norm $\|v\|'_r$ in (5.16)–(5.17) and (2.6), we have

$$
\begin{aligned}
\|v\|'_r &= \int_r^{+\infty} \|U(\sigma,r)v\| e^{a'(\sigma-r)}\,d\sigma \\
&\leq \int_r^{+\infty} D_1 e^{\overline{a}(\sigma-r)+ar} \|v\| e^{(-\overline{a}-\varsigma)(\sigma-r)}\,d\sigma \\
&= D_1 e^{ar}\|v\| \int_r^{+\infty} e^{-\varsigma(\sigma-r)}\,d\sigma = \frac{D_1}{\varsigma} e^{ar}\|v\|.
\end{aligned}
$$

For the other inequality in (5.19) we write

$$
\begin{aligned}
\|v\|'_r &\geq \int_r^{r+1} \|U(\sigma,r)v\| e^{a'(\sigma-r)}\,d\sigma \geq \int_r^{r+1} \|U(\sigma,r)^{-1}\|^{-1}\|v\| e^{a'(\sigma-r)}\,d\sigma \\
&\geq \int_r^{r+1} D_1^{-1} e^{\underline{a}(\sigma-r)-a\sigma} \|v\| e^{(-\underline{a}-\varsigma)(\sigma-r)}\,d\sigma \\
&= D_1^{-1}\|v\| e^{-(\underline{a}-\overline{a}-\varsigma)r} \int_r^{r+1} e^{(\underline{a}-\overline{a}-\varsigma-a)\sigma}\,d\sigma = C_1 e^{-ar}\|v\|.
\end{aligned}
$$

In a similar manner, using the definition of the norm $\|w\|'_r$ in (5.16)–(5.17) and (2.6), we have

$$
\begin{aligned}
\|w\|'_r &= \int_{s-\varrho}^r \|V(r,\sigma)^{-1}w\| e^{-b'(\sigma-r)}\,d\sigma \\
&\leq \int_{s-\varrho}^r D_2 e^{-\underline{b}(r-\sigma)+br} \|w\| e^{(-\underline{b}-\varsigma)(\sigma-r)}\,d\sigma \\
&= D_2 e^{br}\|w\| \int_{s-\varrho}^r e^{-\varsigma(\sigma-r)}\,d\sigma \leq \frac{D_2}{\varsigma} e^{br}\|w\| (e^{\varsigma(r+\varrho-s)}-1).
\end{aligned}
$$

The remaining inequality follows from

$$
\begin{aligned}
\|w\|'_r &\geq \int_{s-\varrho}^r \|V(r,\sigma)\|^{-1}\|w\| e^{-b'(\sigma-r)}\,d\sigma \\
&\geq \int_{r-\varrho}^r D_2^{-1} e^{-\overline{b}(r-\sigma)-b\sigma} \|w\| e^{(-\underline{b}-\varsigma)(\sigma-r)}\,d\sigma \\
&= D_2^{-1}\|w\| e^{(\underline{b}-\overline{b}+\varsigma)r} \int_{r-\varrho}^r e^{-(\underline{b}-\overline{b}+\varsigma+b)\sigma}\,d\sigma \\
&= \frac{\|w\|}{D_2(\underline{b}-\overline{b}+\varsigma+b)} e^{-br}(e^{(\underline{b}-\overline{b}+\varsigma+b)\varrho}-1) = C_2 e^{-br}\|w\|.
\end{aligned}
$$

This completes the proof of the lemma.   □

Note that in view of (5.18) the constant C_2 is well-defined. Although the norms $\|\cdot\|$ and $\|\cdot\|'$ are equivalent (for each fixed $\varrho > 0$, $\varsigma > 0$, and $s \geq \varrho$), by Lemma 5.7 their ratio may deteriorate with exponential speed along each orbit (see (5.19) and (5.20)). Nevertheless, this deterioration, essentially given by the exponents a and b, that is, by the nonuniformity in the exponential dichotomy, is small when compared to the values of the Lyapunov exponents (in view of (5.1)).

5.4.3 Existence of an invariant family of cones

The next step in the proof of Theorem 5.1 is to establish the existence of an invariant family of cones along each orbit of the semiflow Ψ_τ. The construction of the cones uses in a decisive manner the Lyapunov norms introduced in Section 5.4.2. The invariant family of cones is the main element towards the construction of an invariant distribution that later will be shown to coincide with the tangent bundle of $\mathcal{V}$. This procedure will also allow us to discuss the continuity of the distribution, and thus of the tangent spaces, in terms of the cones.

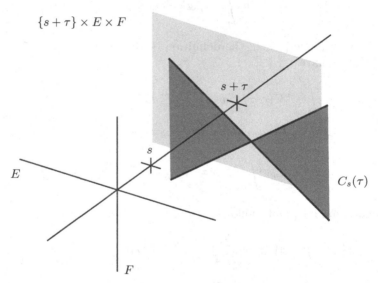

Fig. 5.1. The cone $C_s(\tau)$ at the point $\Psi_\tau(s,\xi,\eta)$. Here the subspaces $E = E(t)$ and $F = F(t)$ are assumed to be independent of t.

For each $s \geq 0$ and $\tau \geq 0$, we consider the cone

$$C_s(\tau) = \{(v,w) \in E(s+\tau) \times F(s+\tau) \colon \|w\|'_{s+\tau} < \|v\|'_{s+\tau}\} \cup \{(0,0)\}.$$

We emphasize that $C_s(\tau)$ is defined in terms of the new norms $\|\cdot\|'$ given by (5.16) and not in terms of the original norm in $\mathbb{R}^n$. Given $(s,\xi,\eta) \in \mathbb{R}_0^+ \times E(s+\tau) \times F(s+\tau)$, we may think of the cone $C_s(\tau)$ as a subset of the plane

$$\{s + \tau\} \times E(s + \tau) \times F(s + \tau)$$

or, alternatively, of the tangent space

$$T_{\Psi_\tau(s,\xi,\eta)}(\{s + \tau\} \times E(s + \tau) \times F(s + \tau)).$$

This is a cone around the space $E(s + \tau)$ at time $s + \tau$. See Figure 5.1.

Note that by the first inequality in (5.1), and (5.9), we can choose $\delta > 0$ sufficiently small so that

$$c_1 = \overline{b} + \theta + (q - 1)\overline{a} + a < 0 \quad \text{and} \quad c_2 = \overline{b} + \theta + q\overline{a} + b < 0. \qquad (5.22)$$

The following lemma shows that the above family of cones is indeed invariant under the differential of Ψ_τ. More precisely, we consider the partial derivatives of the second and third components of Ψ_τ (see (4.7)) with respect to (ξ, η) at the point $(s, \xi, \eta) \in \mathbb{R}^+ \times E(s) \times F(s)$, and we denote it by $\partial_{(s,\xi,\eta)}\Psi_\tau$.

Lemma 5.8. *Given $\delta > 0$ sufficiently small, for each $(s, \xi) \in X_\gamma$ with $s \geq \varrho$, and $\tau > t \geq 0$ we have*

$$(\partial_{\Psi_\tau(s,\xi,\varphi(s,\xi))}\Psi_{t-\tau})\overline{C_s(\tau)} \subset C_s(t). \qquad (5.23)$$

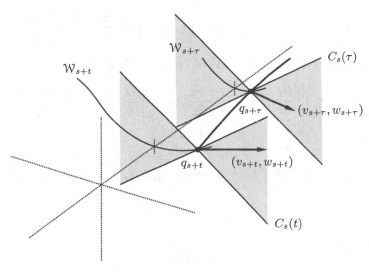

Fig. 5.2. Preimages of vectors inside the cones along a given orbit at the times $s+t$ and $s + \tau$.

Proof. Given $(v_{s+\tau}, w_{s+\tau}) \in \overline{C_s(s + \tau)}$ and $r \in [s, s + \tau]$, we define the vector

$$(v_r, w_r) = (\partial_{\Psi_\tau(s,\xi,\varphi(s,\xi))} \Psi_{r-(s+\tau)})(v_{s+\tau}, w_{s+\tau}) \in E(r) \times F(r). \qquad (5.24)$$

See Figure 5.2 for an illustration. Let now $\tau > t \geq 0$. By (5.24), we have

$$(v_r, w_r) = (\partial_{\Psi_t(s,\xi,\varphi(s,\xi))} \Psi_{r-(s+t)})(v_{s+t}, w_{s+t})$$

for each $r \in [s+t, s+\tau]$. In a somewhat more explicit form, we can write

$$\begin{pmatrix} v_r \\ w_r \end{pmatrix} = \begin{pmatrix} \frac{\partial x}{\partial \xi} & \frac{\partial x}{\partial \eta} \\ \frac{\partial y}{\partial \xi} & \frac{\partial y}{\partial \eta} \end{pmatrix} \begin{pmatrix} v_{s+t} \\ w_{s+t} \end{pmatrix},$$

with the partial derivatives of x and y in (4.5) computed at p_r (see (5.6)). In particular,

$$v_r = \frac{\partial x}{\partial \xi}\Big|_{p_r} v_{s+t} + \frac{\partial x}{\partial \eta}\Big|_{p_r} w_{s+t}, \quad w_r = \frac{\partial y}{\partial \xi}\Big|_{p_r} v_{s+t} + \frac{\partial y}{\partial \eta}\Big|_{p_r} w_{s+t}. \qquad (5.25)$$

We introduce the notation

$$G(r) = \frac{\partial f}{\partial x} v_r + \frac{\partial f}{\partial y} w_r,$$

with the partial derivatives of f computed at q_r (see (5.6)). Then

$$G(r) = \left(\frac{\partial f}{\partial x}\frac{\partial x}{\partial \xi}\Big|_{p_r} + \frac{\partial f}{\partial y}\frac{\partial y}{\partial \xi}\Big|_{p_r} \right) v_{s+t} + \left(\frac{\partial f}{\partial x}\frac{\partial x}{\partial \eta}\Big|_{p_r} + \frac{\partial f}{\partial y}\frac{\partial y}{\partial \eta}\Big|_{p_r} \right) w_{s+t},$$

and in view of (5.10) and (5.11) in Lemma 5.5, we have

$$\begin{aligned}
\|G(r)\| &\leq \left(\left\| \frac{\partial f}{\partial x}\frac{\partial x}{\partial \xi}\Big|_{p_r} + \frac{\partial f}{\partial y}\frac{\partial y}{\partial \xi}\Big|_{p_r} \right\| + \left\| \frac{\partial f}{\partial x}\frac{\partial x}{\partial \eta}\Big|_{p_r} + \frac{\partial f}{\partial y}\frac{\partial y}{\partial \eta}\Big|_{p_r} \right\| \right) \\
&\quad \times (\|v_{s+t}\| + \|w_{s+t}\|) \\
&\leq \theta' e^{q\overline{a}(r-s)-ds}(S_r + T_r)(\|v_{s+t}\| + \|w_{s+t}\|)\| \\
&\leq \theta e^{(\overline{b}+\theta+q\overline{a})(r-s-t)} e^{q\overline{a}t} m(s) \max\{e^{at}, e^{bt}\}(\|v_{s+t}\| + \|w_{s+t}\|),
\end{aligned}$$

where

$$m(s) = e^{-aqs - \max\{a,b\}s} \max\{e^{-(b+2a)s}, e^{-(a+2b)s}\}.$$

Note that

$$e^{as+bs+\max\{a,b\}s} m(s) \leq 1. \qquad (5.26)$$

By the first inequality in (5.1) we obtain

$$\|G(r)\| \leq \theta e^{(\overline{b}+\theta+q\overline{a})(r-s-t)} e^{-(a+b)t-\max\{a,b\}t} m(s)(\|v_{s+t}\| + \|w_{s+t}\|). \qquad (5.27)$$

It follows from (5.13) and (5.25) with $\rho = s + \tau$ that

$$v_{s+\tau} = U(s+\tau, s+t)v_{s+t} + \int_{s+t}^{s+\tau} U(s+\tau, r)G(r)\, dr,$$

$$w_{s+t} = V(s+\tau, s+t)^{-1}w_{s+\tau} - \int_{s+t}^{s+\tau} V(r, s+t)^{-1}G(r)\, dr.$$

Therefore, by (2.6) and since $a' + \bar{a} = -\varsigma$ (see (5.17)), we have

$$\|v_{s+\tau}\|'_{s+\tau} = \int_{s+\tau}^{+\infty} \|U(\sigma, s+\tau)v_{s+\tau}\|e^{a'(\sigma-s-\tau)}\, d\sigma$$

$$= \int_{s+\tau}^{+\infty} \left\| U(\sigma, s+t)v_{s+t} + \int_{s+t}^{s+\tau} U(\sigma, r)G(r)\, dr \right\| e^{a'(\sigma-s-\tau)}\, d\sigma$$

$$\leq \int_{s+t}^{+\infty} \|U(\sigma, s+t)v_{s+t}\|e^{a'(\sigma-s-\tau)}\, d\sigma$$

$$+ e^{-a'(s+\tau)} \int_{s+\tau}^{+\infty} \left(\int_{s+t}^{s+\tau} \|U(\sigma, r)\| \cdot \|G(r)\|\, dr \right) e^{a'\sigma}\, d\sigma$$

$$\leq e^{-a'(\tau-t)}\|v_{s+t}\|'_{s+t}$$

$$+ e^{-a'(s+\tau)}D_1 \int_{s+t}^{s+\tau} \left(\|G(r)\|e^{-\bar{a}r+ar} \int_{s+\tau}^{+\infty} e^{(a'+\bar{a})\sigma}\, d\sigma \right) dr$$

$$= e^{-a'(\tau-t)}\|v_{s+t}\|'_{s+t} + \frac{D_1}{\varsigma}e^{\bar{a}(s+\tau)} \int_{s+t}^{s+\tau} \|G(r)\|e^{-\bar{a}r+ar}\, dr.$$

Set now $C_3 = \max\{C_1^{-1}, C_2^{-1}\}$. By (5.27) and Lemma 5.7 (see (5.19)) we obtain

$$\|v_{s+\tau}\|'_{s+\tau} \leq e^{(\bar{a}+\varsigma)(\tau-t)}\|v_{s+t}\|'_{s+t} + \frac{\theta D_1}{\varsigma}C_3 e^{\max\{a,b\}s}\|(v_{s+t}, w_{s+t})\|'_{s+t}$$

$$\times e^{-(a+b)t}m(s) \int_{s+t}^{s+\tau} e^{\bar{a}(s+\tau-r)}e^{(\bar{b}+\theta+q\bar{a})(r-s-t)}e^{ar}\, dr$$

$$\leq e^{(\bar{a}+\varsigma)(\tau-t)}\|v_{s+t}\|'_{s+t} + \frac{\theta D_1}{\varsigma}C_3 e^{\max\{a,b\}s}\|(v_{s+t}, w_{s+t})\|'_{s+t}$$

$$\times e^{as}e^{-bt}m(s) \int_{s+t}^{s+\tau} e^{\bar{a}(s+\tau-r)}e^{(\bar{b}+\theta+q\bar{a}+a)(r-s-t)}\, dr$$

$$\leq e^{(\bar{a}+\varsigma)(\tau-t)}\|v_{s+t}\|'_{s+t} + \frac{\theta D_1}{\varsigma}C_3 e^{\max\{a,b\}s}\|(v_{s+t}, w_{s+t})\|'_{s+t}$$

$$\times e^{as}e^{-bt}m(s)e^{\bar{a}(\tau-t)} \int_{s+t}^{s+\tau} e^{[\bar{b}+\theta+(q-1)\bar{a}+a](r-s-t)}\, dr.$$

Using the constant $c_1 < 0$ in (5.22), we conclude from (5.26) that

$$\|v_{s+\tau}\|'_{s+\tau} \leq e^{(\bar{a}+\varsigma)(\tau-t)}\|v_{s+t}\|'_{s+t}$$

$$+ \frac{\theta D_1}{\varsigma|c_1|}C_3 e^{-bt}(1 - e^{c_1(\tau-t)})e^{\bar{a}(\tau-t)}\|(v_{s+t}, w_{s+t})\|'_{s+t}. \tag{5.28}$$

In a similar manner, using (2.6) we obtain

$$
\|w_{s+t}\|'_{s+t} = \int_{s-\varrho}^{s+t} \|V(s+t,\sigma)^{-1} w_{s+t}\| e^{-b'(\sigma-s-t)}\, d\sigma
$$

$$
= \int_{s-\varrho}^{s+t} \left\| V(s+\tau,\sigma)^{-1} w_{s+\tau} - \int_{s+t}^{s+\tau} V(r,\sigma)^{-1} G(r)\, dr \right\| e^{-b'(\sigma-s-t)}\, d\sigma
$$

$$
\le e^{-b'(\tau-t)} \int_{s-\varrho}^{s+\tau} \|V(s+\tau,\sigma)^{-1} w_{s+\tau}\| e^{-b'(\sigma-s-\tau)}\, d\sigma
$$

$$
+ e^{b'(s+t)} \int_{s-\varrho}^{s+t} \left(\int_{s+t}^{s+\tau} \|V(r,\sigma)^{-1}\| \cdot \|G(r)\|\, dr \right) e^{-b'\sigma}\, d\sigma
$$

$$
\le e^{-b'(\tau-t)} \|w_{s+\tau}\|'_{s+\tau}
$$

$$
+ e^{b'(s+t)} D_2 \int_{s+t}^{s+\tau} \left(\|G(r)\| e^{-\underline{b}r+br} \int_{s-\varrho}^{s+t} e^{(-b'+\underline{b})\sigma}\, d\sigma \right) dr
$$

$$
\le e^{-b'(\tau-t)} \|w_{s+\tau}\|'_{s+\tau} + e^{b'(s+t)} e^{-\varsigma s} e^{\varsigma\varrho} \frac{D_2}{\varsigma} \int_{s+t}^{s+\tau} \|G(r)\| e^{-\underline{b}r+br}\, dr,
$$

where in the last inequality we have used the identity $b' = \underline{b} + \varsigma$ (see (5.17)). It follows from (5.27) and Lemma 5.7 that

$$
\|w_{s+t}\|'_{s+t} \le e^{-(\underline{b}+\varsigma)(\tau-t)} \|w_{s+\tau}\|'_{s+\tau}
$$

$$
+ e^{(\underline{b}+\varsigma)(s+t)} e^{-\varsigma s} e^{\varsigma\varrho} \frac{\theta D_2}{\varsigma} C_3 e^{\max\{a,b\}s} \|(v_{s+t}, w_{s+t})\|'_{s+t}
$$

$$
\times e^{-(a+b)t} m(s) \int_{s+t}^{s+\tau} e^{-\underline{b}r+br} e^{(\overline{b}+\theta+q\overline{a})(r-s-t)}\, dr
$$

$$
\le e^{-(\underline{b}+\varsigma)(\tau-t)} \|w_{s+\tau}\|'_{s+\tau}
$$

$$
+ e^{(\underline{b}+\varsigma)(s+t)} e^{-\varsigma s} e^{\varsigma\varrho} \frac{\theta D_2}{\varsigma} C_3 \|(v_{s+t}, w_{s+t})\|'_{s+t}
$$

$$
\times e^{-\underline{b}(s+t)} e^{bs} e^{-at} e^{\max\{a,b\}s} m(s) \int_{s+t}^{s+\tau} e^{(\overline{b}+\theta+q\overline{a}+b)(r-s-t)}\, dr
$$

$$
\le e^{-(\underline{b}+\varsigma)(\tau-t)} \|w_{s+\tau}\|'_{s+\tau} + e^{(\varsigma-a)t} e^{\varsigma\varrho} \frac{\theta D_2}{\varsigma} C_3 \|(v_{s+t}, w_{s+t})\|'_{s+t}
$$

$$
\times e^{bs+\max\{a,b\}s} m(s) \int_{s+t}^{s+\tau} e^{(\overline{b}+\theta+q\overline{a}+b)(r-s-t)}\, dr.
$$

Using the constant $c_2 < 0$ in (5.22), we conclude from (5.26) that

$$
\|w_{s+t}\|'_{s+t} \le e^{-(\underline{b}+\varsigma)(\tau-t)} \|w_{s+\tau}\|'_{s+\tau}
$$
$$
+ \frac{\theta D_2}{\varsigma|c_2|} C_3 e^{\varsigma\varrho} e^{(\varsigma-a)t} (1 - e^{c_2(\tau-t)}) \|(v_{s+t}, w_{s+t})\|'_{s+t}. \tag{5.29}
$$

Since $(v_{s+\tau}, w_{s+\tau}) \in \overline{C_s(\tau)}$ we have $\|w_{s+\tau}\|'_{s+\tau} \le \|v_{s+\tau}\|'_{s+\tau}$. It follows from (5.28) and (5.29) that

$$\|w_{s+t}\|'_{s+t} \le e^{-(\underline{b}+\varsigma)(\tau-t)}\left[e^{(\overline{a}+\varsigma)(\tau-t)}\|v_{s+t}\|'_{s+t}\right.$$

$$+\frac{\theta D_1}{\varsigma|c_1|}C_3 e^{-bt}(1-e^{c_1(\tau-t)})e^{\overline{a}(\tau-t)}\|(v_{s+t},w_{s+t})\|'_{s+t}\Big]$$

$$+\frac{\theta D_2}{\varsigma|c_2|}C_3 e^{\varsigma\varrho}e^{(\varsigma-a)t}(1-e^{c_2(\tau-t)})\|(v_{s+t},w_{s+t})\|'_{s+t}$$

$$\le e^{(\overline{a}-\underline{b})(\tau-t)}\|v_{s+t}\|'_{s+t}$$

$$+\frac{\theta D_1}{\varsigma|c_1|}C_3(1-e^{c_1(\tau-t)})\|(v_{s+t},w_{s+t})\|'_{s+t}$$

$$+\frac{\theta D_2}{\varsigma|c_2|}C_3 e^{\varsigma\varrho}e^{(\varsigma-a)t}(1-e^{c_2(\tau-t)})\|(v_{s+t},w_{s+t})\|'_{s+t}.$$

Set now

$$\nu = 2C_3\frac{\theta}{\varsigma}\left(\frac{D_1}{|c_1|}+\frac{D_2}{|c_2|}e^{\varsigma\varrho}\right). \tag{5.30}$$

In view of (5.17)–(5.18) we have $\varsigma - a \le 0$. Thus, setting $c = \min\{c_1, c_2\} < 0$ we obtain

$$\|w_{s+t}\|'_{s+t} \le e^{(\overline{a}-\underline{b})(\tau-t)}\|v_{s+t}\|'_{s+t} + \nu(1-e^{c(\tau-t)})(\|v_{s+t}\|'_{s+t}+\|w_{s+t}\|'_{s+t}).$$

Therefore,

$$[1-\nu(1-e^{c(\tau-t)})]\cdot\|w_{s+t}\|'_{s+t} \le [e^{(\overline{a}-\underline{b})(\tau-t)}+\nu(1-e^{c(\tau-t)})]\cdot\|v_{s+t}\|'_{s+t}.$$

We now consider two cases:

1. if $\overline{a} - \underline{b} \le c$, then

$$\|w_{s+t}\|'_{s+t} \le F(\tau-t)\|v_{s+t}\|'_{s+t},$$

where

$$F(\tau-t) = \frac{e^{c(\tau-t)}+\nu(1-e^{c(\tau-t)})}{1-\nu(1-e^{c(\tau-t)})};$$

2. if $\overline{a} - \underline{b} > c$, then

$$\|w_{s+t}\|'_{s+t} \le F(\tau-t)\|v_{s+t}\|'_{s+t},$$

where

$$F(\tau-t) = \frac{e^{(\overline{a}-\underline{b})(\tau-t)}+\nu(1-e^{c(\tau-t)})}{1-\nu(1-e^{c(\tau-t)})}.$$

When δ is sufficiently small (and thus when θ is also sufficiently small, in view of (5.9)), it follows from (5.30) that ν can be made arbitrarily small. By Lemma 5.6 we conclude that $F(\tau-t) < 1$ for every $\tau > t$. This completes the proof of the lemma. □

5.4.4 Construction and continuity of the stable spaces

Given $(s, \xi) \in X_\gamma$, we set

$$E(s, \xi, \varphi(s, \xi)) = \bigcap_{\tau \geq 0} (\partial_{\Psi_\tau(s,\xi,\varphi(s,\xi))} \Psi_{-\tau}) \overline{C_s(\tau)}. \tag{5.31}$$

It is shown in Lemma 5.11 that $E(s, \xi, \varphi(s, \xi))$ is a vector space, with the same dimension as $E(s)$ in the direct sum $E(s) \oplus F(s)$. We will call it the *stable space* at the point $(s, \xi, \varphi(s, \xi))$. We first establish some auxiliary results concerning the speed at which the norms of vectors inside and outside the cones vary along a given orbit. We start with vectors inside the cones.

Lemma 5.9. *Given $\delta > 0$ sufficiently small, for each $(s, \xi) \in X_\gamma$ with $s \geq \varrho$, $\tau \geq 0$, and $(v, w) \in \overline{C_s(\tau)}$ we have*

$$\|(\partial_{\Psi_\tau(s,\xi,\varphi(s,\xi))} \Psi_{-\tau})(v, w)\|_s' \geq \frac{1}{4} e^{-(\overline{a}+\varsigma)\tau} \|(v, w)\|_{s+\tau}'. \tag{5.32}$$

Proof. We use the same notation as in the proof of Lemma 5.8. Let $\tau > t = 0$ and set $(v_{s+\tau}, w_{s+\tau}) = (v, w)$. We consider the vector (v_s, w_s) given by (5.24) with $r = s$. By Lemma 5.8 (see (5.23)), we have $(v_s, w_s) \in C_s(0)$, and thus $\|w_s\|_s' \leq \|v_s\|_s'$. It follows from (5.28) that

$$\|v_{s+\tau}\|_{s+\tau}' \leq e^{(\overline{a}+\varsigma)\tau} \|v_s\|_s' + \frac{\theta D_1}{\varsigma |c_1|} 2C_3 (1 - e^{c_1\tau}) e^{\overline{a}\tau} \|v_s\|_s'$$

$$\leq e^{(\overline{a}+\varsigma)\tau} \left(1 + \frac{\theta D_1}{\varsigma |c_1|} 2C_3\right) \|v_s\|_s'.$$

In view of (5.9), for each δ sufficiently small we have

$$\frac{\theta D_1}{\varsigma |c_1|} 2C_3 \leq 1, \tag{5.33}$$

and thus

$$\|v_s\|_s' \geq \frac{1}{2} e^{-(\overline{a}+\varsigma)\tau} \|v_{s+\tau}\|_{s+\tau}'. \tag{5.34}$$

Since $(v, w) \in \overline{C_s(\tau)}$, we have $\|(v_{s+\tau}, w_{s+\tau})\|_{s+\tau}' \leq 2\|v_{s+\tau}\|_{s+\tau}'$. It follows from (5.34) that

$$\|(v_s, w_s)\|_s' \geq \|v_s\|_s' \geq \frac{1}{2} e^{-(\overline{a}+\varsigma)\tau} \|v_{s+\tau}\|_{s+\tau}'$$

$$\geq \frac{1}{4} e^{-(\overline{a}+\varsigma)\tau} \|(v_{s+\tau}, w_{s+\tau})\|_{s+\tau}'.$$

This completes the proof of the lemma.   □

We now establish an analogous result to that in Lemma 5.9 for vectors outside the cones.

Lemma 5.10. *Given $\delta > 0$ sufficiently small, for each $(s, \xi) \in X_\gamma$ with $s \geq \varrho$, $\tau \geq 0$, and $z \in F(s)$ we have*

$$\|(\partial_{(s,\xi,\varphi(s,\xi))}\Psi_\tau)(0, z)\|'_{s+\tau} \geq \frac{1}{2}e^{(\underline{b}+\varsigma)\tau}\|(0, z)\|'_s. \tag{5.35}$$

Proof. Let

$$(z_1, z_2) = (\partial_{(s,\xi,\varphi(s,\xi))}\Psi_\tau)(0, z) = \begin{pmatrix} \frac{\partial x}{\partial \xi} & \frac{\partial x}{\partial \eta} \\ \frac{\partial y}{\partial \xi} & \frac{\partial y}{\partial \eta} \end{pmatrix}\begin{pmatrix} 0 \\ z \end{pmatrix},$$

with the partial derivatives computed at the point $p_{s+\tau}$ (see (5.6) and (5.7)). Since $\|(z_1, z_2)\|'_{s+\tau} \geq \|z_2\|'_{s+\tau}$, it is sufficient to find a lower bound for $\|z_2\|'_{s+\tau}$. We use the notation

$$F(r) = \frac{\partial f}{\partial x}\Big|_{q_r}\frac{\partial x}{\partial \eta}\Big|_{p_r}z + \frac{\partial f}{\partial y}\Big|_{q_r}\frac{\partial y}{\partial \eta}\Big|_{p_r}z,$$

with p_r and q_r as in (5.6). It follows from (5.13) with $t = 0$ and $\rho = s + \tau$ that

$$V(s + \tau, s)z = z_2 - \int_s^{s+\tau} V(r, s + \tau)^{-1}\left(\frac{\partial f}{\partial x}\Big|_{q_r}\frac{\partial x}{\partial \eta}\Big|_{p_r}z + \frac{\partial f}{\partial y}\Big|_{q_r}\frac{\partial y}{\partial \eta}\Big|_{p_r}z\right)dr.$$

Using (5.29) with $t = 0$ and since $c_2 < 0$, we obtain

$$\|z\|'_s \leq e^{-(\underline{b}+\varsigma)\tau}\|z_2\|'_{s+\tau} + \frac{\theta D_2}{\varsigma|c_2|}C_3 e^{\varsigma\varrho}\|z\|'_s.$$

In view of (5.9), for each δ sufficiently small we have

$$\frac{\theta D_2}{\varsigma|c_2|}C_3 e^{\varsigma\varrho} \leq 1/2, \tag{5.36}$$

and thus $\|z\|'_s \leq 2e^{-(\underline{b}+\varsigma)\tau}\|z_2\|'_{s+\tau}$. Therefore, since $\|z\|'_s = \|(0, z)\|'_s$, we conclude that

$$\|(z_1, z_2)\|'_{s+\tau} \geq \|z_2\|'_{s+\tau} \geq \frac{1}{2}e^{(\underline{b}+\varsigma)\tau}\|(0, z)\|'_s.$$

This completes the proof of the lemma. $\qquad\square$

We recall that $-(\overline{a}+\varsigma) > 0$ and $\underline{b}+\varsigma > 0$ (see (5.17)). Thus, the inequalities (5.32) and (5.35) say respectively that vectors inside the cones expand as time goes to the past, and that vectors in the subspaces $F(s)$ (which are outside the cones) expand as time goes to the future. With the help of Lemmas 5.9 and 5.10 we can now establish that the set $E(s, \xi, \varphi(s, \xi))$ defined by (5.31) is a vector space varying continuously with the pair (s, ξ).

Lemma 5.11. *Given $\delta > 0$ sufficiently small, the following properties hold:*

1. for every $(s, \xi) \in X_\gamma$ with $s \geq \varrho$, the set $E(s, \xi, \varphi(s, \xi))$ is a subspace with $\dim E(s, \xi, \varphi(s, \xi)) = \dim E(s)$;
2. the map $X_\gamma \cap ([\varrho, +\infty) \times \mathbb{R}^n) \ni (s, \xi) \mapsto E(s, \xi, \varphi(s, \xi))$ is continuous.

Proof. Set

$$D(\tau) = (\partial_{\Psi_\tau(s, \xi, \varphi(s, \xi))} \Psi_{-\tau}) \overline{C_s(\tau)}.$$

It follows readily from Lemma 5.8, applying $\partial_{\Psi_t(s, \xi, \varphi(s, \xi))} \Psi_{-t}$ to both sides of (5.23), that for every $\tau \geq t \geq 0$,

$$D(\tau) \subset \text{int } D(t) \subset \text{int } D(0) = C_s(0) \setminus \{(0, 0)\}. \tag{5.37}$$

Therefore, $(D(\tau))_{\tau \geq 0}$ is a strictly decreasing family of closed sets inside the cone $C_s(0)$, and thus, $E(s, \xi, \varphi(s, \xi))$ is a nonempty closed subset of $C_s(0)$. Furthermore, for each $k \in \mathbb{N}$ the set $D(k)$ contains a subspace E_k of dimension $\dim E(s)$. Therefore, by the compactness of the unit ball in $\mathbb{R}^n$, there exists a subspace $E' \subset E(s, \xi, \varphi(s, \xi))$ of dimension $\dim E(s)$ (consider the subspaces E_k and an orthonormal basis for each of them; the compactness allows us to find a subsequence for which each of the components of the orthonormal basis converges).

Given $v \in E(s, \xi, \varphi(s, \xi))$, we write $v = v_1 + v_2$ with $v_1 \in E'$ and $v_2 \in F(s)$. We note that E' is inside the cone $C_s(0)$ while $F(s)$ is outside this cone, and thus, we can always write v in this form. Since $v, v_1 \in E(s, \xi, \varphi(s, \xi))$ it follows from the definition of $E(s, \xi, \varphi(s, \xi))$ in (5.31) that

$$(\partial_{(s, \xi, \varphi(s, \xi))} \Psi_\tau) v, (\partial_{(s, \xi, \varphi(s, \xi))} \Psi_\tau) v_1 \in \overline{C_s(\tau)}.$$

By Lemmas 5.9 and 5.10 we obtain

$$\|v_2\|'_s \leq 2 e^{(-\underline{b} - \varsigma) \tau} \|(\partial_{(s, \xi, \varphi(s, \xi))} \Psi_\tau) v_2\|'_{s+\tau}$$
$$= 2 e^{(-\underline{b} - \varsigma) \tau} \|(\partial_{(s, \xi, \varphi(s, \xi))} \Psi_\tau)(v - v_1)\|'_{s+\tau}$$
$$\leq 8 e^{(\overline{a} - \underline{b}) \tau} (\|v\|'_s + \|v_1\|'_s).$$

By (2.6), letting $\tau \to 0$ we obtain $v_2 = 0$ and thus $v \in E'$. Therefore, $E' = E(s, \xi, \varphi(s, \xi))$ and $E(s, \xi, \varphi(s, \xi))$ is a subspace of dimension $\dim E(s)$.

It remains to establish the continuity in the last property. Recall that $v \in E(s, \xi, \varphi(s, \xi))$ if and only if (see (5.31))

$$(\partial_{(s, \xi, \varphi(s, \xi))} \Psi_\tau) v \in \overline{C_s(\tau)} \text{ for every } \tau \geq 0.$$

Consider a sequence $(s_k, \xi_k)_k \in X_\gamma \cap ([\varrho, +\infty) \times \mathbb{R}^n)$ converging to a point (s, ξ) in the same set as $k \to +\infty$, and a sequence $v_k \in E(s_k, \xi_k, \varphi(s_k, \xi_k))$ such that $\|v_k\| = 1$ for each $k \in \mathbb{N}$. Then

$$(\partial_{(s_k, \xi_k, \varphi(s_k, \xi_k))} \Psi_\tau) v_k \in \overline{C_{s_k}(\tau)} \text{ for every } k \in \mathbb{N} \text{ and } \tau \geq 0 \tag{5.38}$$

(we stress that the cone in (5.38) is computed with respect to the norms $\|\cdot\|'_{s_k+\tau}$ in (5.16)). We first assume that $(v_k)_k$ converges and let $v \in \mathbb{R}^n$ be the limit of the sequence. It follows from the C^1 regularity of Ψ_τ (which is an immediate consequence of the C^1 regularity of $(t, v) \mapsto A(t)v + f(t, v)$) and the Lipschitz property of φ in (4.10) that $(s, \xi) \mapsto \partial_{(s,\xi,\varphi(s,\xi))}\Psi_\tau$ is continuous, and hence,

$$(\partial_{(s_k,\xi_k,\varphi(s_k,\xi_k))}\Psi_\tau)v_k \to (\partial_{(s,\xi,\varphi(s,\xi))}\Psi_\tau)v \text{ as } k \to +\infty \text{ for every } \tau \geq 0.$$

Furthermore, since the norms in (5.16) are independent of ξ and vary continuously with s, we conclude from (5.38) that

$$(\partial_{(s,\xi,\varphi(s,\xi))}\Psi_\tau)v \in \overline{C_s(\tau)} \text{ for every } \tau \geq 0.$$

Therefore, $v \in E(s, \xi, \varphi(s, \xi))$ (see (5.31)). When $(v_k)_k$ does not converge, let $(m_k)_k$ be some subsequence for which $(v_{m_k})_k$ converges, say to a vector $v \in \mathbb{R}^n$ (recall that $\|v_k\| = 1$ for each k, and thus there are always sublimits). Proceeding in a similar manner, it follows from (5.38) that

$$(\partial_{(s,\xi,\varphi(s,\xi))}\Psi_\tau)v \in \overline{C_s(\tau)} \text{ for every } \tau \geq 0.$$

Therefore, $v \in E(s, \xi, \varphi(s, \xi))$, that is, any sublimit of the sequence $(v_k)_k$ is in $E(s, \xi, \varphi(s, \xi))$. It follows from the property

$$\dim E(s_k, \xi_k, \varphi(s_k, \xi_k)) = \dim E(s, \xi, \varphi(s, \xi)) = \dim E(s) \text{ for each } k \in \mathbb{N},$$

that any sublimit of a sequence of orthonormal bases (with respect to the original norm $\|\cdot\|$) of the vector spaces $E(s_k, \xi_k, \varphi(s_k, \xi_k))$ (obtained by considering any subsequence $(m_k)_k$ such that every component of the orthonormal bases converges) is also an orthonormal basis of $E(s, \xi, \varphi(s, \xi))$. Therefore,

$$E(s_k, \xi_k, \varphi(s_k, \xi_k)) \to E(s, \xi, \varphi(s, \xi)) \text{ as } k \to \infty,$$

and the map $\xi \mapsto E(s, \xi, \varphi(s, \xi))$ is continuous. $\qquad\square$

It follows readily from (5.31) and the inclusions in (5.37) that for every increasing sequence $\tau_k \to +\infty$ as $k \to +\infty$ we have

$$E(s, \xi, \varphi(s, \xi)) = \bigcap_{k \in \mathbb{N}} (\partial_{\Psi_{\tau_k}(s,\xi,\varphi(s,\xi))}\Psi_{-\tau_k})\overline{C_s(\tau_k)}.$$

5.4.5 Behavior of the tangent sets

We now introduce sets that at each point of $\mathcal{V}$ contain all possible tangential behavior (with respect to $\mathcal{V}$). Given $s \geq 0$ and $\xi, \bar{\xi} \in R_s(\delta e^{-\gamma s})$ with $\xi \neq \bar{\xi}$, we set

$$\Delta_{\xi,\bar{\xi}}\varphi = \frac{(\xi, \varphi(s, \xi)) - (\bar{\xi}, \varphi(s, \bar{\xi}))}{\|(\xi, \varphi(s, \xi)) - (\bar{\xi}, \varphi(s, \bar{\xi}))\|},$$

and

$$t_{(s,\xi)}\varphi = \{v \in \mathbb{R}^n : \Delta_{\xi,\xi_m}\varphi \to v \text{ for some sequence } \xi_m \to \xi\}.$$

We define the *tangent set* of the graph of φ at $(s, \xi, \varphi(s, \xi))$ (when restricted to $\{s\} \times \mathbb{R}^n$) by

$$V(s, \xi, \varphi(s, \xi)) = \{\lambda v : v \in t_{(s,\xi)}\varphi \text{ and } \lambda \in \mathbb{R}\}.$$

One can easily verify that the function φ is differentiable at (s, ξ) when restricted to $\{s\} \times \mathbb{R}^n$ if and only if $V(s, \xi, \varphi(s, \xi))$ is a subspace of dimension $\dim E(s)$. This is precisely the basis in the present approach to establish the smoothness of $\mathcal{V}$. In order to effect this approach we first establish a relation between the tangent sets and the invariant family of cones constructed in the former section.

For each $r \geq s$, we write

$$x(r) = x(r, s, \xi, \varphi(s, \xi)) \quad \text{and} \quad \overline{x}(r) = x(r, s, \overline{\xi}, \varphi(s, \overline{\xi})). \tag{5.39}$$

We also set

$$\zeta = c2^{q+2}D^q\delta^q D_1. \tag{5.40}$$

We start with some auxiliary results.

Lemma 5.12. *Given $\delta \in (0,1)$ sufficiently small, for each $s \geq 0$, $\xi, \overline{\xi} \in R_s(\delta e^{-\gamma s})$, and $\tau > t \geq 0$ we have*

$$\|x(s + \tau) - \overline{x}(s + \tau)\| \leq D_1 e^{(\overline{a}+\zeta)(\tau-t)+a(s+t)}\|x(s + t) - \overline{x}(s + t)\|. \tag{5.41}$$

Proof. For each $r \in [s + t, s + \tau]$, it follows from (4.10) that

$$\|(x(r), \varphi(r, x(r)))\| = \|(x(r), \varphi(r, x(r)) - \varphi(r, 0))\| \leq 2\|x(r)\|,$$
$$\|(\overline{x}(r), \varphi(r, \overline{x}(r)))\| = \|(\overline{x}(r), \varphi(r, \overline{x}(r)) - \varphi(r, 0))\| \leq 2\|\overline{x}(r)\|,$$

and

$$\|(x(r), \varphi(r, x(r))) - (\overline{x}(r), \varphi(r, \overline{x}(r)))\| \leq 2\|x(r) - \overline{x}(r)\|.$$

By (4.2), we obtain

$$\|f(r, x(r), \varphi(r, x(r))) - f(r, \overline{x}(r), \varphi(r, \overline{x}(r)))\|$$
$$\leq c2^{q+1}\|x(r) - \overline{x}(r)\|(\|x(r)\|^q + \|\overline{x}(r)\|^q).$$

Using Theorem 4.1 (see (4.16)), this yields

$$\|f(r, x(r), \varphi(r, x(r))) - f(r, \overline{x}(r), \varphi(r, \overline{x}(r)))\|$$
$$\leq c2^{q+2}\|x(r) - \overline{x}(r)\|D^q e^{q\overline{a}(r-s)+aqs}(\|\xi\|^q + \|\overline{\xi}\|^q) \tag{5.42}$$
$$\leq \eta e^{q\overline{a}(r-s)-(2a+2b+aq+\max\{a,b\})s}\|x(r) - \overline{x}(r)\|,$$

where

$$\eta = c2^{q+2}D^q\delta^q < c2^{q+2}D^q, \tag{5.43}$$

since $\delta < 1$. Note that the last constant in (5.43) is independent of δ. Therefore, setting

$$\rho(r) = \|x(r) - \overline{x}(r)\|$$

and using (2.6), it follows from the identities

$$x(s + \tau) = U(s + \tau, s + t)x(s + t) + \int_{s+t}^{s+\tau} U(s + \tau, r)f(r, x(r), \varphi(r, x(r)))\, dr$$

$$\overline{x}(s + \tau) = U(s + \tau, s + t)\overline{x}(s + t) + \int_{s+t}^{s+\tau} U(s + \tau, r)f(r, \overline{x}(r), \varphi(r, \overline{x}(r)))\, dr$$

that

$$
\begin{aligned}
\rho(s + \tau) &\leq \|U(s + \tau, s + t)\|\rho(s + t) \\
&\quad + \int_{s+t}^{s+\tau} \|U(s + \tau, r)\|\eta e^{q\overline{a}(r-s) - (2a + 2b + aq + \max\{a,b\})s}\rho(r)\, dr \\
&\leq D_1 e^{\overline{a}(\tau - t) + a(s+t)}\rho(s + t) + D_1 \eta e^{-(a + 2b + aq + \max\{a,b\})s} \\
&\quad \times \int_{s+t}^{s+\tau} e^{\overline{a}(\tau - t)}e^{T_1(r-s)}e^{-\overline{a}(r-t-s)}\rho(r)\, dr,
\end{aligned}
$$
(5.44)

with $T_1 = q\overline{a} + a$. By the first inequality in (5.1), we have $T_1 < 0$. Setting

$$\Gamma(\sigma) = e^{-\overline{a}(\sigma - t - s)}\rho(\sigma),$$

it follows from (5.44) that for every $\tau > t$,

$$\Gamma(s + \tau) \leq D_1 e^{a(s+t)}\rho(s + t) + \zeta \int_{s+t}^{s+\tau} \Gamma(r)\, dr.$$

Using Gronwall's lemma for the function $\tau \mapsto \Gamma(s + \tau)$, we obtain

$$\rho(s + \tau) \leq D_1 e^{a(s+t) + \zeta(\tau - t)}\rho(s + t)e^{\overline{a}(\tau - t)}$$

for every $\tau > t$, which is the same as (5.41). This completes the proof.   □

Note now that by (2.6) and (5.1), we have

$$-\underline{b} + \overline{a} + q\overline{a} + b < 0 \quad \text{and} \quad \overline{a} - \underline{b} + b < 0,$$

and hence, in view of (5.40), we can choose $\delta > 0$ sufficiently small so that

$$T_2 = -\underline{b} + \overline{a} + \zeta + q\overline{a} + b < 0 \quad \text{and} \quad \overline{a} + \zeta - \underline{b} + b < 0.$$
(5.45)

We will use the notations

$$\chi(r) = \|x(r) - \overline{x}(r)\| \quad \text{and} \quad \rho(r) = \|\varphi(r, x(r)) - \varphi(r, \overline{x}(r))\|,$$
(5.46)

with $x(r)$ and $\overline{x}(r)$ as in (5.39). The following is another auxiliary result.

Lemma 5.13. *Given $\delta \in (0,1)$ sufficiently small, for each $s \geq 0$, $\xi, \bar{\xi} \in R_s(\delta e^{-\gamma s})$, and $\tau > t \geq 0$ we have*

$$\rho(s+t) \leq D_2 e^{-(\underline{b}-b)(\tau-t)} e^{b(s+t)} \rho(s+\tau)$$
$$+ \frac{D_1 D_2 \eta}{|T_2|} e^{(q\bar{a}+a+b)t-(a+b+aq+\max\{a,b\})s} \chi(s+t).$$

Proof. It follows from (4.17) and (4.17) that

$$\varphi(s+t, x(s+t)) = V(s+\tau, s+t)^{-1} \varphi(s+\tau, x(s+\tau))$$
$$- \int_{s+t}^{s+\tau} V(r, s+t)^{-1} f(r, x(r), \varphi(r, x(r))) \, dr, \tag{5.47}$$

$$\varphi(s+t, \bar{x}(s+t)) = V(s+\tau, s+t)^{-1} \varphi(s+\tau, \bar{x}(s+\tau))$$
$$- \int_{s+t}^{s+\tau} V(r, s+t)^{-1} f(r, \bar{x}(r), \varphi(r, \bar{x}(r))) \, dr. \tag{5.48}$$

By (5.42), we have

$$\|f(r, x(r), \varphi(r, x(r))) - f(r, \bar{x}(r), \varphi(r, \bar{x}(r)))\|$$
$$\leq \eta e^{q\bar{a}(r-s)-(2a+2b+aq+\max\{a,b\})s} \chi(r),$$

with η as in (5.43). It follows from Lemma 5.12, setting $s + \tau = r$ in (5.41), that for every $r \geq s+t$ we have

$$\|f(r, x(r), \varphi(r, x(r))) - f(r, \bar{x}(r), \varphi(r, \bar{x}(r)))\|$$
$$\leq D_1 \eta e^{q\bar{a}(r-s)-(2a+2b+aq+\max\{a,b\})s} e^{(\bar{a}+\zeta)(r-s-t)+a(s+t)} \chi(s+t).$$

Subtracting (5.47) and (5.48), and using (2.6) we obtain

$$\rho(s+t) \leq \|V(s+\tau, s+t)^{-1}\| \rho(s+\tau) + D_1 \eta e^{at} \chi(s+t)$$
$$\times \int_{s+t}^{s+\tau} \|V(r, s+t)^{-1}\| e^{q\bar{a}(r-s)-(a+2b+aq+\max\{a,b\})s} e^{(\bar{a}+\zeta)(r-t-s)} \, dr$$
$$\leq D_2 e^{-\underline{b}(\tau-t)+b(s+\tau)} \rho(s+\tau) + D_1 D_2 \eta e^{(q\bar{a}+a+b)t}$$
$$\times e^{-(a+b+aq+\max\{a,b\})s} \chi(s+t) \int_{s+t}^{s+\tau} e^{T_2(r-s-t)} \, dr$$
$$\leq D_2 e^{-\underline{b}(\tau-t)+b(s+\tau)} \rho(s+\tau)$$
$$+ \frac{D_1 D_2 \eta}{|T_2|} e^{(q\bar{a}+a+b)t-(a+b+aq+\max\{a,b\})s} \chi(s+t),$$

with T_2 as in (5.45). This completes the proof of the lemma.   □

We can now establish a relation between the tangent sets and the invariant family of cones along each orbit.

Lemma 5.14. *Given $\delta \in (0,1)$ sufficiently small, for each $(s,\xi) \in X_\gamma$ with $s \geq \varrho$, and $t \geq 0$ we have $V(\Psi_t(s,\xi,\varphi(s,\xi))) \subset \overline{C_s(t)}$.*

Proof. We proceed by contradiction. Namely, assume that there exists $t \geq s$ such that

$$V(\Psi_t(s,\xi,\varphi(s,\xi))) \setminus \overline{C_s(t)} \neq \varnothing. \tag{5.49}$$

Then, there exists $\bar{\xi} \in R_s(\delta e^{-\gamma s})$ arbitrarily close to ξ for which

$$\|\varphi(s+t, x(s+t)) - \varphi(s+t, \overline{x}(s+t))\|'_{s+t} > \|x(s+t) - \overline{x}(s+t)\|'_{s+t}, \tag{5.50}$$

where x and $\overline{x}$ are the functions in (5.39). Using the same notation as in (5.46), it follows from (5.50) and Lemma 5.7 that

$$\chi(s+t) < C_1 e^{a(s+t)} \|\varphi(s+t, x(s+t)) - \varphi(s+t, \overline{x}(s+t))\|'_{s+t}$$
$$\leq \frac{D_2 C_1}{\varsigma} e^{(a+b)(s+t)+\varsigma(\varrho+t)} \rho(s+t).$$

By Lemma 5.13, we obtain

$$\chi(s+t) \leq \frac{D_2^2 C_1}{\varsigma} e^{-(\underline{b}-b)(\tau-t)} e^{(a+2b)(s+t)+\varsigma(\varrho+t)} \rho(s+\tau)$$
$$+ \frac{D_1 D_2^2 C_1 \eta}{\varsigma |T_2|} e^{(q\overline{a}+a+b)t-(a+b+aq+\max\{a,b\})s} e^{(a+b)(s+t)} \chi(s+t)$$
$$\leq \frac{D_2^2 C_1}{\varsigma} e^{-(\underline{b}-b)(\tau-t)} e^{(a+2b)(s+t)+\varsigma(\varrho+t)} \rho(s+\tau)$$
$$+ \frac{D_1 D_2^2 C_1 \eta}{\varsigma |T_2|} e^{(q\overline{a}+2a+2b)t} e^{-aqs-\max\{a,b\}s} \chi(s+t). \tag{5.51}$$

By the first inequality in (5.1), we have $q\overline{a} + 2a + 2b < 0$. In view of (5.43), we can choose $\delta > 0$ sufficiently small so that

$$\frac{D_1 D_2^2 C_1 \eta}{\varsigma |T_2|} e^{(q\overline{a}+2a+2b)t} e^{-aqs-\max\{a,b\}s} \leq \frac{D_1 D_2^2 C_1 \eta}{\varsigma |T_2|} \leq \frac{1}{2}.$$

Hence, it follows from (5.51) that

$$\chi(s+t) \leq 2\frac{D_2^2 C_1}{\varsigma} e^{-(\underline{b}-b)(\tau-t)} e^{(a+2b)(s+t)+\varsigma(\varrho+t)} \rho(s+\tau).$$

By (5.41) in Lemma 5.12, we conclude that

$$\chi(s+\tau) \leq \frac{2D_1 D_2^2 C_1}{\varsigma} e^{(\overline{a}+\varsigma-\underline{b}+b)(\tau-t)} e^{(2a+2b)(s+t)+\varsigma(\varrho+t)} \rho(s+\tau).$$

Recall that s and t are fixed (see (5.49)). Therefore, by (5.45) (see also (5.46)), there exists $\tau > t$ such that

$$\|x(s+\tau) - \overline{x}(s+\tau)\| < \|\varphi(s+\tau, x(s+\tau)) - \varphi(s+\tau, \overline{x}(s+\tau))\|.$$

But this contradicts the fact that the points

$$(x(s+\tau), \varphi(s+\tau, x(s+\tau))) \quad \text{and} \quad (\overline{x}(s+\tau), \varphi(s+\tau, \overline{x}(s+\tau)))$$

belong to the stable manifold, since φ possesses the Lipschitz property in (4.10). This completes the proof of the lemma. $\square$

5.4.6 C^1 regularity of the stable manifolds

We have now all the tools that are needed to prove that the subset $\mathcal{V} \subset \mathcal{W}$ (see (5.4)) of the Lipschitz manifold $\mathcal{W}$ is in fact a smooth manifold of class C^1.

Proof of Theorem 5.1. We note that $\Delta_{\xi,\xi_m}\varphi \to v$ as $m \to \infty$ (with $\xi_m \to \xi$ as $m \to \infty$) if and only if for every $\tau \geq 0$,

$$\lim_{m \to \infty} \frac{\Psi_\tau(s,\xi_m,\varphi(s,\xi_m)) - \Psi_\tau(s,\xi,\varphi(s,\xi))}{\|\Psi_\tau(s,\xi_m,\varphi(s,\xi_m)) - \Psi_\tau(s,\xi,\varphi(s,\xi))\|} = \frac{(\partial_{(s,\xi,\varphi(s,\xi))}\Psi_\tau)v}{\|(\partial_{(s,\xi,\varphi(s,\xi))}\Psi_\tau)v\|}.$$

This implies that

$$(\partial_{(s,\xi,\varphi(s,\xi))}\Psi_\tau)V(s,\xi,\varphi(s,\xi)) = V(\Psi_\tau(s,\xi,\varphi(s,\xi))). \tag{5.52}$$

Let now $(s,\xi) \in X_\gamma$ with $s > \varrho$. By Lemma 5.14, we have

$$V(\Psi_\tau(s,\xi,\varphi(s,\xi))) \subset \overline{C_s(\tau)} \text{ for every } \tau \geq 0.$$

Therefore, in view of (5.52),

$$V(s,\xi,\varphi(s,\xi)) \subset (\partial_{\Psi_\tau(s,\xi,\varphi(s,\xi))}\Psi_{-\tau})\overline{C_s(\tau)}$$

for every $\tau \geq 0$, and hence, by (5.31),

$$V(s,\xi,\varphi(s,\xi)) \subset E(s,\xi,\varphi(s,\xi)).$$

On the other hand, for each $v \in E(s) \setminus \{0\}$ there exists a sequence $t_m \to 0$ such that $\Delta_{\xi,\xi+t_m v}\varphi$ converges as $m \to +\infty$ (due to the compactness of the unit ball in $\mathbb{R}^n$). This implies that the first $\dim E(s)$ components of $V(s,\xi,\varphi(s,\xi))$ project onto $E(s)$. On the other hand, by Lemma 5.11, the space $E(s,\xi,\varphi(s,\xi))$ has dimension $\dim E(s)$ and hence

$$V(s,\xi,\varphi(s,\xi)) = E(s,\xi,\varphi(s,\xi)). \tag{5.53}$$

In particular, $V(s,\xi,\varphi(s,\xi))$ is a subspace of dimension $\dim E(s)$. Therefore (see the discussion in the beginning of Section 5.4.5), the function φ is differentiable at each point $(s,\xi,\varphi(s,\xi))$ when restricted to $\{s\} \times \mathbb{R}^n$. Furthermore, it follows from the continuity of the map

$$(s, \xi) \mapsto E(s, \xi, \varphi(s, \xi))$$

(see Lemma 5.11) and (5.53) that φ is of class C^1 on each plane $\{s\} \times \mathbb{R}^n$ (since the tangent set varies continuously). This shows that the set $\mathcal{V} \cap (\{s\} \times \mathbb{R}^n)$ is a C^1 manifold for each $s > \varrho$, of dimension $\dim E(s)$.

We now consider some $\varepsilon = \varepsilon(s) > 0$ such that $s - \varepsilon > \varrho$, and we define the map

$$F_s \colon \{(t, \xi) \colon t \in (-\varepsilon, \varepsilon) \text{ and } \xi \in R_{s+t}(\delta e^{-\gamma(s+t)})\} \to \mathbb{R}^+ \times \mathbb{R}^n$$

by

$$F_s(t, \xi) = \Psi_t(s, \xi, \varphi(s, \xi)). \tag{5.54}$$

We showed above that $\xi \mapsto \varphi(s, \xi)$ is of class C^1 (for each fixed s). Furthermore, it follows from B1 and B2 that the map $(t, s, \xi, \eta) \mapsto \Psi_t(s, \xi, \eta)$ is of class C^1. Therefore, F_s is also of class C^1 (for each fixed s). In addition, one can verify that the map F_s is injective: if $F_s(t, \xi) = F_s(t', \xi')$ then the first component of F_s gives $s + t = s + t'$ and hence $t = t'$; therefore,

$$\Psi_t(s, \xi, \varphi(s, \xi)) = \Psi_t(s, \xi', \varphi(s, \xi'))$$

and applying Ψ_{-t} to both sides of the identity yields $\xi = \xi'$. This shows that F_s is a parametrization of class C^1 of an open subset of $\mathcal{V}$ containing $(s, 0)$. Since this procedure can be effected for every s, and ε is arbitrarily small, we conclude that $\mathcal{V}$ is a smooth manifold of class C^1 of dimension $\dim E(s) + 1$.

For the remaining properties, note that by Theorem 4.1 (see (4.15)) we have

$$\varphi(s, \xi) = -\int_s^{+\infty} V(\tau, s)^{-1} f(\Psi_{\tau-s}(s, \xi, \varphi(s, \xi))) \, d\tau.$$

Taking derivatives with respect to ξ, we obtain

$$\frac{\partial \varphi}{\partial \xi}(s, 0) = -\int_s^{+\infty} V(\tau, s)^{-1} \left(\frac{\partial f}{\partial x}(\tau, 0) \frac{\partial x}{\partial \xi} + \frac{\partial f}{\partial y}(\tau, 0) \frac{\partial y}{\partial \xi} \right) d\tau,$$

with $\frac{\partial x}{\partial \xi}$ and $\frac{\partial y}{\partial \xi}$ computed at $(\tau, s, 0) \in \mathbb{R}^+ \times \mathbb{R}^+ \times \mathbb{R}^n$. By Lemma 5.4 we have

$$\frac{\partial f}{\partial x}(\tau, 0) = \frac{\partial f}{\partial y}(\tau, 0) = 0,$$

and hence, $\frac{\partial \varphi}{\partial \xi}(s, 0) = 0$. This implies that

$$(T_{(s,0)}\mathcal{V}) \cap (\{s\} \times \mathbb{R}^n) = \{s\} \times E(s) \text{ for each } s > \varrho. \tag{5.55}$$

Furthermore, since $\varphi(s, 0) = 0$ for every $s > \varrho$, we have $(\varrho, +\infty) \times \{0\} \subset \mathcal{V}$ and thus, $\mathbb{R} \times \{0\} \subset T_{(s,0)}\mathcal{V}$ for every $s > \varrho$. Together with the identities in (5.55) and the fact that $\dim \mathcal{V} = \dim E(s) + 1$, we conclude that

$$T_{(s,0)}\mathcal{V} = \mathbb{R} \times E(s).$$

This completes the proof of the theorem. $\square$

It follows immediately from (5.52) and (5.53) in the proof of the theorem that given $(s, \xi) \in X_\gamma$ with $s > \varrho$ we have

$$(\partial_{(s,\xi,\varphi(s,\xi))}\Psi_\tau)E(s, \xi, \varphi(s,\xi)) = E(\Psi_\tau(s, \xi, \varphi(s,\xi))) \qquad (5.56)$$

for every $\tau \geq 0$. However, a priori (without considering the tangent sets), the identity in (5.56) must be considered nontrivial, due to the fact that the cones which are used to define the space $E(\Psi_\tau(s, \xi, \varphi(s,\xi)))$ in (5.56) are obtained from the norms $\|\cdot\|'$ in (5.16) with s replaced by $s+\tau$. In particular, it follows easily from the definition of the stable spaces in (5.31) that

$$(\partial_{(s,\xi,\varphi(s,\xi))}\Psi_\tau)E(s, \xi, \varphi(s,\xi)) \subset E(\Psi_\tau(s, \xi, \varphi(s,\xi)))$$

and a priori this inclusion could be proper. We note that when we consider cones instead of tangent spaces, the corresponding inclusion is indeed proper for the cones at the points $(s, \xi, \varphi(s,\xi))$ and $\Psi_\tau(s, \xi, \varphi(s,\xi))$, that is,

$$(\partial_{(s,\xi,\varphi(s,\xi))}\Psi_\tau)C_s(0) \subset C_{s+\tau}(\tau)$$

and this inclusion is proper (since $\|w\|'_{s+\tau} < \|w\|'_s$ whenever $w \neq 0$).

We now explain why Theorem 5.1 requires the extra parameter ϱ which is absent in Theorem 4.1 concerning the existence of a Lipschitz manifold. This has to do with the expressions in (5.30), (5.33), and (5.36) which involve the product

$$\theta C_3 = \theta \max\{C_1^{-1}, C_2^{-1}\},$$

with θ as in (5.9), and with C_1 and $C_2 = C_2(\varrho)$ as in (5.21). Indeed, it follows from (5.21) that as $\varrho \to 0$ the constant C_2 approaches infinity, and thus θ and consequently $\delta = \delta(\varrho)$ must approach zero. On the other hand, we can fix an arbitrarily small positive ϱ, and choose a corresponding δ in Theorem 5.1.

5.5 C^k stable manifolds

Given $k \in \mathbb{N}$, we now replace the conditions B1 and B2 in Section 5.1 by the new conditions:

C1. the function $A\colon \mathbb{R}_0^+ \to M_n(\mathbb{R})$ is of class C^k and satisfies (2.2);
C2. the function $f\colon \mathbb{R}_0^+ \times \mathbb{R}^n \to \mathbb{R}^n$ is of class C^k, satisfies (4.1), and there exist $c > 0$ and $q > k$ such that

$$\left\| \frac{\partial^j f}{\partial v^j}(t, u) - \frac{\partial^j f}{\partial v^j}(t, v) \right\| \leq c\|u - v\|(\|u\|^{q-j} + \|v\|^{q-j}) \qquad (5.57)$$

for every $j = 0, \ldots, k-1$, $t \geq 0$, and $u, v \in \mathbb{R}^n$.

When $j \geq 1$, the norm considered in the left-hand side of (5.57) is the norm of a multilinear (j-linear) operator. One can easily verify that the condition C2 is equivalent (with the same constant q) to the condition: $f(t, 0) = 0$ for every $t \geq 0$, and there exist $c > 0$ and $q > k$ such that

$$\left\| \frac{\partial^j f}{\partial v^j}(t, v) \right\| \leq c \|v\|^{q-j+1} \text{ for every } j = 1, \ldots, k, \ t \geq 0, \text{ and } v \in \mathbb{R}^n.$$

That this implies the condition C2 is immediate from the mean value theorem. The converse follows readily from the definitions (see (5.73)).

We continue to set $\vartheta = \max\{a, b\}$ (see (2.6)). We consider the conditions

$$q\bar{a} + 4\vartheta < \min\{\bar{a} - \bar{b}, (2 - q)\vartheta\} \quad \text{and} \quad \bar{a} + 5\vartheta(1 + 1/q) \leq 0. \tag{5.58}$$

Note that the second inequality in (5.58) is always satisfied for ϑ sufficiently small. Again we slightly reduce the size of the neighborhood $R_s(\delta e^{-(\beta + 3\vartheta)s/q})$, with β as in (4.12). Let

$$\omega = \gamma + a = a(3 + 1/q) + b/q + 3\vartheta/q, \tag{5.59}$$

with γ as in (5.3). Note that

$$\omega \geq \gamma \geq \beta \geq \alpha. \tag{5.60}$$

We now formulate the higher regularity result. We write $p_{s,\xi} = (s, \xi, \varphi(s, \xi))$.

Theorem 5.15 ([5]). *Assume that C1–C2 hold for some functions A and f of class C^k, for some $k \in \mathbb{N}$. If the equation $v' = A(t)v$ admits a strong nonuniform exponential dichotomy in $\mathbb{R}^+$ and the conditions in (5.58) hold, then for each $\varrho > 0$ there exist $\delta > 0$ and a unique function $\varphi \in \mathfrak{X}_{\omega-a}$ such that the graph*

$$\mathcal{V} = \{p_{s,\xi} : (s, \xi) \in X_\omega \text{ and } s > \varrho\} \subset \mathbb{R}^+ \times \mathbb{R}^n$$

is forward invariant under the semiflow Ψ_τ, in the sense that

$$\text{if } (s, \xi) \in X_{\omega+a} \text{ then } \Psi_\tau(p_{s,\xi}) \in \mathcal{V} \text{ for every } \tau \geq 0. \tag{5.61}$$

In addition, the following properties hold:

1. *$\mathcal{V}$ is a smooth manifold of class C^k containing the line $(\varrho, +\infty) \times \{0\}$ and satisfying $T_{(s,0)}\mathcal{V} = \mathbb{R} \times E(s)$ for every $s > \varrho$;*
2. *for every $(s, \xi) \in X_\omega$ we have*

$$\varphi(s, \xi) = -\int_s^{+\infty} V(\tau, s)^{-1} f(\Psi_{\tau-s}(p_{s,\xi})) \, d\tau;$$

3. *there exists $D > 0$ such that for every $s \geq 0$, ξ, $\bar{\xi}$, u, $\bar{u} \in R_s(\delta e^{-\omega s})$, and $\tau \geq 0$ we have*

$$\|\Psi_\tau(p_{s,\xi}) - \Psi_\tau(p_{s,\bar{\xi}})\| \leq De^{\bar{a}\tau + as}\|\xi - \bar{\xi}\|, \tag{5.62}$$

and setting $z_{s,\xi,u} = (u, \frac{\partial \varphi}{\partial \xi}(s, \xi)u)$,

$$\left\|\frac{\partial \Psi_\tau}{\partial v}(p_{s,\xi})z_{s,\xi,u} - \frac{\partial \Psi_\tau}{\partial v}(p_{s,\bar{\xi}})z_{s,\bar{\xi},\bar{u}}\right\| \leq De^{\bar{a}\tau + as}(\|\xi - \bar{\xi}\| + \|u - \bar{u}\|). \tag{5.63}$$

The proof of Theorem 5.15 is given in Section 5.6, and proceeds by induction on k. When $k = 1$, the statement in Theorem 5.15 is a consequence of Theorem 5.1. Indeed, note that the second condition in (5.58) implies the second condition in (5.1). Theorem 5.1 is the first step in the induction process in the proof of Theorem 5.15. We observe that for technical reasons in general we are not able to take $\varrho = 0$ in Theorem 5.1 and thus also in Theorem 5.15.

We emphasize that we also establish the exponential decay on the tangent bundle of the stable manifold of the derivatives of Ψ_τ with respect to the initial condition (see (5.63)). Since $u \mapsto z_{s,\xi,u}$ is linear, setting $u = \bar{u}$ in (5.63) we obtain

$$\left\|\frac{\partial \Psi_\tau}{\partial v}(p_{s,\xi})z_{s,\xi,u} - \frac{\partial \Psi_\tau}{\partial v}(p_{s,\bar{\xi}})z_{s,\bar{\xi},u}\right\| \leq D\delta^{-1}e^{\bar{a}\tau + (a+\omega)s}\|\xi - \bar{\xi}\| \cdot \|u\| \tag{5.64}$$

for every $s \geq 0$, ξ, $\bar{\xi} \in R_s(\delta e^{-\omega s})$, $u \in E(s)$, and $\tau \geq 0$. Furthermore, setting $\bar{\xi} = 0$ in (5.64) we obtain

$$\left\|\frac{\partial \Psi_\tau}{\partial v}(p_{s,\xi})|T_{p_{s,\xi}}\mathcal{V} \cap (\{s\} \times \mathbb{R}^n)\right\| \leq De^{\bar{a}\tau + as} \sup_{u \neq 0} \frac{\|u\|}{\|z_{s,\xi,u}\|}$$

for every $s \geq 0$, $\xi \in R_s(\delta e^{-\omega s})$, and $\tau \geq 0$. When we consider higher-order jets (other than the first, which amounts to the linear variational equation, that is used in the proof of Theorem 5.15; see Section 5.6.1), the corresponding higher-order linearized vector fields maintain the linear part $A(t)$ of the original linear variational equation. This ensures that the higher-order jets possess essentially the same nonuniform exponential dichotomies as the linear variational equation, although in higher-dimensional spaces. Thus, we can expect each of the corresponding higher-order derivatives to exhibit an exponential decay similar to that in (5.62) and (5.63), up to order $k - 1$. That this is indeed the case is established in Chapter 6 with another approach (see Theorem 6.1). The main "computational" difficulty in using the approach in the present chapter to deal with the higher-order derivatives is that when $j \geq 2$ the derivative $\partial^j \Psi_\tau / \partial v^j$ never occurs alone in any component of the semiflows associated with the higher-order linearized vector fields. For example, to obtain the second-order linearized vector field we can consider the first-order vector field in (5.66) and take derivatives with respect to v and z, which gives the "second order" semiflow on $\mathbb{R}_0^+ \times (\mathbb{R}^n)^4$,

$$\widetilde{\Theta}_\tau(s, v, z, \widetilde{v}, \widetilde{z})$$

$$= \left(\Psi_\tau(s, v), \frac{\partial \Psi_\tau}{\partial v}(s, v)z, \frac{\partial \Psi_\tau}{\partial v}(s, v)\widetilde{v}, \left(\frac{\partial^2 \Psi_\tau}{\partial v^2}(s, v) \right) (z, \widetilde{v}) + \frac{\partial \Psi_\tau}{\partial v}(s, v)\widetilde{z} \right).$$

In particular, one can establish similar inequalities to those in (5.62) and (5.63), or equivalently to (5.78), that is, for every $s \geq 0$ and $\tau \geq 0$,

$$\|\widetilde{\Theta}_\tau(s, p_1) - \widetilde{\Theta}_\tau(s, p_2)\| \leq De^{\bar{a}\tau + as}\|p_1 - p_2\|$$

whenever $(s, p_i) = (s, v_i, z_i, \widetilde{v}_i, \widetilde{z}_i) \in T(T\mathcal{V}) \cap (\{s\} \times (\mathbb{R}^n)^4)$ for $i = 1, 2$.

We now make some comments on the assumption $q > k$ in the condition C2. This is caused by the nonuniformity in the exponential dichotomy, that is, the constants a and b in (2.17). Namely, the norms of the operators $U(t, s)$ and $V(t, s)$ may increase with exponential speed along a given orbit, and in its turn this may cause that one is not able to control the solutions for an arbitrary perturbation f. More precisely, we require some *a priori control* of the perturbation. In the present section, the corresponding condition $q > k$ is a consequence of the nonuniformity. Namely, in the proof we need that f and its derivatives up to order k are Lipschitz, with sufficiently small Lipschitz constants, in appropriate neighborhoods of the origin with size decreasing exponentially to zero along each orbit (due to the nonuniformity). Unfortunately, when $q \geq k$ this is in general not true for the k-th derivative unless $q > k$.

A somewhat related possible approach could be to require an extra amount of differentiability, such as in the nonuniform hyperbolicity theory in which it is customary to assume the dynamics to be of class $C^{1+\varepsilon}$. Incidentally, we note that in [77] Pugh gave an explicit example of a C^1 diffeomorphism which is not $C^{1+\varepsilon}$ for any $\varepsilon \in (0, 1)$, for which there exists no stable manifold theorem (see also Section 5.3). Of course that this not mean that all C^1 diffeomorphisms and flows with nonuniformly hyperbolic trajectories which lack higher regularity have no stable or unstable invariant manifolds. In fact, we showed in Section 5.3 that replacing the $C^{1+\varepsilon}$ hypothesis by the above condition C2, now with $q > 1$, there exist classes of C^1 vector fields which need not be $C^{1+\varepsilon}$ for any $\varepsilon \in (0, 1)$ but for which each nonuniformly hyperbolic trajectory possesses an invariant stable manifold.

We note that if in Theorem 5.15 we still have the condition in (5.57) but now with $q \leq k$ (where k continues to be the degree of differentiability of A and f), then we can show the existence of an invariant stable manifold $\mathcal{V}$ of class C^r where r is the largest integer smaller than q. In another direction, we would like to comment that in some situations, such as in dimension one, it is easy to reduce the perturbation to one satisfying the condition C2. Namely, it is sufficient to make the change of variables $v = u^p$ for a sufficiently large odd integer $p \in \mathbb{N}$. We also point out that in the case of *uniform exponential dichotomies*, that is, when $a = b = 0$, we do not require the condition $q > k$ and it can be replaced by $q \geq 1$ both in Theorems 5.1 and 5.15. This follows from a simple inspection of the proofs, although since in the uniform case the corresponding results are well known we do not include the corresponding discussion.

5.6 Proof of the C^k regularity

5.6.1 Method of proof

The proof of Theorem 5.15 is based on the extension of the vector field of the original equation by its differential. The regularity of the stable manifold of the original equation will be obtained "integrating" the corresponding second component of the stable manifold of the extended vector field, thus gaining one additional derivative in the process. This allows us to proceed by induction on k, the regularity of the vector field. We briefly describe the main elements of the argument:

1. The induction step is based on the linear extension of the equation (4.4) by the linear variational equation along a solution $v(t)$ of (4.4), that is, of the equation

$$z' = A(t)z + \frac{\partial f}{\partial v}(t, v(t))z. \tag{5.65}$$

   The extended vector field is of class C^{k-1}. Thus, provided that all the conditions in Theorem 5.15 are satisfied with k replaced by $k - 1$, we obtain by induction a stable manifold $\widetilde{\mathcal{V}}$ of class C^{k-1} for the equation defined by the extended vector field (see Section 5.6.2). We recall that the case when $k = 1$ was already considered in Section 5.1.
2. The invariant manifold $\widetilde{\mathcal{V}}$ has essentially two components, namely the stable manifold $\mathcal{V}$ for the original equation (4.4), since the first component of the extended vector field coincides with the original vector field (the right-hand side of (4.4)), and a vector bundle B over this first component. This follows from the fact that the equation (5.65) is linear in z. Furthermore, both manifolds $\mathcal{V}$ and B are of class C^{k-1}. If we can show that the second component B of $\widetilde{\mathcal{V}}$ coincides with the tangent bundle $T\mathcal{V}$ of the stable manifold, then $\mathcal{V}$ is of class C^k (since the tangent bundle is of class C^{k-1}). The details of this claim are given in Section 5.6.5.
3. We show that indeed $B = T\mathcal{V}$ by carefully studying the behavior of solutions along the stable manifold of the extension. More precisely, we first obtain a characterization of the vectors in $T\mathcal{V}$ in terms of the norms of the action of the differential of Ψ_τ (see Section 5.6.3). We then establish an exponential decay of the tangential component of the solutions of the extended equation (see Section 5.6.4) or equivalently of the solutions of the equation (5.65) on the stable manifold $\widetilde{\mathcal{V}}$. Combining the two results we are finally able to show that $B \subset T\mathcal{V}$ (see Section 5.6.4). Since the vector bundles B and $T\mathcal{V}$ have the same dimension they must be equal. This completes the proof of Theorem 5.15.

We assume from now on that A and f are of class C^k, for some $k \geq 2$, and that C1 and C2 hold. We also assume that the equation $v' = A(t)v$ admits a nonuniform exponential dichotomy in $\mathbb{R}^+$ and that the conditions in (5.58) hold.

We will refer to the statement in Theorem 5.15 as the *induction hypothesis for* k. We divide the proof into several steps.

5.6.2 Linear extension of the vector field

We consider the new vector field $X \colon \mathbb{R}_0^+ \times \mathbb{R}^n \times \mathbb{R}^n \to \mathbb{R}^n \times \mathbb{R}^n$ given by

$$X(t, v, z) = \left(A(t)v + f(t, v), A(t)z + \frac{\partial f}{\partial v}(t, v)z \right). \tag{5.66}$$

We note that X is a linear extension of its first component. The corresponding nonautonomous differential equation is

$$(v', z') = X(t, v, z). \tag{5.67}$$

Observe that the first component $v(t)$ of a solution of (5.67) satisfies the decoupled equation

$$v' = A(t)v + f(t, v). \tag{5.68}$$

The second component $z(t)$ of a solution of (5.67) satisfies the linear variational equation obtained from (5.68), that is,

$$z' = A(t)z + \frac{\partial f}{\partial v}(t, v(t))z. \tag{5.69}$$

Therefore, if Ψ_τ is the semiflow in (4.7), the autonomous equation

$$(t', v', z') = (1, X(t, v, z)) \tag{5.70}$$

generates the semiflow Θ_τ on $\mathbb{R}_0^+ \times \mathbb{R}^n \times \mathbb{R}^n$ given by

$$\Theta_\tau(s, v, z) = \left(\Psi_\tau(s, v), \frac{\partial \Psi_\tau}{\partial v}(s, v)z \right). \tag{5.71}$$

Note that writing

$$\widetilde{A}(t) = \begin{pmatrix} A(t) & 0 \\ 0 & A(t) \end{pmatrix} \quad \text{and} \quad \widetilde{f}(t, v, z) = \left(f(t, v), \frac{\partial f}{\partial v}(t, v)z \right),$$

and using the new variable $p = (v, z)$, we can rewrite (5.70) in the form

$$t' = 1, \quad p' = \widetilde{A}(t)p + \widetilde{f}(t, p).$$

Clearly, the matrix $\widetilde{A}(t)$ satisfies the condition C1. The evolution operator associated with $\widetilde{A}(t)$ is given by

$$\begin{pmatrix} U(t,s) & 0 \\ 0 & V(t,s) \end{pmatrix} \oplus \begin{pmatrix} U(t,s) & 0 \\ 0 & V(t,s) \end{pmatrix}.$$

We now verify that the extended vector field satisfies the remaining hypotheses of Theorem 5.1. We consider the norm $\|(v,z)\| = \|v\| + \|z\|$ for $(v,z) \in \mathbb{R}^n \times \mathbb{R}^n$.

Lemma 5.16. *The following properties hold:*

1. $\widetilde{f}$ *is of class* C^{k-1} *and* $\widetilde{f}(t,0) = 0$ *for every* $t \geq 0$;
2. *there exists* $C > 0$ *such that for every* $j = 0, \ldots, k-2$, $t \geq 0$, *and* $p_1, p_2 \in \mathbb{R}^n \times \mathbb{R}^n$ *we have*

$$\left\| \frac{\partial^j \widetilde{f}}{\partial p^j}(t,p_1) - \frac{\partial^j \widetilde{f}}{\partial p^j}(t,p_2) \right\| \leq C\|p_1 - p_2\|(\|p_1\|^{q-j} + \|p_2\|^{q-j}). \tag{5.72}$$

Proof. The first property follows immediately from the definitions. We now establish the second property. Since by hypothesis f is of class C^k, it follows from (5.57) that for every $j = 0, \ldots, k-1$ and $(t,v,z) \in \mathbb{R}_0^+ \times \mathbb{R}^n \times \mathbb{R}^n$,

$$\begin{aligned}
\left\| \frac{\partial^{j+1} f}{\partial v^{j+1}}(t,v)z \right\| &= \lim_{h \to 0} \frac{1}{|h|} \left\| \frac{\partial^j f}{\partial v^j}(t,v+hz) - \frac{\partial^j f}{\partial v^j}(t,v) \right\| \\
&\leq c \lim_{h \to 0} \frac{\|v + hz - v\|}{|h|}(\|v + hz\|^{q-j} + \|v\|^{q-j}) \\
&\leq 2c\|v\|^{q-j}\|z\|.
\end{aligned} \tag{5.73}$$

Therefore, $\|(\partial^{j+1} f/\partial v^{j+1})(t,v)\| \leq 2c\|v\|^{q-j}$. Set $p_i = (v_i, z_i)$ for $i = 1, 2$. Again by (5.57), for $j = 0, \ldots, k-2$ we obtain

$$\begin{aligned}
& \left\| \frac{\partial^{j+1} f}{\partial v^{j+1}}(t,v_1)z_1 - \frac{\partial^{j+1} f}{\partial v^{j+1}}(t,v_2)z_2 \right\| \\
&\leq \left\| \frac{\partial^{j+1} f}{\partial v^{j+1}}(t,v_1) - \frac{\partial^{j+1} f}{\partial v^{j+1}}(t,v_2) \right\| \cdot \|z_1\| + \left\| \frac{\partial^{j+1} f}{\partial v^{j+1}}(t,v_2) \right\| \cdot \|z_1 - z_2\| \\
&\leq c\|v_1 - v_2\|(\|v_1\|^{q-j-1} + \|v_2\|^{q-j-1})\|z_1\| + 2c\|v_2\|^{q-j}\|z_1 - z_2\| \\
&\leq c\|p_1 - p_2\|(\|p_1\|^{q-j-1} + \|p_2\|^{q-j-1})\|p_1\| + 2c\|p_2\|^{q-j}\|p_1 - p_2\| \\
&\leq 2c\|p_1 - p_2\|(\|p_1\| + \|p_2\|)^{q-j-1}(\|p_1\| + \|p_2\|) \\
&\quad + 2c(\|p_1\| + \|p_2\|)^{q-j}\|p_1 - p_2\| \\
&= 4c\|p_1 - p_2\|(\|p_1\| + \|p_2\|)^{q-j}.
\end{aligned}$$

Since all norms in $\mathbb{R}^2$ are equivalent and $q - j > 1$ (since $q > k \geq j+2$), there exists a universal constant $d = d(q,k)$ such that

$$\|p_1\| + \|p_2\| \leq d(\|p_1\|^{q-j} + \|p_2\|^{q-j})^{1/(q-j)}.$$

Therefore,

$$\left\| \frac{\partial^{j+1} f}{\partial v^{j+1}}(t, v_1)z_1 - \frac{\partial^{j+1} f}{\partial v^{j+1}}(t, v_2)z_2 \right\| \le 4cd^{q-j}\|p_1 - p_2\|(\|p_1\|^{q-j} + \|p_2\|^{q-j}).$$
$$(5.74)$$

In order to establish (5.72), it suffices to observe that for $j = 0, \ldots, k-2$ each component of the derivative

$$\frac{\partial^j \tilde{f}}{\partial p^j}(t, p) = \frac{\partial^j}{\partial p^j}\left(f(t, v), \frac{\partial f}{\partial v}(t, v)z\right) \equiv \left(\frac{\partial^j f}{\partial v^j}(t, v), \frac{\partial^j}{\partial p^j}\left(\frac{\partial f}{\partial v}(t, v)z\right)\right)$$

is a linear combination of components of the derivatives $(\partial^{j+1} f/\partial v^{j+1})(t, v)z$ and $(\partial^j f/\partial v^j)(t, v)$. Therefore, by (5.57) and (5.74) we obtain (5.72) (for some constant C depending only on q and k). $\qquad\square$

We now consider the set

$$Y_\omega = \{(s, \xi, u) : s \ge 0 \text{ and } \xi, u \in R_s(\delta e^{-\omega s})\},$$

and let $\mathcal{Y}_\omega$ be the space of continuous functions $\Phi \colon Y_\omega \to \mathbb{R}^n$ such that:

1. $\Phi(s, \xi, u) \in F(s)$ for each $s \ge 0$ and $\xi, u \in R_s(\delta e^{-\omega s})$;
2. the function $u \mapsto \Phi(s, \xi, u)$ is linear for each $(s, \xi) \in X_\omega$;
3. for each $s \ge 0$, and $\xi, u, \bar{\xi}, \bar{u} \in R_s(\delta e^{-\omega s})$ we have

$$\|\Phi(s, \xi, u) - \Phi(s, \bar{\xi}, \bar{u})\| \le \|\xi - \bar{\xi}\| + \|u - \bar{u}\|. \qquad (5.75)$$

Note that in particular, given $\Phi \in \mathcal{Y}_\omega$ we have

$$\Phi(s, \xi, 0) = 0 \text{ for every } s \ge 0 \text{ and } \xi \in R_s(\delta e^{-\omega s}). \qquad (5.76)$$

The induction hypothesis allows us to obtain a stable manifold of the origin for the equation (5.67). We continue to write $p_{s,\xi} = (s, \xi, \varphi(s, \xi))$.

Lemma 5.17. *If the induction hypothesis holds for $k-1$, then for each $\varrho > 0$ there exist $\delta > 0$ and unique functions $\varphi \in X_{\omega - a}$ and $\Phi \in \mathcal{Y}_{\omega - a}$ such that the set*

$$\tilde{\mathcal{V}} = \{(p_{s,\xi}, u, \Phi(s, \xi, u)) : (s, \xi, u) \in Y_{\omega - a} \text{ and } s > \varrho\}, \qquad (5.77)$$

is a $(2 \dim E(s) + 1)$-dimensional stable manifold of class C^{k-1} which is forward invariant under the semiflow Θ_τ, in the sense that if $(s, \xi, u) \in Y_\omega$ then

$$\Theta_\tau(p_{s,\xi}, u, \Phi(s, \xi, u)) \in \tilde{\mathcal{V}} \text{ for every } \tau \ge 0.$$

In addition, there exists a constant $D > 0$ such that for every $s \ge 0$, $\xi, \bar{\xi}, u, \bar{u} \in R_s(\delta e^{-\omega s})$, and $\tau \ge 0$ we have

$$\|\Theta_\tau(p_{s,\xi}, u, \Phi(s, \xi, u)) - \Theta_\tau(p_{s,\bar{\xi}}, \bar{u}, \Phi(s, \bar{\xi}, \bar{u}))\| \le De^{\bar{a}\tau + as}(\|\xi - \bar{\xi}\| + \|u - \bar{u}\|). \qquad (5.78)$$

Proof. Since the first component of (5.67) is the decoupled equation (5.68) (and the first component of Θ_τ in (5.71) is the semiflow Ψ_τ; see (4.7)), the first components (s, ξ, η) of a solution in a stable manifold of the origin for the equation (5.67) must satisfy the forward invariance property of the manifold $\mathcal{V}$ in Theorem 5.15 (for the equation in (4.4)). Therefore, in view of Lemma 5.16 (and of the discussion before the lemma), the existence of a unique stable manifold $\widetilde{\mathcal{V}}$ of class C^{k-1} in (5.77) follows immediately from Theorem 5.15, although with the linearity assumption in the definition of $\mathcal{Y}_{\omega-a}$ replaced by the condition $\Phi(s, 0, 0) = 0$ for every $s \geq 0$.

Using the fact that the equation in (5.69) is linear in z, we can show that for each $(s, \xi) \in X_{\omega-a}$ the section

$$\widetilde{\mathcal{V}}(s, \xi) = \{(p_{s,\xi}, u, \Phi(s, \xi, u)) : u \in R_s(\delta e^{-(\omega-a)s})\}$$

is an open subset of a linear space. This can be seen in the following manner. Since $\widetilde{\mathcal{V}}(s, \xi)$ projects onto the space $E(s)$ (on the component u), given (s, ξ) we can choose linearly independent vectors $u_1, \ldots, u_\ell \in R_s(\delta e^{-(\omega-a)s})$, where $\ell = \dim E(s)$, and define the ℓ-dimensional space

$$\widetilde{\mathcal{V}}^*(s, \xi) = \{p_{s,\xi}\} \times E(s, \xi),$$

where

$$E(s, \xi) = \mathrm{span}\{(u_i, \Phi(s, \xi, u_i)) : i = 1, \ldots, \ell\}.$$

Since the second component of Θ_τ is linear in the variable z, the set $\widetilde{\mathcal{V}}^* = \{\Theta_\tau(\widetilde{\mathcal{V}}^*(s, \xi)) : \tau \geq 0\}$ is a smooth manifold with the property that

$$(\{\Psi_\tau(p_{s,\xi})\} \times \mathbb{R}^n) \cap \widetilde{\mathcal{V}}^* = \{\Psi_\tau(p_{s,\xi})\} \times \left(\frac{\partial \Psi_\tau}{\partial v}(p_{s,\xi})\right) E(s, \xi)$$

is a linear space of dimension ℓ for each $\tau \geq 0$. Furthermore, the manifold $\widetilde{\mathcal{V}}^*$ is forward invariant and projects over $\mathbb{R}^+ \times E(s)$. In particular, $\widetilde{\mathcal{V}}^*$ and $\widetilde{\mathcal{V}}$ have the same dimension. In view of the uniqueness of the manifold $\widetilde{\mathcal{V}}$ among those with the same dimension (and over the same vector space) which are forward invariant, we conclude that $\widetilde{\mathcal{V}}$ coincides with an open subset of a linear space. Therefore, the function $u \mapsto \Phi(s, \xi, u)$ must be linear for each (s, ξ). This completes the proof of the lemma. $\qquad \square$

5.6.3 Characterization of the stable spaces

For each $s \geq \varrho$, we consider the tangent set

$$V(p_{s,\xi}) = T_{p_{s,\xi}} \mathcal{V} \cap (\{s\} \times \mathbb{R}^n) \tag{5.79}$$

(note that by Theorem 5.1 the set $\mathcal{V}$ is a smooth manifold of class C^1). We will refer to each of these sets as a *stable space*. We give a characterization of the vectors in the stable spaces $V(p_{s,\xi})$ in (5.79). We consider the Lyapunov norms introduced in Section 5.4.2 (see (5.16)). For simplicity of the notation, from now on we will also write $\partial_{(s,v)} \Psi_\tau$ for the partial derivative $(\partial \Psi_\tau / \partial v)(s, v)$.

Lemma 5.18. *Given $(s, \xi) \in X_\omega$ and $(u, w) \in E(s) \times F(s)$, if the function*

$$\mathbb{R}_0^+ \ni \tau \mapsto \|(\partial_{p_{s,\xi}}\Psi_\tau)(u, w)\|'_{s+\tau} \text{ is bounded,} \qquad (5.80)$$

then $(u, w) \in V(p_{s,\xi})$.

Proof. We proceed by contradiction. For each $t \geq s$, we write $(u(t), w(t)) = (\partial_{p_{s,\xi}}\Psi_{t-s})(u, w)$. Note that $(u(s), w(s)) = (u, w)$. The space $V(p_{s,\xi})$ satisfies the invariance property

$$(\partial_{p_{s,\xi}}\Psi_\tau)V(p_{s,\xi}) = V(\Psi_\tau(p_{s,\xi})).$$

Therefore, it is sufficient to prove that $(u(t), w(t)) \in V(\Psi_{t-s}(p_{s,\xi}))$ for some $t \geq s$. We now write $(u(s), w(s))$ in the form

$$(u(s), w(s)) = (\bar{u}(s), \bar{w}(s)) + (0, z(s)) \qquad (5.81)$$

with $(\bar{u}(s), \bar{w}(s)) \in V(p_{s,\xi})$. This is always possible since the "vertical" vectors $(0, z) \in F(s)$ and the tangent vectors in $V(p_{s,\xi})$ (which are obtained from the tangent vectors to the graph of the Lipschitz function φ in (4.10)) generate the whole space $\mathbb{R}^n$. We want to prove that $z(s) = 0$. Applying $(\partial_{p_{s,\xi}}\Psi_{t-s})$ to both sides in (5.81) we obtain

$$(u(t), w(t)) = (\bar{u}(t), \bar{w}(t)) + (\partial_{p_{s,\xi}}\Psi_{t-s})(0, z(s)), \qquad (5.82)$$

where

$$(\bar{u}(t), \bar{w}(t)) = (\partial_{p_{s,\xi}}\Psi_{t-s})(\bar{u}(s), \bar{w}(s)). \qquad (5.83)$$

Replacing (v, w) by $(\bar{u}(t), \bar{w}(t))$ in (5.32) in Lemma 5.9 (note that by (5.60) we have $X_\omega \subset X_\gamma$), and using (5.83) we obtain

$$\|(\bar{u}(t), \bar{w}(t))\|'_t \leq 4e^{(\bar{a}+\varsigma)\tau}\|(u, w)\|'_s.$$

In particular, the pair $(\bar{u}(t), \bar{w}(t))$ is bounded in t. By the hypothesis in the lemma, the pair $(u(t), w(t))$ is also bounded in t, and hence, by (5.82), the derivative $(\partial_{p_{s,\xi}}\Psi_{t-s})(0, z(s))$ must also be bounded in t. But by (5.35) in Lemma 5.10 the norm of this derivative increases exponentially with t whenever $z(s) \neq 0$. Therefore, we must have $z(s) = 0$ and the desired statement follows from (5.81). $\qquad \square$

5.6.4 Tangential component of the extension

We will write from now on, for each $t \geq s$,

$$\Theta_{t-s}(s, \xi, \varphi(s, \xi), u(s), \Phi(s, \xi, u(s))) = (t, x(t), y(t), u(t), w(t)). \qquad (5.84)$$

By Lemma 5.17, for each $(s, \xi, u(s)) \in Y_\omega$ we have

$$y(t) = \varphi(t, x(t)) \quad \text{and} \quad w(t) = \Phi(t, x(t), u(t)). \tag{5.85}$$

We want to show that for each $t > s$ the component $u(t)$ in (5.84) can be estimated in terms of $u(s)$ solely, without needing the component ξ of the initial condition. The stable component of (5.69) can be written in the form

$$u(t) = U(t, s)u(s) + \int_s^t U(t, \tau)\left[\frac{\partial f}{\partial x}u(\tau) + \frac{\partial f}{\partial y}\Phi(\tau, x(\tau), u(\tau))\right] d\tau \tag{5.86}$$

with the partial derivatives of f computed at the point $(\tau, x(\tau), y(\tau))$, where $y(\tau) = \varphi(\tau, x(\tau))$ for each τ.

Lemma 5.19. *There exists $M > 0$ such that for every $\delta > 0$ sufficiently small, given $(s, \xi, u(s)) \in Y_\omega$ we have*

$$\|u(t)\|_t' \le Me^{(\bar{a}+\varsigma)(t-s)}\|u(s)\|_s' \text{ for every } t \ge s.$$

Proof. We consider the space

$$\mathcal{B} = \{u \colon [s, +\infty) \to \mathbb{R}^n \text{ continuous: } u(t) \in E(t) \text{ for } t \ge s \text{ and } \|u\|' \le N\},$$

where $N = \delta e^{-(\omega+a)s}/C_1'$ (we recall that δ is fixed, and the constant $C_1' = C_1^{-1}$ is given by Lemma 5.7), with the norm

$$\|u\|' = \sup\{\|u(t)\|_t' e^{-(\bar{a}+\varsigma)(t-s)} : t \ge s\}.$$

One can easily verify, with the help of Lemma 5.7, that with the distance induced by this norm $\mathcal{B}$ is a complete metric space. By (5.17) and (5.58) we have

$$\bar{a} + a + \varsigma + \omega \le \bar{a} + 5\vartheta(1 + 1/q) \le 0.$$

Therefore, for each $u \in \mathcal{B}$, it follows from Lemma 5.7 that

$$\|u(t)\| \le C_1' e^{at}\|u(t)\|_t' \le \delta e^{-\omega s}e^{(\bar{a}+a+\varsigma)(t-s)} \le \delta e^{-\omega t}e^{(\bar{a}+a+\varsigma+\omega)(t-s)},$$

and hence $u(t) \in R_t(\delta e^{-\omega t})$ for every $t \ge s$. On the other hand, by Lemma 4.5, provided that δ is sufficiently small, and since $\xi \in R_s(\delta e^{-\omega s})$, we obtain

$$\|x(t)\| \le K_1 e^{\bar{a}(t-s)+as}\|\xi\| \le K_1\delta e^{-\omega s}e^{\bar{a}(t-s)+as} \le K_1\delta e^{-(\omega-a)s}. \tag{5.87}$$

Therefore, in view of (5.75) and since $\Phi \in \mathcal{Y}_{\omega-a}$, for each $u, \bar{u} \in \mathcal{B}$ we have

$$\|\Phi(t, x(t), u(t)) - \Phi(t, x(t), \bar{u}(t))\| \le \|u(t) - \bar{u}(t)\|. \tag{5.88}$$

We now define an operator J for each $u \in \mathcal{B}$ and $t \ge s$ by

$$(Ju)(t) = \int_s^t U(t, \tau)G(\tau, x(\tau), u(\tau)) d\tau,$$

where (see (5.86))

$$G(\tau, x(\tau), u(\tau)) = \frac{\partial f}{\partial x} u(\tau) + \frac{\partial f}{\partial y} \Phi(\tau, x(\tau), u(\tau)) \qquad (5.89)$$

with the partial derivatives of f computed at the point $(\tau, x(\tau), \varphi(\tau, x(\tau))) = (\tau, x(\tau), y(\tau))$. Proceeding as in (5.73) with $j = 0$, it follows from (5.57) that for every $u \in E(t)$,

$$\begin{aligned}
\left\| \frac{\partial f}{\partial x}(t, x, y) u \right\| &= \lim_{h \to 0} \frac{\|f(t, x + hu, y) - f(t, x, y)\|}{|h|} \\
&\leq c \lim_{h \to 0} \frac{\|x + uh - x\|(\|(x + uh, y)\|^q + \|(x, y)\|^q)}{|h|} \\
&\leq 2c\|(x, y)\|^q \|u\|.
\end{aligned}$$

Therefore, $\|\partial f/\partial x\| \leq 2c\|(x, y)\|^q$. Since

$$\|(x(\tau), y(\tau))\| \leq \|x(\tau)\| + \|y(\tau)\|, \quad \|y(\tau)\| = \|y(\tau) - y(0)\| \leq \|x(\tau)\|,$$

it follows from Theorem 4.1 (more precisely, from the inequality (4.16)) and (5.87) that

$$\begin{aligned}
\left\| \frac{\partial f}{\partial x}(\tau, x(\tau), y(\tau)) \right\| &\leq 2c(\|x(\tau)\| + \|y(\tau)\|)^q \leq c2^{q+1}\|x(\tau)\|^q \\
&\leq c2^{q+1} D^q e^{q\bar{a}(t-s)+qas} \|\xi\|^q \qquad (5.90) \\
&= c2^{q+1} D^q \delta^q e^{q\bar{a}(t-s)} e^{-q(\omega-a)s}.
\end{aligned}$$

The same bound can be obtained for the derivative $\partial f/\partial y$ at the point $(\tau, x(\tau), y(\tau))$. Furthermore, given $u, \bar{u} \in \mathcal{B}$ we have

$$\begin{aligned}
&G(\tau, x(\tau), u(\tau)) - G(\tau, x(\tau), \bar{u}(\tau)) \\
&= \frac{\partial f}{\partial x}(u(\tau) - \bar{u}(\tau)) + \frac{\partial f}{\partial y}[\Phi(\tau, x(\tau), u(\tau)) - \Phi(\tau, x(\tau), \bar{u}(\tau))].
\end{aligned}$$

By (5.88), (5.90) (and the corresponding inequality for the derivative $\partial f/\partial y$), and Lemma 5.7, we obtain

$$\begin{aligned}
a(\tau) :&= \|G(\tau, x(\tau), u(\tau)) - G(\tau, x(\tau), \bar{u}(\tau))\| \\
&\leq c2^{q+2} D^q \delta^q e^{q\bar{a}(\tau-s)} e^{-q(\omega-a)s} \|u(\tau) - \bar{u}(\tau)\| \\
&\leq c2^{q+2} D^q \delta^q e^{q\bar{a}(\tau-s)} C_1' e^{a\tau} e^{-q(\omega-a)s} \|u(\tau) - \bar{u}(\tau)\|_\tau' \\
&\leq c2^{q+2} D^q \delta^q C_1' e^{((q+1)\bar{a}+a+\varsigma)(\tau-s)} e^{-q(\omega-a)s+as} \|u - \bar{u}\|'.
\end{aligned}$$

Therefore,

$$\|(Ju)(t) - (J\bar{u})(t)\|'_t$$

$$= \int_t^{+\infty} \|U(r,t)((Ju)(t) - (J\bar{u})(t))\|e^{-(\bar{a}+\varsigma)(r-t)}\,dr$$

$$\leq \int_t^{+\infty} \left(\int_s^t \|U(r,\tau)\|a(\tau)\,d\tau \right) e^{-(\bar{a}+\varsigma)(r-t)}\,dr$$

$$\leq c2^{q+2}D^q\delta^q C'_1 e^{-q(\omega-a)s+as}\|u-\bar{u}\|'$$

$$\times \int_t^{+\infty} \left(\int_s^t D_1 e^{\bar{a}(r-\tau)+a\tau} e^{((q+1)\bar{a}+a+\varsigma)(\tau-s)}\,d\tau \right) e^{-(\bar{a}+\varsigma)(r-t)}\,dr$$

$$= cD_1 2^{q+2}D^q\delta^q C'_1 e^{-q(\omega-a)s+2as}\|u-\bar{u}\|' e^{(\bar{a}+\varsigma)(t-s)}$$

$$\times \int_s^t \left(e^{(q\bar{a}+2a)(\tau-s)} \int_t^{+\infty} e^{-\varsigma(r-\tau)}\,dr \right) d\tau$$

$$\leq \frac{cD_1 2^{q+2}D^q\delta^q C'_1}{\varsigma} e^{-q(\omega-a)s+2as}\|u-\bar{u}\|' e^{\bar{a}(t-s)} \int_s^t e^{T_3(\tau-s)}\,d\tau,$$

where $T_3 = q\bar{a} + 2a + \varsigma$. Note that in view of the first inequality in (5.18) and (5.58), we have

$$T_3 \leq q\bar{a} + 3a \leq q\bar{a} + 4\vartheta < 0.$$

Furthermore, in view of (5.59) we have $e^{-q(\omega-a)s+2as} \leq 1$. Therefore,

$$\|Ju - J\bar{u}\|' \leq \theta\|u - \bar{u}\|', \tag{5.91}$$

where

$$\theta = cD_1 2^{q+2}D^q\delta^q C'_1/(\varsigma|T_3|).$$

For $u = 0$ we obtain $Ju = 0$ (note that by (5.76) we have that $\Phi(s,\xi,0) = 0$ for every $(s,\xi) \in X_\omega$, and thus, by (5.91), $\|Ju\|' \leq \theta\|u\|'$.

We now choose $\delta > 0$ sufficiently small, independently of s, so that $\theta < 1/2$. Given $u(s) \in R_s(Ne^{-as}/(2C'_1))$, we consider the operator $\bar{J}$ on the space $\mathcal{B}$ defined by

$$(\bar{J}u)(t) = T(t) + (Ju)(t),$$

where $T(t) = U(t,s)u(s)$. For each $t \geq s$, we have

$$\|T(t)\|'_t = \int_t^{+\infty} \|U(r,t)T(t)\|e^{-(\bar{a}+\varsigma)(r-t)}\,dr$$

$$= e^{(\bar{a}+\varsigma)(t-s)} \int_t^{+\infty} \|U(r,s)u(s)\|e^{-(\bar{a}+\varsigma)(r-s)}\,dr$$

$$\leq e^{(\bar{a}+\varsigma)(t-s)}\|u(s)\|'_s.$$

Since $u(s) \in R_s(Ne^{-as}/(2C'_1))$, it follows from Lemma 5.7 that

$$\|T\|' \leq \|u(s)\|'_s \leq C'_1 e^{as}\|u(s)\| \leq \frac{N}{2}.$$

Therefore,

$$\|\bar{J}u\|' \leq \|T\|' + \|Ju\|' \leq \|u(s)\|_s' + \theta\|u\|' \leq \frac{N}{2} + \frac{N}{2} = N, \qquad (5.92)$$

and $\bar{J} \colon \mathcal{B} \to \mathcal{B}$ is a well-defined operator. In view of (5.91),

$$\|\bar{J}u - \bar{J}\bar{u}\|' = \|Ju - J\bar{u}\|' \leq \theta\|u - \bar{u}\|',$$

and $\bar{J}$ is a contraction in the complete metric space $\mathcal{B}$. Hence, there exists a unique function $u \in \mathcal{B}$ such that $\bar{J}u = u$. It follows from (5.92) that $\|u\|' \leq \|u(s)\|_s' + \theta\|u\|'$, and hence,

$$\|u\|' \leq \frac{\|u(s)\|_s'}{1 - \theta}.$$

Therefore, for every $u(s) \in R_s(Ne^{-as}/(2C_1'))$ and $t \geq s$,

$$\|u(t)\|_t' \leq e^{(\bar{a}+\varsigma)(t-s)}\frac{\|u(s)\|_s'}{1 - \theta}.$$

To establish this inequality for $u(s)$ outside $R_s(Ne^{-as}/(2C_1'))$ it suffices to observe that in view of (5.86) (see also (5.89)) the function $u(s) \mapsto u(t)$ is linear (for each s and t), and thus it suffices to consider initial conditions $u(s)$ in an arbitrarily small ball. Thus, we obtain the desired result setting $M = 1/(1 - \theta)$. $\qquad \square$

We now show that each tangential component of the extension belongs to a stable space.

Lemma 5.20. *For every $s \geq 0$ and $\xi, u \in R_s(\delta e^{-\omega s})$ we have*

$$(u, \Phi(s, \xi, u)) \in V(p_{s,\xi}).$$

Proof. In view of Lemma 5.18, it is enough to prove that for the vector (u, w) with $w = \Phi(s, \xi, u)$ the function

$$(u(t), w(t)) = (\partial_{p_{s,\xi}}\Psi_{t-s})(u, w) \qquad (5.93)$$

in (5.80) is bounded for $\tau \geq 0$. Since $w(t) = \Phi(t, x(t), u(t))$ (see (5.85)), it follows from (5.75) that $\|w(t)\| \leq \|u(t)\|$. By Lemma 5.7, setting $C_2' = D_2 e^{\varsigma(\rho-s)}/\varsigma$, we obtain

$$\|w(t)\|_t' \leq C_2'e^{(b+\varsigma)t}\|w(t)\| \leq C_2'e^{(b+\varsigma)t}\|u(t)\| \leq C_1'C_2'e^{(a+b+\varsigma)t}\|u(t)\|_t',$$

for every $t \geq s$, and thus, since $C_1', C_2' \geq 1$,

$$\|(u(t), w(t))\|_t' \leq (1 + C_1'C_2'e^{(a+b+\varsigma)t})\|u(t)\|_t' \leq 2C_1'C_2'e^{(a+b+\varsigma)t}\|u(t)\|_t'.$$

It follows from Lemma 5.19 that

$$
\begin{aligned}
\|(u(t), w(t))\|'_t &\le 2C'_1 C'_2 M e^{(\bar{a}+a+b+2\varsigma)(t-s)} e^{(a+b+\varsigma)s} \|u(s)\|'_s \\
&\le 2C'_1 C'_2 M e^{(\bar{a}+a+b+2\varsigma)(t-s)} e^{(a+b+\varsigma)s} \|(u(s), w(s))\|'_s.
\end{aligned}
\tag{5.94}
$$

By (5.58), provided that ς is sufficiently small we have $\bar{a} + a + b + 2\varsigma < 0$. Thus, for each fixed $s \ge 0$ the quantities in (5.94) are bounded for $\tau \ge 0$. By (5.93), the function in (5.80) is also bounded for $\tau \ge 0$. This completes the proof of the lemma. $\qquad\square$

5.6.5 C^k regularity of the stable manifolds

We have now all the necessary tools to establish that the stable manifold $\mathcal{V}$ of the origin for the equation (4.4) is of class C^k when the functions A and f are of class C^k.

Proof of Theorem 5.15. When $k = 1$, the statement is a simple consequence of Theorems 4.1 and 5.1 (note that the second condition in (5.58) implies the second condition in (5.1)). We now proceed by induction on k. Assume that the induction hypothesis (see the initial paragraph of Section 5.6) holds for $k - 1$, for some $k \ge 2$. We consider the extended vector field $\tilde{f}$, and by Lemma 5.17 we obtain the C^{k-1} manifold $\tilde{\mathcal{V}}$ given by (5.77). By Lemma 5.20 and the fact that $\dim V(p_{s,\xi}) = \dim E(s)$ (see (5.79)) we obtain

$$
V(p_{s,\xi}) = \{(u, \Phi(s, \xi, u)) : u \in E(s)\},
\tag{5.95}
$$

where we continue to denote by $\Phi(s, \xi, \cdot)$ the unique linear extension of $R_s(\delta e^{-\omega s}) \ni u \mapsto \Phi(s, \xi, u)$ to the whole space $E(s)$. Therefore, in view of (5.77),

$$
\tilde{\mathcal{V}} \cap (\{p_{s,\xi}\} \times \mathbb{R}^n) = \{p_{s,\xi}\} \times V_\delta(p_{s,\xi})
$$

for every $s > \varrho$ and $\xi \in R_s(\delta e^{-\omega s})$, where

$$
V_\delta(p_{s,\xi}) = \{(u, \Phi(s, \xi, u)) : u \in R_s(\delta e^{-\omega s})\}
$$

is an open subset of $V(p_{s,\xi})$ containing the origin. That is,

$$
\tilde{\mathcal{V}} \cap (\{p\} \times \mathbb{R}^n) = \{p\} \times V_\delta(p) \text{ for every } p \in \mathcal{V}.
\tag{5.96}
$$

By (5.96) and the C^{k-1} regularity of the manifold $\tilde{\mathcal{V}}$, the map $\mathcal{V} \ni p \mapsto V(p)$ is of class C^{k-1}. Therefore,

$$
\xi \mapsto \varphi(s, \xi) \text{ is of class } C^k \text{ for each fixed } s.
\tag{5.97}
$$

We now consider, for each $s > 0$, a constant $\varepsilon = \varepsilon(s) > 0$ such that $s - \varepsilon > \varrho$, and the parametrization F_s in the proof of Theorem 5.1 (see (5.54)) of an open subset of $\mathcal{V}$ containing $(s, 0)$. Since A and f are of class C^k, the map

$(t, s, \xi, \eta) \mapsto \Psi_t(s, \xi, \eta)$ is also of class C^k (see for example [46]). By (5.97), we conclude that F_s is of class C^k (for each fixed s), and $\mathcal{V}$ is a smooth manifold of class C^k.

With the exception of the exponential decay of the derivatives in (5.63), the remaining properties are already given by Theorem 4.1. In particular, the manifold $\mathcal{V}$ satisfies the forward invariance property in (5.61). To establish the exponential decay of the derivatives we note that by (5.78), the second component of Θ_τ in (5.71) gives

$$\|(\partial_{p_{s,\xi}} \Psi_\tau) z_{s,\xi,u} - (\partial_{p_{s,\bar{\xi}}} \Psi_\tau) z_{s,\bar{\xi},\bar{u}}\| \leq De^{\bar{a}\tau + as}(\|\xi - \bar{\xi}\| + \|u - \bar{u}\|),$$

where $z_{s,\xi,u} = (u, \Phi(s, \xi, u))$, for every $s \geq 0$, ξ, $\bar{\xi}$, u, $\bar{u} \in R_s(\delta e^{-\omega s})$, and $\tau \geq 0$. By (5.97), the tangent vectors in $V(p_{s,\xi})$ (see (5.79)) are of the form

$$\frac{\partial}{\partial \xi}(\xi, \varphi(s, \xi))u = \left(u, \frac{\partial \varphi}{\partial \xi}(s, \xi)u\right) \quad \text{for some } u \in E(s).$$

But by the uniqueness of the function Φ in Lemma 5.17, it follows from (5.95) that $\Phi(s, \xi, u) = \frac{\partial \varphi}{\partial \xi}(s, \xi)u$. This completes the proof of the theorem. $\square$

6

Smooth stable manifolds in Banach spaces

We establish in this chapter the existence of smooth stable manifolds for semiflows defined by nonautonomous differential equations in a Banach space. One can obtain unstable manifolds simply by reversing the time. We also establish the exponential decay on the stable manifold of the derivatives of the semiflow with respect to the initial condition (see (6.8) and (6.9)). We are not aware of any similar result in the literature even in the case of uniform exponential dichotomies. Our approach to the proof of the stable manifold theorem consists again in using the differential equation and the invariance of the stable manifold under the dynamics to conclude that it must be the graph of a function satisfying a certain fixed point problem. However, the extra small exponentials led us to consider two fixed-point problems—one to obtain an a priori estimate for the speed of decay of the stable component of solutions along a given graph, and the other to obtain the graph which is the stable manifold. To obtain the estimates in the fixed point problems, we need sharp bounds for the derivatives of the stable component of the solutions, and for the derivatives of the vector field along a given graph. For this, we use a multivariate version of the Faà di Bruno formula in [30] for the derivatives of a composition (see (6.15) for a particular case). This formula allows us to estimate the norms of the derivatives of the composition in terms of the norms of the derivatives of the original functions (see (6.16)). Although several special cases were treated before, the general formula for the derivatives of a composition was first obtained by Faà di Bruno in [36]. We recommend [56] for the history of the problem and for many related references. We also use a result in [34] (see Proposition 6.3), that goes back to a lemma of Henry in [46]. This allows us to establish the existence and *simultaneously* the regularity of the stable manifolds using a single fixed point problem, instead of one for each of the successive higher-order derivatives. We follow closely [11], although now considering arbitrary stable and unstable subspaces.

6.1 Existence of smooth stable manifolds

We present here the result on the existence of a stable manifold for the origin in equation (4.4) assuming the existence of a nonuniform exponential dichotomy in $\mathbb{R}^+$. Let again $R_s(\delta) \subset E(s)$ be the open ball of radius $\delta \in (0,1)$ centered at zero. Given $k \in \mathbb{N}$ and $q > k$ we set

$$\alpha = \frac{a(q+k+1)+b}{q-k} > a \quad \text{and} \quad \beta = \alpha + a = \frac{a(2q+1)+b}{q-k}, \qquad (6.1)$$

with a and b as in (2.6). Notice that in the particular case when $k = 0$ the numbers α and β coincide respectively with α and β given by (4.8) and (4.12). The number α specifies the size of the neighborhood $R_s(\delta e^{-\alpha s})$ in which we will take the initial condition. We continue to consider the set $X_\alpha = X_\alpha(\delta)$ defined by (4.9). We also denote by ∂ the partial derivative with respect to the second variable of any given function of two variables. We assume that for some $k \in \mathbb{N}$ the following conditions hold:

D1. the function $A \colon \mathbb{R}_0^+ \to B(X)$ is of class C^k and satisfies (2.2);
D2. the function $f \colon \mathbb{R}_0^+ \times X \to X$ is of class C^k, and there exist $c > 0$, $q > k$, and $\varepsilon \in (0,1]$ such that for every $t \geq 0$ and $u, v \in X$ we have

$$f(t,0) = 0, \quad \partial f(t,0) = 0, \quad f|Y_\beta \equiv 0, \qquad (6.2)$$

where $Y_\beta = \{(s,v) \in \mathbb{R}_0^+ \times X : \|v\| \geq \delta e^{-\beta s}/(2D_1)\}$,

$$\|\partial^j f(t,u)\| \leq c\|u\|^{q+1-j} \text{ for } j = 1,\ldots,k, \qquad (6.3)$$

$$\|\partial^k f(t,u) - \partial^k f(t,v)\| \leq c\|u-v\|^\varepsilon (\|u\| + \|v\|)^{q-k}. \qquad (6.4)$$

The last condition in (6.2) may sometimes be obtained with an appropriate cut-off of the function f. We will verify that for the stable manifold of the origin in equation (4.4) constructed in Theorem 6.1, the solutions with sufficiently small initial condition starting on the manifold never intersect the region Y_β.

We denote by $\mathcal{Z}_\alpha$ the space of continuous functions $\varphi \colon X_\alpha \to X$ of class C^k in ξ such that for each $s \geq 0$ and $x, y \in R_s(\delta e^{-\alpha s})$ we have:

1. $\varphi(s, R_s(\delta e^{-\alpha s})) \subset F(s)$;
2. $\varphi(s,0) = 0$ and $\partial\varphi(s,0) = 0$;
3. $\|\partial^j \varphi(s,x)\| \leq 1$ for $j = 1,\ldots,k$;
4. $\|\partial^k \varphi(s,x) - \partial^k \varphi(s,y)\| \leq \|x-y\|^\varepsilon$.

Given a function $\varphi \in \mathcal{Z}_\alpha$ we consider its graph

$$\mathcal{V} = \{(s,\xi,\varphi(s,\xi)) : (s,\xi) \in X_\alpha\} \subset \mathbb{R}_0^+ \times X. \qquad (6.5)$$

We look for the stable manifold in this form. We will show that there exists $\varphi \in \mathcal{Z}_\alpha$ such that for every $(s,\xi) \in X_\beta \subset X_\alpha$ the corresponding solution of (4.4) is entirely contained in $\mathcal{V}$. We will assume the conditions

$$\bar{a} + \alpha < 0 \quad \text{and} \quad \bar{a} + b < \underline{b}. \tag{6.6}$$

Note that both inequalities in (6.6) hold when a and b are sufficiently small.

We now formulate the stable manifold theorem. We continue to write $p_{s,\xi} = (s, \xi, \varphi(s, \xi))$.

Theorem 6.1 ([11]). *Assume that D1–D2 hold. If the equation $v' = A(t)v$ in the Banach space X admits a nonuniform exponential dichotomy in $\mathbb{R}^+$ and the conditions in (6.6) hold, then there exist $\delta > 0$ and a unique function $\varphi \in \mathcal{Z}_\alpha$ such that the set $\mathcal{V}$ in (6.5) is forward invariant under the semiflow Ψ_τ, that is,*

$$\text{if } (s, \xi) \in X_\beta \text{ then } \Psi_\tau(p_{s,\xi}) \in \mathcal{V} \text{ for every } \tau \geq 0. \tag{6.7}$$

Furthermore:

1. *$\mathcal{V}$ is a manifold of class C^k that contains the line $\mathbb{R}_0^+ \times \{0\}$ and satisfies $T_{(s,0)}\mathcal{V} = \mathbb{R} \times E(s)$ for each $s \geq 0$;*
2. *for every $(s, \xi) \in X_\beta$ we have*

$$\varphi(s, \xi) = -\int_s^{+\infty} V(\tau, s)^{-1} f(\Psi_{\tau-s}(p_{s,\xi})) \, d\tau;$$

3. *there exists $D > 0$ such that for every $s \geq 0$, $\xi, \bar{\xi} \in R_s(\delta e^{-\beta s})$, and $\tau \geq 0$ we have*

$$\|\partial_\xi^j(\Psi_\tau(p_{s,\xi})) - \partial_\xi^j(\Psi_\tau(p_{s,\bar{\xi}}))\| \leq D e^{\bar{a}\tau + a(j+1)s}\|\xi - \bar{\xi}\| \tag{6.8}$$

for $j = 0, \ldots, k - 1$, and

$$\|\partial_\xi^k(\Psi_\tau(p_{s,\xi})) - \partial_\xi^k(\Psi_\tau(p_{s,\bar{\xi}}))\| \leq D e^{\bar{a}\tau + a(k+1)s}\|\xi - \bar{\xi}\|^\varepsilon. \tag{6.9}$$

The exponential decay on the stable manifold, of the derivatives up to order $k - 1$ (see (6.9)), can be understood in the following manner. When we consider higher-order jets (other than the first, with the linear variational equation), the corresponding higher-order linearized vector fields maintain the linear part $A(t)$ of the original linear variational equation. Thus, the higher-order jets possess essentially the same nonuniform exponential dichotomies as the linear variational equation, although in higher-dimensional spaces, and we can expect each of the corresponding higher-dimensional dynamics to possess a similar exponential behavior. Since the lower-dimensional parts of these jets coincide with the lower-order jets, the corresponding initial components of the higher-order jets maintain the exponential behavior along the stable manifolds of the lower-order jets.

6.2 Nonuniformly hyperbolic trajectories

We now use Theorem 6.1 to obtain smooth stable manifolds for nonuniformly hyperbolic solutions of a differential equation.

Consider a function $F\colon \mathbb{R}_0^+ \times X \to X$ of class C^k $(k \in \mathbb{N})$. We assume that $A(t) = \partial F(t, v_0(t))$ satisfies (2.2), that is, all solutions of $v' = A(t)v$ are global. We also consider the function $G(t, y)$ in (4.22).

Theorem 6.2. *Let F be of class C^k $(k \in \mathbb{N})$ and let $v_0(t)$ be a nonuniformly hyperbolic solution of (4.18). We assume that $G|Y_\beta \equiv 0$ and that there exist $c > 0$, $q > 1$, and $\varepsilon \in (0, 1]$ such that for every $t \geq 0$ and $u, v \in X$ we have*

$$\|\partial^j G(t, u) - \partial^j G(t, v)\| \leq c\|u - v\|(\|u\|^{q-j} + \|v\|^{q-j}) \qquad (6.10)$$

for $j = 0, \ldots, k - 1$, and

$$\|\partial^k G(t, u) - \partial^k G(t, v)\| \leq c\|u - v\|^\varepsilon (\|u\|^{q-k} + \|v\|^{q-k}). \qquad (6.11)$$

If the conditions in (6.6) hold, then there exist $\delta > 0$ and a unique function $\varphi \in \mathcal{Z}_\alpha$ such that the set

$$\mathcal{V} = \{(s, \xi, \varphi(s, \xi)) + (0, v_0(s)) : (s, \xi) \in X_\alpha\}$$

is a manifold of class C^k with the following properties:

1. *$(s, v_0(s)) \in \mathcal{V}$ and $T_{(s,0)}\mathcal{V} = \mathbb{R}_0^+ \times E(s)$ for every $s \geq 0$;*
2. *$\mathcal{V}$ is forward invariant under solutions of $t' = 1$, $v' = F(t, v)$, that is, if*

$$(s, v_s) \in \mathcal{V}' = \{(r, \xi, \varphi(r, \xi)) + (0, v_0(r)) : (r, \xi) \in X_\beta\}$$

   then $(t, v(t)) \in \mathcal{V}$ for every $t \geq s$, where $v(t) = v(t, v_s)$ is the unique solution of (4.18) for $t \geq s$ with $v(s) = v_s$;
3. *there exists $D > 0$ such that for every $s \geq 0$, (s, v_s), $(s, \bar{v}_s) \in \mathcal{V}'$, and $t \geq s$ we have*

$$\|v(t, v_s) - v(t, \bar{v}_s)\| \leq De^{\bar{a}(t-s)+as}\|v_s - \bar{v}_s\|.$$

Proof. Proceeding as in the proof of Theorem 4.2, and using the same notation, we obtain the equation $y' = A(t)y + G(t, y)$, with $G(t, y)$ as in (4.22). By hypothesis $A(t)$ satisfies the assumption D1 in Section 6.1. It follows from (4.22) that G is of class C^k. Furthermore, $G(t, 0) = 0$ and $\partial G(t, 0) = \partial F(t, v_0(t)) - A(t) = 0$, using the definition of $A(t)$. By (6.10), for every $(t, y, u) \in \mathbb{R}_0^+ \times X \times X$,

$$\|\partial G(t, y)u\| = \left\| \lim_{h \to 0} \frac{F(t, y + v_0(t) + hu) - F(t, y + v_0(t))}{h} - A(t)u \right\|$$

$$= \lim_{h \to 0} \frac{1}{|h|} \|F(t, y + v_0(t) + hu) - F(t, y + v_0(t)) - A(t)hu\|$$

$$\leq \lim_{h \to 0} \frac{1}{|h|} c\|hu\|(\|y + hu\|^q + \|y\|^q) \leq 2c\|u\| \cdot \|y\|^q,$$

and thus, $\|\partial G(t, y)\| \leq 2c\|y\|^q$. For $j = 2, \ldots, k - 1$, it follows from (6.10) that

$$\|\partial^j G(t,y)u\| = \left\| \lim_{h \to 0} \frac{\partial^{j-1} F(t, y + v_0(t) + hu) - \partial^{j-1} F(t, y + v_0(t))}{h} \right\|$$

$$\leq \lim_{h \to 0} \frac{c}{|h|} \|hu\| (\|y + hu\|^{q-j+1} + \|y\|^{q-j+1}) = 2c\|u\| \cdot \|y\|^{q-j+1},$$

and thus, $\|\partial^j G(t,y)\| \leq 2c\|y\|^{q-j+1}$. Together with (6.11) this shows that the function G satisfies the assumption D2 in Section 6.1. We can now apply Theorem 6.1 to obtain the desired statement. $\qquad \square$

6.3 Proof of the existence of smooth stable manifolds

6.3.1 Functional spaces

In view of the required forward invariance of $\mathcal{V}$ under solutions of the equation in (4.4) (see (6.7)), any solution with initial condition in $\mathcal{V}$ at time $s \geq 0$ must remain in $\mathcal{V}$ for every $t \geq s$, and thus must be of the form $(x(t), \varphi(t, x(t)))$ for every $t \geq s$. In particular, on such a manifold $\mathcal{V}$ the equations in (4.6) can be written in the form (4.23)–(4.24).

We equip the space $\mathcal{Z}_\alpha$ in Section 6.1 with the norm in (4.25). We want to verify that $\mathcal{Z}_\alpha$ is a complete metric space with this norm, or in other words, that $\mathcal{Z}_\alpha$ is closed in the complete metric space $\mathcal{X}_\alpha$ in Section 4.4.1. Let X, Y be Banach spaces and let $U \subset X$ be an open set. Given constants $\varepsilon \in (0,1]$, $k \in \mathbb{N} \cup \{0\}$, and $c > 0$ we define the set

$$C_c^{k,\varepsilon}(U,Y) = \left\{ u \in C^{k,\varepsilon}(U,Y) : \|u\|_{k,\varepsilon} \leq c \right\},$$

where $C^{k,\varepsilon}(U,Y)$ is the space of functions $u \colon U \to Y$ of class C^k with Hölder continuous k-th derivative with Hölder exponent ε. Furthermore, we set

$$\|u\|_{k,\varepsilon} = \max\{\|u\|_\infty, \|du\|_\infty, \ldots, \|d^k u\|_\infty, H_\varepsilon(d^k u)\},$$

where $\|\cdot\|_\infty$ denotes the supremum norm and

$$H_\varepsilon(u) = \sup\left\{ \frac{\|u(x) - u(y)\|}{\|x - y\|^\varepsilon} : x, y \in X \text{ and } x \neq y \right\}.$$

The following result shows that $C_c^{k,\varepsilon}(U,Y)$ is closed in the space of continuous functions $C(U,Y)$ with the supremum norm.

Proposition 6.3 ([34]). *Let X, Y be Banach spaces and let $U \subset X$ be an open subset. If $u_n \in C_c^{k,\varepsilon}(U,Y)$ for each $n \in \mathbb{N}$ and $u \colon U \to Y$ is a function such that $\|u_n - u\|_\infty \to 0$ as $n \to \infty$, then $u \in C_c^{k,\varepsilon}(U,Y)$ and for each $x \in U$ we have $d^k u_n(x) \to d^k u(x)$ as $n \to \infty$.*

The proposition says that the closed unit ball of the space $C^{k,\varepsilon}$ of functions of class C^k between two Banach spaces with ε-Hölder continuous k-th derivative is closed with respect to the C^0-topology. This allows us to consider contraction maps solely using the supremum norm instead of any norm involving also the derivatives. When $k = 1$, the statement in Proposition 6.3 was first established by Henry in [46, Lemma 6.1.6]. A similar result was obtained by Lanford in [54, Lemma 2.5] (with all limits, in the hypothesis and in the conclusion, pointwise in the weak topology).

Proposition 6.4. *With the norm in* (4.25), $\mathcal{Z}_\alpha$ *is a complete metric space.*

Proof. Given $\varphi \in \mathcal{Z}_\alpha$, $t \geq 0$, and $x \in R_t(\delta e^{-\alpha t})$ we have

$$\|\varphi(t, x)\| \leq \delta e^{-\alpha t} \frac{\|\varphi(t, x)\|}{\|x\|} \leq \delta e^{-\alpha t} \|\varphi\| \leq \delta.$$

Thus, if $(\varphi_n)_n \subset \mathcal{Z}_\alpha$ is a Cauchy sequence with respect to the norm in (4.25), then, for each $t \geq 0$, $(\varphi_n(t, \cdot))_n \subset C_\delta^{k,\varepsilon}(R_t(\delta e^{-\alpha t}), F(t))$ is a Cauchy sequence in the supremum norm. A simple application of Proposition 6.3 now yields the desired result. $\qquad\square$

For technical reasons, we also consider the space $\mathcal{Z}_\alpha^*$ of continuous functions $\varphi \colon X_\alpha^* \to X$, with X_α^* as in (4.26), such that $\varphi|X_\alpha \in \mathcal{Z}_\alpha$ (see (4.9)) and

$$\varphi(s, \xi) = \varphi(s, \delta e^{-\alpha s} \xi / \|\xi\|) \text{ whenever } s \geq 0 \text{ and } \xi \notin R_s(\delta e^{-\alpha s}).$$

Clearly, there is a one-to-one correspondence between functions in $\mathcal{Z}_\alpha$ and functions in $\mathcal{Z}_\alpha^*$. In particular we have the following.

Proposition 6.5. *With the norm* $\varphi \mapsto \|\varphi|X_\alpha\|$, $\mathcal{Z}_\alpha^*$ *is a complete metric space.*

As in Section 6.1, we denote by ∂ the partial derivative with respect to the second variable. For each fixed $s \geq 0$, set

$$\rho(t) = \bar{a}(t - s) + as, \tag{6.12}$$

and let $\mathcal{B}$ be the space of continuous functions

$$x \colon \{(t, \xi) : t \geq s \text{ and } \xi \in R_s(\delta e^{-\alpha s})\} \to X$$

of class C^k in the second variable such that for some constant $C > 0$ we have:

1. $x(t, \xi) \in E(t)$ for every $t \geq s$ and $\xi \in R_s(\delta e^{-\alpha s})$;
2. $x(s, \xi) = \xi$ for every $\xi \in R_s(\delta e^{-\alpha s})$;
3.

$$\|x\|' \leq C\delta e^{-\alpha s} \text{ and } \|\partial^j x\|' \leq C \text{ for } j = 1, \ldots, k, \tag{6.13}$$

where

$$\|x\|' := \sup \left\{ \|x(t, \xi)\| e^{-\rho(t)} : t \geq s \text{ and } \xi \in R_s(\delta e^{-\alpha s}) \right\}; \tag{6.14}$$

4.

$$|x|'_k := \sup\left\{\frac{\|\partial^k x(t,\xi) - \partial^k x(t,\bar{\xi})\|}{\|\xi - \bar{\xi}\|^\varepsilon}e^{-\rho(t)}\right\} \leq C$$

with the supremum taken over $t \geq s$ and $\xi, \bar{\xi} \in R_s(\delta e^{-\alpha s})$ with $\xi \neq \bar{\xi}$.

An application of Proposition 6.3 now yields the following result.

Proposition 6.6. *With the norm in* (6.14), $\mathcal{B}$ *is a complete metric space.*

6.3.2 Derivatives of compositions

We recall the Faà di Bruno formula for the n-th derivative of a composition. Consider open sets Y, Z, and W of Banach spaces. Let $g\colon Y \to Z$ be defined in a neighborhood of $x \in Y$ with derivatives up to order n at x. Let also $f\colon Z \to W$ be defined in a neighborhood of $y = g(x) \in Z$ with derivatives up to order n at y. Then the n-th derivative of the composition $h = f \circ g$ at x is given by

$$d_x^n h = \sum_{k=1}^n d_y^k f \sum_{\substack{0 \leq r_1,\ldots,r_k \leq n \\ r_1 + \cdots + r_k = n}} c_{r_1 \cdots r_k} d_x^{r_1} g \cdots d_x^{r_k} g, \tag{6.15}$$

for some nonnegative integers $c_{r_1 \cdots r_k}$. Collecting derivatives of equal order, one can show that for each $n \in \mathbb{N}$ there exists $c = c(n) > 0$ such that (see [30])

$$\|d_x^n h\| \leq c \sum_{k=1}^n \|d_y^k f\| \sum_{p(n,k)} \prod_{j=1}^n \|d_x^j g\|^{k_j}, \tag{6.16}$$

where

$$p(n,k) = \left\{(k_1,\ldots,k_n) \in \mathbb{N}_0^n : \sum_{j=1}^n k_j = k \text{ and } \sum_{j=1}^n jk_j = n\right\} \tag{6.17}$$

(here $\mathbb{N}_0$ is the set of nonnegative integers). Furthermore, using (6.15) and the triangular inequality one can show that for $y = g(x)$ and $\bar{y} = g(\bar{x})$,

$$\|d_x^n h - d_{\bar{x}}^n h\| \leq c \sum_{k=1}^n \|d_y^k f - d_{\bar{y}}^k f\| \sum_{p(n,k)} \prod_{j=1}^n \|d_x^j g\|^{k_j} + c' \sum_{k=1}^n \|d_{\bar{y}}^k f\| S_k, \tag{6.18}$$

for some constant $c' = c'(n) > 0$, where

$$S_k := \sum_{p(n,k)} \sum_{j=1}^n T_j \prod_{m=1}^{j-1} \|d_{\bar{x}}^m g\|^{k_m} \prod_{m=j+1}^n \|d_x^m g\|^{k_m},$$

and

$$T_j := \|d_x^j g - d_{\bar{x}}^j g\| \sum_{k=0}^{k_j-1} \|d_x^j g\|^{k_j-1-k} \|d_{\bar{x}}^j g\|^k.$$

A multivariate extension of the Faà di Bruno formula was established in [30]. It can be readily generalized to transformations in Banach spaces as follows. Let $g = (g_1, g_2)$ be defined in a neighborhood of x with derivatives up to order n at x. Let also $f(y)$ be defined in a neighborhood of $(y_1, y_2) = (g_1(x), g_2(x))$ with derivatives up to order n at (y_1, y_2). Then for each $n \in \mathbb{N}$ there exists $c = c(n) > 0$ such that the n-th derivative of $h = f \circ (g_1, g_2)$ at x satisfies

$$\|d_x^n h\| \le c \sum_{q(n)} \|\partial_{y_1, y_2}^{\lambda_1, \lambda_2} f\| \sum_{\sigma=1}^{n} \sum_{p_\sigma(n,\lambda)} \prod_{j=1}^{\sigma} \|d_x^{l_j} g_1\|^{k_{j1}} \|d_x^{l_j} g_2\|^{k_{j2}}, \qquad (6.19)$$

with the notations

$$\partial_{y_1, y_2}^{\lambda_1, \lambda_2} f = \frac{\partial^{\lambda_1+\lambda_2} f(y_1, y_2)}{\partial y_1^{\lambda_1} \partial y_2^{\lambda_2}}, \quad q(n) = \{(\lambda_1, \lambda_2) : \lambda_1 + \lambda_2 \in \{1, \dots, n\}\},$$

and, setting $\lambda = (\lambda_1, \lambda_2)$,

$$p_\sigma(n, \lambda) = \left\{ (k_{11}, k_{12}, \dots, k_{\sigma 1}, k_{\sigma 2}; l_1, \dots, l_\sigma) \in \mathbb{N}_0^{2\sigma} \times \mathbb{N}^\sigma : \right.$$
$$(k_{j1}, k_{j2}) \ne (0,0) \text{ for } 1 \le j \le \sigma, \, l_1 < \cdots < l_\sigma, \qquad (6.20)$$
$$\left. \sum_{j=1}^{\sigma} k_{jl} = \lambda_l \text{ for } l = 1, 2, \text{ and } \sum_{j=1}^{\sigma} l_j(k_{j1} + k_{j2}) = n \right\}.$$

Furthermore, in a similar manner to that in (6.18) one can show that for $(y_1, y_2) = (g_1(x), g_2(x))$ and $(\bar{y}_1, \bar{y}_2) = (g_1(\bar{x}), g_2(\bar{x}))$,

$$\|d_x^n h - d_{\bar{x}}^n h\| \le c \sum_{q(n)} \|\partial_{y_1, y_2}^{\lambda_1, \lambda_2} f - \partial_{\bar{y}_1, \bar{y}_2}^{\lambda_1, \lambda_2} f\| \sum_{\sigma=1}^{n} \sum_{p_\sigma(n,\lambda)} \prod_{j=1}^{\sigma} \|d_x^{l_j} g_1\|^{k_{j1}} \|d_x^{l_j} g_2\|^{k_{j2}}$$
$$+ c' \sum_{q(n)} \|\partial_{\bar{y}_1, \bar{y}_2}^{\lambda_1, \lambda_2} f\| \sum_{\sigma=1}^{n} \widetilde{S}_\sigma,$$
$$(6.21)$$

for some constant $c' = c'(n) > 0$, where

$$\widetilde{S}_\sigma := \sum_{p_\sigma(n,\lambda)} \sum_{j=1}^{\sigma} \widetilde{T}_{k_{j1}, k_{j2}, l_j} \prod_{i=1}^{j-1} \|d_{\bar{x}}^{l_i} g_1\|^{k_{i1}} \|d_{\bar{x}}^{l_i} g_2\|^{k_{i2}} \prod_{i=j+1}^{\sigma} \|d_x^{l_i} g_1\|^{k_{i1}} \|d_x^{l_i} g_2\|^{k_{i2}},$$
$$(6.22)$$

and

$$\widetilde{T}_{k_{j1},k_{j2},l_j} := \|d_x^{l_j} g_2\|^{k_{j2}} \|d_x^{l_j} g_1 - d_{\bar x}^{l_j} g_1\| \sum_{k=0}^{k_{j1}-1} \|d_{\bar x}^{l_j} g_1\|^{k_{j1}-1-k} \|d_x^{l_j} g_1\|^k$$
$$+ \|d_{\bar x}^{l_j} g_1\|^{k_{j1}} \|d_x^{l_j} g_2 - d_{\bar x}^{l_j} g_2\| \sum_{k=0}^{k_{j2}-1} \|d_x^{l_j} g_2\|^{k_{j2}-1-k} \|d_{\bar x}^{l_j} g_2\|^k. \tag{6.23}$$

6.3.3 A priori control of the derivatives

We now use the inequalities in Section 6.3.2 to obtain several bounds for the derivatives that are needed in the proof of Theorem 6.1. Given $\varphi \in Z_\alpha^*$ and $x \in \mathcal{B}$ we write

$$\varphi^*(t,\xi) = \varphi(t, x(t,\xi)). \tag{6.24}$$

Lemma 6.7. *There exists $A > 0$ such that for each $j = 1, \ldots, k$, $\varphi \in Z_\alpha^*$, $(s,\xi) \in X_\alpha$, and $t \geq s$ satisfying $x(t,\xi) \in R_t(\delta e^{-\alpha t})$ we have*

$$\|\partial^j \varphi^*(t,\xi)\| \leq A e^{\rho(t)+(j-1)as}.$$

Proof. Using (6.16) for the derivative $\partial^j \varphi^*$ we obtain

$$\|\partial^j \varphi^*(t,\xi)\| \leq d \sum_{m=1}^{j} \|\partial^m \varphi(t, x(t,\xi))\| \sum_{p(j,m)} \prod_{l=1}^{j} \|\partial^l x(t,\xi)\|^{k_l}$$

with $p(j,m)$ as in (6.17). Since $\sum_{l=1}^{j} k_l = m$ (see (6.17)), using (6.13) we obtain

$$\|\partial^j \varphi^*(t,\xi)\| \leq d \sum_{m=1}^{j} \sum_{p(j,m)} \prod_{l=1}^{j} (C e^{\rho(t)})^{k_l} \leq c_1 \sum_{m=1}^{j} \sum_{p(j,m)} e^{\rho(t) \sum_{l=1}^{j} k_l}$$
$$\leq c_2 \sum_{m=1}^{j} e^{m\rho(t)} \leq c_3 e^{\rho(t)+(j-1)as},$$

for some constants $c_1, c_2, c_3 > 0$, where we have used (6.12) in the last inequality (note that $\bar a < 0$ and $a \geq 0$). $\qquad\qquad\square$

Given $\varphi \in Z_\alpha^*$ and $x \in \mathcal{B}$ we write

$$f^*(t,\xi) = f(t, x(t,\xi), \varphi(t, x(t,\xi))). \tag{6.25}$$

Lemma 6.8. *There exists $B > 0$ such that for each $j = 1, \ldots, k$, $\varphi \in Z_\alpha^*$, $(s,\xi) \in X_\alpha$, and $t \geq s$ we have*

$$\|\partial^j f^*(t,\xi))\| \leq B\delta^{q+1-j} e^{-\alpha(q+1-j)s} e^{(q+1)\rho(t)+jas}.$$

Proof. We note that the derivative $\partial^j f^*(t, \xi)$ is defined for every $t \geq s$. This is due to the fact that, by (6.2), whenever $t \geq 0$ is such that $x(t, \xi) \in E(t) \setminus R_t(\delta e^{-\alpha t})$ we have $f^*(t, \xi) = 0$. Indeed, setting $t = s$ in (2.5) we obtain $\|P(s)\| \leq D_1 e^{as}$. Therefore, since

$$P(t)(x(t, \xi), \varphi(t, x(t, \xi))) = x(t, \xi)$$

we have

$$\|(x(t, \xi), \varphi(t, x(t, \xi)))\| \geq D_1^{-1} e^{-at} \|x(t, \xi)\| \geq \delta D_1^{-1} e^{-\beta t}, \qquad (6.26)$$

and by (6.2) we conclude that $f^*(t, \xi) = 0$ whenever $\|x(t, \xi)\| \geq \delta e^{-\alpha t}$.

Assume now that $x(t, \xi) \in R_t(\delta e^{-\alpha t})$. Using (6.19) for the derivative $\partial^j f^*$ we obtain

$$\|\partial^j f^*(t, \xi)\| \leq c \sum_{q(j)} \|\partial_{x(t, \xi), \varphi(t, x(t, \xi))}^{\lambda_1, \lambda_2} f(t, \cdot)\| \sum_{s=1}^{j} \sum_{p_s(j, \lambda)}$$
$$\times \prod_{m=1}^{s} \|\partial^{l_m} x(t, \xi)\|^{k_{m1}} \|\partial^{l_m} \varphi^*(t, \xi)\|^{k_{m2}},$$

with $\varphi^*(t, \xi)$ as in (6.24), and $p_s(j, \lambda)$ as in (6.20).

Since $\varphi \in \mathcal{Z}_\alpha^*$, using (6.3), (6.13), and the fact that in $p_s(j, \lambda)$ we have

$$\sum_{j=1}^{s} (k_{j1} + k_{j2}) = \lambda_1 + \lambda_2 \quad \text{and} \quad \sum_{j=1}^{s} l_j k_{j2} \leq \sum_{j=1}^{s} l_j (k_{j1} + k_{j2}) = j$$

(see (6.20)), it follows from Lemma 6.7 that

$$\|\partial^j f^*(t, \xi)\| \leq c_1 \sum_{q(j)} (\|x(t, \xi)\| + \|\varphi(t, x(t, \xi))\|)^{q+1-\lambda_1-\lambda_2} \sum_{s=1}^{j} \sum_{p_s(j, \lambda)}$$
$$\times \prod_{m=1}^{s} (Ce^{\rho(t)})^{k_{m1}} (Ae^{\rho(t)+(l_m-1)as})^{k_{m2}}$$
$$\leq c_2 \sum_{q(j)} (2\|x(t, \xi)\|)^{q+1-\lambda_1-\lambda_2} e^{\rho(t)(\lambda_1+\lambda_2)+jas}$$
$$\leq c_3 e^{(q+1)\rho(t)+jas} \sum_{q(j)} (\delta e^{-\alpha s})^{q+1-(\lambda_1+\lambda_2)},$$

for some constants $c_1, c_2, c_3 > 0$. Since $\delta e^{-\alpha s} < 1$ and $1 \leq \lambda_1 + \lambda_2 \leq j$ we obtain the desired statement. $\qquad \square$

Lemma 6.9. *For each* $j = 0, \ldots, k-1$, $(s, \xi), (s, \bar{\xi}) \in X_\alpha$, *and* $t \geq s$ *we have*

$$\|\partial^j x(t, \xi) - \partial^j x(t, \bar{\xi})\| \leq Ce^{\rho(t)} \|\xi - \bar{\xi}\|. \qquad (6.27)$$

Proof. To prove the lemma it suffices to observe that by (6.13),

$$\|\partial^j x(t,\xi) - \partial^j x(t,\bar\xi)\| \leq \sup_{r\in[0,1]} \|\partial^{j+1} x(t,\xi+r(\bar\xi-\xi))\| \cdot \|\xi-\bar\xi\| \leq Ce^{\rho(t)}\|\xi-\bar\xi\|,$$

applying the mean value theorem.   □

6.3.4 Hölder regularity of the top derivatives

Lemma 6.10. *There exists $C' > 0$ such that for each $\varphi \in \mathcal{Z}_\alpha^*$, $(s,\xi), (s,\bar\xi) \in X_\alpha$, and $t \geq s$ satisfying $x(t,\xi), x(t,\bar\xi) \in R_t(\delta e^{-\alpha t})$ we have*

$$\|\partial^k \varphi^*(t,\xi) - \partial^k \varphi^*(t,\bar\xi)\| \leq C' e^{\rho(t)+kas}\|\xi-\bar\xi\|^\varepsilon.$$

Proof. By (6.18) we obtain

$$\|\partial^k \varphi^*(t,\xi) - \partial^k \varphi^*(t,\bar\xi)\|$$

$$\leq c \sum_{m=1}^k \|\partial^m \varphi(t,x(t,\xi)) - \partial^m \varphi(t,x(t,\bar\xi))\| \sum_{p(k,m)} \prod_{l=1}^k \|\partial^l x(t,\xi)\|^{k_l} \tag{6.28}$$

$$+ c' \sum_{m=1}^k \|\partial^m \varphi(t,x(t,\bar\xi))\| S_m,$$

with $p(k,m)$ as in (6.17), and where

$$S_m := \sum_{p(k,m)} \sum_{l=1}^k T_l \prod_{i=1}^{l-1} \|\partial^i x(t,\bar\xi)\|^{k_i} \prod_{i=l+1}^k \|\partial^i x(t,\xi)\|^{k_i}, \tag{6.29}$$

and

$$T_l := \|\partial^l x(t,\xi) - \partial^l x(t,\bar\xi)\| \sum_{k=0}^{k_l-1} \|\partial^l x(t,\xi)\|^{k_l-1-k}\|\partial^l x(t,\bar\xi)\|^k.$$

Since $\varphi \in \mathcal{Z}_\alpha^*$, it follows from Lemma 6.9 that for $i = 1,\dots,k-1$,

$$\|\partial^i \varphi(t,x(t,\xi)) - \partial^i \varphi(t,x(t,\bar\xi))\|$$
$$\leq \sup_{r\in[0,1]} \|\partial^{i+1}\varphi(t,x(t,\xi) + r(x(t,\bar\xi) - x(t,\xi)))\| \cdot \|x(t,\xi) - x(t,\bar\xi)\|$$
$$\leq \|x(t,\xi) - x(t,\bar\xi)\| \leq Ce^{\rho(t)}\|\xi-\bar\xi\|.$$

Furthermore, again using Lemma 6.9,

$$\|\partial^k \varphi(t,x(t,\xi)) - \partial^k \varphi(t,x(t,\bar\xi))\| \leq \|x(t,\xi) - x(t,\bar\xi)\|^\varepsilon$$
$$\leq C^\varepsilon e^{\varepsilon\rho(t)}\|\xi-\bar\xi\|^\varepsilon. \tag{6.30}$$

Since $\delta < 1$, we have $\|\xi - \bar{\xi}\| \leq \|\xi - \bar{\xi}\|^{\varepsilon}$, and using (6.13) with $j = l$ we obtain

$$T_l \leq C e^{\rho(t)} \|\xi - \bar{\xi}\|^{\varepsilon} \sum_{k=0}^{k_l-1} (C e^{\rho(t)})^{k_l-1} \leq C^{k_l} k_l e^{k_l \rho(t)} \|\xi - \bar{\xi}\|^{\varepsilon}. \qquad (6.31)$$

By (6.29) together with (6.31), (6.13), and the fact that in $p(k,m)$ we have $\sum_{i=1}^{k} k_i = m$ (see (6.17)) we conclude that

$$\begin{aligned}
S_m &\leq \sum_{p(k,m)} \sum_{l=1}^{k} T_l \prod_{i=1, i \neq l}^{k} (C e^{\rho(t)})^{k_i} \\
&\leq c_1 \sum_{p(k,m)} \sum_{l=1}^{k} C^{k_l} k_l e^{k_l \rho(t)} \|\xi - \bar{\xi}\|^{\varepsilon} \prod_{i=1, i \neq l}^{k} e^{k_i \rho(t)} \\
&\leq c_2 e^{m\rho(t)} \|\xi - \bar{\xi}\|^{\varepsilon},
\end{aligned} \qquad (6.32)$$

for some constants $c_1, c_2 > 0$. By (6.28), (6.30), (6.32), (6.13), the fact that in $p(k,m)$ we have $\sum_{i=1}^{k} k_i = m$ (see (6.17)), and since $\rho(t) = \bar{a}(t-s) + as$ with $\bar{a} < 0$ and $a > 0$, we obtain

$$\begin{aligned}
\|\partial^k \varphi^*(t,\xi) - \partial^k \varphi^*(t,\bar{\xi})\| &\leq c_3 e^{as} \|\xi - \bar{\xi}\|^{\varepsilon} \sum_{m=1}^{k} \sum_{p(k,m)} \prod_{l=1}^{k} (C e^{\rho(t)})^{k_l} \\
&\quad + c_4 \|\xi - \bar{\xi}\|^{\varepsilon} \sum_{m=1}^{k} e^{m\rho(t)} \\
&\leq c_5 e^{as} \|\xi - \bar{\xi}\|^{\varepsilon} \sum_{m=1}^{k} e^{m\rho(t)} \leq c_6 e^{\rho(t)+kas} \|\xi - \bar{\xi}\|^{\varepsilon},
\end{aligned}$$

for some constants $c_3, c_4, c_5, c_6 > 0$. This gives the desired statement. $\qquad \square$

In view of Lemmas 6.7 and 6.10, for each $j = 0, \ldots, k$ we have

$$\|\partial^j \varphi^*(t,\xi) - \partial^j \varphi^*(t,\bar{\xi})\| \leq V e^{\rho(t)+jas} \|\xi - \bar{\xi}\|^{\varepsilon}, \qquad (6.33)$$

for some constant $V > 0$.

Lemma 6.11. *There exists $D' > 0$ such that for each $\varphi \in \mathfrak{Z}_\alpha^*$, $x \in \mathcal{B}$, $(s,\xi), (s,\bar{\xi}) \in X_\alpha$, and $t \geq s$ we have*

$$\|\partial^k f^*(t,\xi) - \partial^k f^*(t,\bar{\xi})\| \leq D' (\delta e^{-\alpha s})^{q-k} e^{q\rho(t)+(k+1)as} \|\xi - \bar{\xi}\|^{\varepsilon}. \qquad (6.34)$$

Proof. The reason why the derivatives in (6.34) are always well-defined is the same as in the proof of Lemma 6.8. We first assume that $x(t,\xi), x(t,\bar{\xi}) \in R_t(\delta e^{-\alpha t})$. We use (6.21) to obtain

$$\|\partial^k f^*(t,\xi) - \partial^k f^*(t,\bar{\xi})\|$$

$$\leq c\sum_{q(k)} G_{\lambda_1,\lambda_2} \sum_{\sigma=1}^{k} \sum_{p_\sigma(k,\lambda)} \prod_{m=1}^{\sigma} \|\partial^{l_m} x(t,\xi)\|^{k_{m1}} \|\partial^{l_m}\varphi^*(t,\xi)\|^{k_{m2}} \tag{6.35}$$

$$+ c'\sum_{q(k)} \|\partial^{\lambda_1,\lambda_2}_{x(t,\bar{\xi}),\varphi^*(t,\bar{\xi})} f(t,\cdot)\| \sum_{\sigma=1}^{k} \widetilde{S}_\sigma,$$

where

$$G_{\lambda_1,\lambda_2} = \|\partial^{\lambda_1,\lambda_2}_{x(t,\xi),\varphi^*(t,\xi)} f(t,\cdot) - \partial^{\lambda_1,\lambda_2}_{x(t,\bar{\xi}),\varphi^*(t,\bar{\xi})} f(t,\cdot)\|,$$

and, in view of (6.22) and (6.23),

$$\widetilde{S}_\sigma := \sum_{p_\sigma(k,\lambda)} \sum_{m=1}^{\sigma} \widetilde{T}_{k_{m1},k_{m2},l_m} \left(\prod_{i=1}^{m-1} \|\partial^{l_i} x(t,\bar{\xi})\|^{k_{i1}} \|\partial^{l_i}\varphi^*(t,\bar{\xi})\|^{k_{i2}} \right.$$

$$\left. \times \prod_{i=m+1}^{\sigma} \|\partial^{l_i} x(t,\xi)\|^{k_{i1}} \|\partial^{l_i}\varphi^*(t,\xi)\|^{k_{i2}} \right), \tag{6.36}$$

and

$$\widetilde{T}_{k_{m1},k_{m2},l_m} := \|\partial^{l_m}\varphi^*(t,\xi)\|^{k_{m2}} \|\partial^{l_m} x(t,\xi) - \partial^{l_m} x(t,\bar{\xi})\|$$

$$\times \sum_{k=0}^{k_{m1}-1} \|\partial^{l_m} x(t,\zeta)\|^{k_{m1}-1-k} \|\partial^{l_m} x(t,\bar{\xi})\|^{k}$$

$$+ \|\partial^{l_m} x(t,\bar{\xi})\|^{k_{m1}} \|\partial^{l_m}\varphi^*(t,\xi) - \partial^{l_m}\varphi^*(t,\bar{\xi})\| \tag{6.37}$$

$$\times \sum_{k=0}^{k_{m2}-1} \|\partial^{l_m}\varphi^*(t,\xi)\|^{k_{m2}-1-k} \|\partial^{l_m}\varphi^*(t,\bar{\xi})\|^{k}.$$

By the mean value theorem, for $\lambda_1 + \lambda_2 = 1,\ldots,k-1$ we have

$$G_{\lambda_1,\lambda_2} \leq \sup_{r\in[0,1]} \|\partial^{\lambda_1+1,\lambda_2}_{a(r)} f(t,\cdot)\| \cdot \|x(t,\xi) - x(t,\bar{\xi})\|$$

$$+ \sup_{r\in[0,1]} \|\partial^{\lambda_1,\lambda_2+1}_{b(r)} f(t,\cdot)\| \cdot \|\varphi^*(t,\xi) - \varphi^*(t,\bar{\xi})\|,$$

where

$$a(r) = (x(t,\xi) + r(x(t,\bar{\xi}) - x(t,\xi)), \varphi^*(t,\xi)),$$
$$b(r) = (x(t,\xi), \varphi^*(t,\xi) + r(\varphi^*(t,\bar{\xi}) - \varphi^*(t,\xi))).$$

By (6.3), (6.13), and Lemma 6.7, for $\lambda_1 + \lambda_2 = 1,\ldots,k-1$ we obtain

$$G_{\lambda_1,\lambda_2} \leq c(2C\delta e^{-\alpha s} e^{\rho(t)})^{q-\lambda_1-\lambda_2} Ce^{\rho(t)}\|\xi - \bar{\xi}\|$$

$$+ c(2C\delta e^{-\alpha s} e^{\rho(t)})^{q-\lambda_1-\lambda_2} Ae^{\rho(t)}\|\xi - \bar{\xi}\|$$

$$= c_1(\delta e^{-\alpha s})^{q-\lambda_1-\lambda_2} e^{\rho(t)(q+1-(\lambda_1+\lambda_2))}\|\xi - \bar{\xi}\|,$$

for some constant $c_1 > 0$. Furthermore, since $\varphi \in Z_\alpha^*$, in view of (6.4), (6.13), and Lemma 6.9, for $\lambda_1 + \lambda_2 = k$ we have

$$
\begin{aligned}
G_{\lambda_1, \lambda_2} &\le c_2 e^{\varepsilon \rho(t)} \|\xi - \bar{\xi}\|^\varepsilon (4 C \delta e^{-\alpha s} e^{\rho(t)})^{q-k} \\
&\le c_3 (\delta e^{-\alpha s})^{q-k} e^{\rho(t)(q-k+\varepsilon)} \|\xi - \bar{\xi}\|^\varepsilon,
\end{aligned}
$$

for some constants $c_2, c_3 > 0$. Thus, for each $\lambda_1 + \lambda_2 \in \{1, \ldots, k\}$ we have

$$
G_{\lambda_1, \lambda_2} \le \max\{c_1, c_3\} (\delta e^{-\alpha s})^{q-(\lambda_1+\lambda_2)} e^{\rho(t)(q-(\lambda_1+\lambda_2))} e^{as} \|\xi - \bar{\xi}\|^\varepsilon,
$$

where we have used (6.12).

By Lemma 6.7 together with the fact that in $p_\sigma(k, \lambda)$ we have

$$
\sum_{m=1}^{\sigma} (k_{m1} + k_{m2}) = \lambda_1 + \lambda_2 \quad \text{and} \quad \sum_{m=1}^{\sigma} (l_m - 1) k_{m2} \le \sum_{m=1}^{\sigma} l_m (k_{m1} + k_{m2}) = k,
$$

the first summand in (6.35) can be bounded by

$$
c_4 (\delta e^{-\alpha s})^{q-k} e^{q\rho(t)+(k+1)as} \|\xi - \bar{\xi}\|^\varepsilon, \tag{6.38}
$$

for some constant $c_4 > 0$. To bound the second summand in (6.35) we first bound $\tilde{T}_{k_{m1}, k_{m2}, l_m}$. By (6.37), using Lemmas 6.7 and 6.9, (6.13), and (6.33), we obtain

$$
\begin{aligned}
\tilde{T}_{k_{m1}, k_{m2}, l_m} &\le c_5 e^{\rho(t)(k_{m1}+k_{m2})} \|\xi - \bar{\xi}\| e^{k_{m2}(l_m-1)as} \\
&\quad + c_6 e^{\rho(t)(k_{m1}+k_{m2})} \|\xi - \bar{\xi}\|^\varepsilon e^{((k_{m2}-1)(l_m-1)+l_m)as} \\
&\le c_7 e^{\rho(t)(k_{m1}+k_{m2})} e^{(k_{m2}(l_m-1)+1)as} \|\xi - \bar{\xi}\|^\varepsilon,
\end{aligned}
$$

for some constants $c_5, c_6, c_7 > 0$. In view of (6.36), using Lemma 6.7 and the fact that in $p_\sigma(k, \lambda)$ we have $\sum_{m=1}^{\sigma} (k_{m1} + k_{m2}) = \lambda_1 + \lambda_2$ and

$$
\sum_{m=1}^{\sigma} (l_m - 1)(k_{m1} + k_{m2}) \le \sum_{m=1}^{\sigma} l_m (k_{m1} + k_{m2}) = k,
$$

we conclude that

$$
\begin{aligned}
\tilde{S}_\sigma &\le c_8 e^{as} \|\xi - \bar{\xi}\|^\varepsilon \sum_{p_\sigma(k, \lambda)} \sum_{m=1}^{\sigma} e^{\rho(t)(k_{m1}+k_{m2})} e^{k_{m2}(l_m-1)as} \\
&\quad \times \prod_{i=1, i \neq m}^{\sigma} e^{\rho(t)(k_{i1}+k_{i2})} e^{(k_{i1}+k_{i2})(l_i-1)as} \\
&\le c_9 e^{\rho(t)(\lambda_1+\lambda_2)+(k+1)as} \|\xi - \bar{\xi}\|^\varepsilon,
\end{aligned}
$$

for some constants $c_8, c_9 > 0$. Therefore, the second summand in (6.35) can be bounded by

$$c_{10}e^{(k+1)as}\|\xi - \bar{\xi}\|^\varepsilon \sum_{q(k)}(2\|x(t,\bar{\xi})\|)^{q+1-\lambda_1-\lambda_2}e^{\rho(t)(\lambda_1+\lambda_2)}$$

$$\leq c_{10}e^{(q+1)\rho(t)+(k+1)as}\|\xi - \bar{\xi}\|^\varepsilon \sum_{q(k)}(2C\delta e^{-as})^{q+1-\lambda_1-\lambda_2} \qquad (6.39)$$

$$\leq c_{11}(\delta e^{-as})^{q+1-k}e^{(q+1)\rho(t)+(k+1)as}\|\xi - \bar{\xi}\|^\varepsilon,$$

for some constants $c_{10}, c_{11} > 0$. By (6.35), (6.38), and (6.39) we obtain

$$\|\partial^k f^*(t,\xi) - \partial^k f^*(t,\bar{\xi})\| \leq c_4(\delta e^{-as})^{q-k}e^{q\rho(t)+(k+1)as}\|\xi - \bar{\xi}\|^\varepsilon$$
$$+ c_{11}(\delta e^{-as})^{q+1-k}e^{(q+1)\rho(t)+(k+1)as}\|\xi - \bar{\xi}\|^\varepsilon$$
$$\leq D'(\delta e^{-as})^{q-k}e^{q\rho(t)+(k+1)as}\|\xi - \bar{\xi}\|^\varepsilon,$$

for some constant $D' > 0$, since $\alpha > a$.

When $x(t,\xi), x(t,\bar{\xi}) \notin R_t(\delta e^{-\alpha t})$ we have $\partial^k f^*(t,\xi) = \partial^k f^*(t,\bar{\xi}) = 0$ and there is nothing to show. It remains to consider the case when $x(t,\xi) \in R_t(\delta e^{-\alpha t})$ and $x(t,\bar{\xi}) \notin R_t(\delta e^{-\alpha t})$. Take $c \in (0,1]$ such that the vector $z = c\xi + (1-c)\bar{\xi}$ satisfies $\|x(t,z)\| = \delta e^{-\alpha t}$. Then $\partial^k f^*(t,\bar{\xi}) = \partial^k f(t,z) = 0$ and

$$\|\partial^k f^*(t,\xi) - \partial^k f^*(t,\bar{\xi})\|$$
$$\leq \|\partial^k f^*(t,\xi) - \partial^k f^*(t,z)\| + \|\partial^k f^*(t,z) - \partial^k f^*(t,\bar{\xi})\|$$
$$\leq D'(\delta e^{-as})^{q-k}e^{q\rho(t)+(k+1)as}\|\xi - z\|^\varepsilon$$
$$\leq D'(\delta e^{-as})^{q-k}e^{q\rho(t)+(k+1)as}\|\xi - \bar{\xi}\|^\varepsilon.$$

This completes the proof. □

6.3.5 Solution on the stable direction

We now proceed with the proof of Theorem 6.1. It is obtained in several steps. We first establish the existence of a unique function $x(t) = x_\varphi(t)$ satisfying (4.23) for each given $\varphi \in Z_\alpha^*$.

Lemma 6.12. *For every $\delta > 0$ sufficiently small the following properties hold:*

1. *for each $\varphi \in Z_\alpha^*$, given $s \geq 0$ there exists a unique function $x = x_\varphi \in \mathcal{B}$ satisfying (4.23) for every $t \geq s$;*
2. *we have*

$$\|x(t,\xi)\| \leq Ce^{\rho(t)}\|\xi\| \text{ for every } t \geq s. \qquad (6.40)$$

Proof. Given $x \in \mathcal{B}$, we define the operator

$$(Jx)(t,\xi) = U(t,s)\xi + \int_s^t U(t,\tau)f(\tau, x(\tau,\xi), \varphi(\tau, x(\tau,\xi)))\, d\tau$$

for each $t \geq s$ and $\xi \in R_s(\delta e^{-\alpha s})$. Using the last condition in (6.2) and (6.26), we find that Jx is a continuous function of class C^k in ξ. The fact that $(Jx)(s, \xi) = \xi$ is immediate from $U(s, s)\xi = \xi$. Furthermore, using (6.3) and (6.13),

$$\|f(\tau, x(\tau), \varphi(\tau, x(\tau)))\| \leq c2^{q+1}\|x(\tau)\|^{q+1}$$
$$\leq c2^{q+1}C^{q+1}\delta^{q+1}e^{-\alpha(q+1)s}e^{(q+1)\rho(\tau)}.$$

By (6.1) we have

$$\alpha \geq a(q + k + 1)/(q - k) > a(q + 1)/q.$$

Therefore, using the first inequality in (2.17) and (6.12), we obtain

$$\|(Jx)(t, \xi) - U(t, s)\xi\| \leq \int_s^t \|U(t, \tau)\| \cdot \|f(\tau, x(\tau), \varphi(\tau, x(\tau)))\| \, d\tau$$

$$\leq c2^{q+1}C^{q+1}D_1\delta^{q+1} \int_s^t e^{\bar{a}(t-\tau)+a\tau}e^{\bar{a}(q+1)(\tau-s)}e^{(q+1)(a-\alpha)s} \, d\tau$$

$$\leq c2^{q+1}C^{q+1}D_1\delta^{q+1}e^{\bar{a}(t-s)+as+(q+1)(a-\alpha)s} \int_s^\infty e^{(q\bar{a}+a)(\tau-s)} \, d\tau$$

$$\leq c2^{q+1}C^{q+1}D_1\delta^{q+1}e^{-\alpha s}e^{(q+1)as-q\alpha s}e^{\rho(t)} \int_s^\infty e^{T_1(\tau-s)} \, d\tau,$$

with

$$T_1 = (q - 1)\bar{a} + a < 0. \tag{6.41}$$

Indeed, in view of (6.6) we have

$$T_1 + (1 - k)\bar{a} + a(q + k) + b = (q - k)(\bar{a} + \alpha) < 0,$$

and hence $T_1 < 0$. Therefore

$$\|(Jx)(t, \xi) - U(t, s)\xi\| \leq \theta\delta e^{-\alpha s}e^{\rho(t)},$$

where

$$\theta = c2^{q+1}C^{q+1}D_1\delta^q/|T_1|.$$

Furthermore, by (2.17), (6.12), and since $\xi \in R_s(\delta e^{-\alpha s})$ we have

$$\|U(t, s)\xi\| \leq D_1 e^{\rho(t)}\|\xi\| \leq D_1\delta e^{-\alpha s}e^{\rho(t)}.$$

Thus, choosing $C > D_1$ in the definition of $\mathcal{B}$ independently of s, and taking δ sufficiently small we obtain

$$\|Jx\|' \leq (D_1 + \theta)\delta e^{-\alpha s} \leq C\delta e^{-\alpha s}.$$

We now consider the derivatives $\partial^j(Jx)$. By Lemma 6.8 applied to the function f^* (see (6.25)), for $j = 1, \ldots, k$ we have

$$\|\partial^j f^*(\tau,\xi)\| \le B\delta^{q+1-j}e^{-\alpha(q+1-j)s}e^{(q+1)\rho(\tau)+jas}. \tag{6.42}$$

Therefore, by (6.42) and the first inequality in (2.17), for $j = 2, \ldots, k$,

$$\|\partial^j (Jx)(t,\xi)\| \le \int_s^t \|U(t,\tau)\| \cdot \|\partial^j f^*(\tau,\xi)\|\, d\tau$$

$$\le BD_1\delta^{q+1-j}e^{-\alpha(q+1-j)s}e^{(q+1+j)as}e^{\rho(t)} \int_s^t e^{(q\bar{a}+a)(\tau-s)}\, d\tau$$

$$\le BD_1\delta^{q+1-j}e^{-\alpha(q+1-j)s}e^{(q+1+j)as}e^{\rho(t)} \int_s^\infty e^{T_1(\tau-s)}\, d\tau,$$

with $T_1 < 0$ as in (6.41). Therefore, taking δ sufficiently small, for $j = 2, \ldots, k$ we have

$$\|\partial^j (Jx)\|' \le BD_1\delta^{q+1-j}e^{-\alpha(q+1-j)s}e^{(q+1+j)as}/|T_1| \le C,$$

where we have used that by (6.1) we have $\alpha > a(q + 1 + j)/(q - j)$. When $j = 1$, the term $U(t,s)$ is also present in the derivative, and thus

$$\|\partial(Jx)(t,\xi)\| \le \|U(t,s)\| + \frac{BD_1\delta^q e^{-\alpha qs}e^{(q+2)as}}{|T_1|}e^{\rho(t)}$$

$$\le (D_1 + BD_1\delta^q e^{-\alpha qs}e^{(q+2)as}/|T_1|)e^{\rho(t)}.$$

Taking δ sufficiently small, and using (6.1) we obtain

$$\|\partial(Jx)\|' \le D_1 + BD_1\delta^q e^{-\alpha qs}e^{(q+2)as}/|T_1| \le C.$$

Finally, by Lemma 6.11 and the first inequality in (2.17), for each $t \ge s$ and $\xi, \bar{\xi} \in R_s(\delta e^{-\alpha s})$ we have

$$\|\partial^k (Jx)(t,\xi) - \partial^k (Jx)(t,\bar{\xi})\|$$

$$\le \int_s^t \|U(t,\tau)\| \cdot \|\partial^k f^*(\tau,\xi) - \partial^k f^*(\tau,\bar{\xi})\|\, d\tau$$

$$\le D'\delta^{q-k}e^{-\alpha(q-k)s}e^{(q+k+1)as}\|\xi - \bar{\xi}\|^\varepsilon D_1 \int_s^t e^{\bar{a}(t-\tau)+a\tau}e^{\bar{a}q(\tau-s)}\, d\tau$$

$$\le D'D_1\delta^{q-k}\|\xi - \bar{\xi}\|^\varepsilon e^{\bar{a}(t-s)+as} \int_s^t e^{((q-1)\bar{a}+a)(\tau-s)}\, d\tau$$

$$= D'D_1\delta^{q-k}\|\xi - \bar{\xi}\|^\varepsilon e^{\rho(t)} \int_s^\infty e^{T_1(\tau-s)}\, d\tau = \frac{D'D_1\delta^{q-k}}{|T_1|}\|\xi - \bar{\xi}\|^\varepsilon e^{\rho(t)}.$$

Taking δ sufficiently small we obtain

$$|(Jx)|_k' \le D'D_1\delta^{q-k}/|T_1| \le C.$$

Hence $Jx \in \mathcal{B}$, and $J\colon \mathcal{B} \to \mathcal{B}$ is a well-defined operator.

We now prove that J is a contraction in the norm $\|\cdot\|'$. Given $x, y \in \mathcal{B}$ and $\tau \geq s$, it follows from (6.3), the mean value theorem, and (6.13) that

$$
\begin{aligned}
&\|f(\tau, x(\tau, \xi), \varphi(\tau, x(\tau, \xi))) - f(\tau, y(\tau, \xi), \varphi(\tau, y(\tau, \xi)))\| \\
&\leq \sup_{r \in [0,1]} \|\partial_{a(r)} f(\tau, \cdot)\| \cdot \|x(\tau, \xi) - y(\tau, \xi)\| \\
&\quad + \sup_{r \in [0,1]} \|\partial_{b(r)} f(\tau, \cdot)\| \cdot \|\varphi(\tau, x(\tau, \xi)) - \varphi(\tau, y(\tau, \xi))\| \\
&\leq 2c(2C\delta e^{-\alpha s} e^{\rho(\tau)})^q \|x(\tau, \xi) - y(\tau, \xi)\| \\
&= 2c(2C\delta e^{-\alpha s} e^{\rho(\tau)})^q \|x - y\|' e^{\rho(\tau)},
\end{aligned}
$$

with

$$
\begin{aligned}
a(r) &= (x(\tau, \xi) + r(y(\tau, \xi) - x(\tau, \xi)), \varphi(\tau, x(\tau, \xi))), \\
b(r) &= (x(\tau, \xi), \varphi(\tau, x(\tau, \xi)) + r(\varphi(\tau, y(\tau, \xi)) - \varphi(\tau, x(\tau, \xi)))).
\end{aligned}
$$

By the first inequality in (2.17) and (6.1) we obtain

$$
\begin{aligned}
&\|(Jx)(t, \xi) - (Jy)(t, \xi)\| \\
&\leq \int_s^t \|U(t, \tau)\| \cdot \|f(\tau, x(\tau, \xi), \varphi(\tau, x(\tau, \xi))) - f(\tau, y(\tau, \xi), \varphi(\tau, y(\tau, \xi)))\| \, d\tau \\
&\leq c2^{q+1} C^q \delta^q \|x - y\|' \int_s^t D_1 e^{\bar{a}(t-\tau)+a\tau} e^{\bar{a}(q+1)(\tau-s)} \, d\tau \\
&\leq c2^{q+1} C^q D_1 \delta^q \|x - y\|' e^{\bar{a}(t-s)+as} \int_s^\infty e^{(q\bar{a}+a)(\tau-s)} \, d\tau \\
&\leq c2^{q+1} C^q D_1 \delta^q \|x - y\|' e^{\rho(t)} \int_s^\infty e^{T_1(\tau-s)} \, d\tau \leq \theta_1 \|x - y\|' e^{\rho(t)},
\end{aligned}
$$

where

$$
\theta_1 = c2^{q+1} C^q D_1 \delta^q / |T_1|.
$$

Therefore, taking δ sufficiently small, we obtain

$$
\|Jx - Jy\|' \leq \theta_1 \|x - y\|'
$$

with $\theta_1 < 1$. Hence, J is a contraction. Thus, by Proposition 6.6 there exists a unique function $x = x_\varphi \in \mathcal{B}$ such that $Jx = x$. The inequality in (6.40) is an immediate consequence of Lemma 6.9, by setting $j = 0$ and $\bar{\xi} = 0$ in (6.27); note that it follows readily from (4.23) that $x(t, 0) = 0$ for every t, by the uniqueness of solutions. □

6.3.6 Behavior under perturbations of the data

We need to discuss how the function x_φ varies with φ. Given $\varphi, \psi \in \mathcal{Z}_\alpha^*$ and $(s, \xi) \in X_\alpha$, we denote by x_φ and x_ψ the functions given by Lemma 6.12 such that $x_\varphi(s, \xi) = x_\psi(s, \xi) = \xi$.

Lemma 6.13. *There exists $K > 0$ such that for every $\delta > 0$ sufficiently small, $\varphi, \psi \in Z_\alpha^*$, and $(s, \xi) \in X_\alpha$ we have*

$$\|x_\varphi(t,\xi) - x_\psi(t,\xi)\| \le K e^{\bar{a}(t-s)} \|\xi\| \cdot \|\varphi - \psi\| \text{ for every } t \ge s. \tag{6.43}$$

Proof. Using (6.3) and the mean value theorem we obtain

$$\begin{aligned}
&\|f(\tau, x_\varphi(\tau,\xi), \varphi(\tau, x_\varphi(\tau,\xi))) - f(\tau, x_\psi(\tau,\xi), \psi(\tau, x_\psi(\tau,\xi)))\| \\
&\le c2^q \|(x_\varphi(\tau,\xi) - x_\psi(\tau,\xi), \varphi(\tau, x_\varphi(\tau,\xi)) - \psi(\tau, x_\psi(\tau,\xi)))\| \\
&\quad \times (\|x_\varphi(\tau,\xi)\|^q + \|x_\psi(\tau,\xi)\|^q).
\end{aligned} \tag{6.44}$$

Furthermore,

$$\begin{aligned}
&\|\varphi(\tau, x_\varphi(\tau,\xi)) - \psi(\tau, x_\psi(\tau,\xi))\| \\
&\le \|\varphi(\tau, x_\varphi(\tau,\xi)) - \psi(\tau, x_\varphi(\tau,\xi))\| + \|\psi(\tau, x_\varphi(\tau,\xi)) - \psi(\tau, x_\psi(\tau,\xi))\| \\
&\le \|x_\varphi(\tau,\xi)\| \cdot \|\varphi - \psi\| + 2\|x_\varphi(\tau,\xi) - x_\psi(\tau,\xi)\|.
\end{aligned} \tag{6.45}$$

When $x_\varphi(\tau,\xi) \notin R_\tau(\delta e^{-\alpha\tau})$, the first term after the last inequality in (6.45) appears in this form due to the fact that

$$\begin{aligned}
&\|\varphi(\tau, x_\varphi(\tau,\xi)) - \psi(\tau, x_\varphi(\tau,\xi))\| \\
&= \left\| \varphi\left(\tau, \delta e^{-\alpha\tau} \frac{x_\varphi(\tau,\xi)}{\|x_\varphi(\tau,\xi)\|} \right) - \psi\left(\tau, \delta e^{-\alpha\tau} \frac{x_\varphi(\tau,\xi)}{\|x_\varphi(\tau,\xi)\|} \right) \right\| \\
&\le \delta e^{-\alpha\tau} \|\varphi - \psi\| \le \|x_\varphi(\tau,\xi)\| \cdot \|\varphi - \psi\|.
\end{aligned}$$

By (6.40) in Lemma 6.12 we conclude that

$$\begin{aligned}
&\|f(\tau, x_\varphi(\tau,\xi), \varphi(\tau, x_\varphi(\tau,\xi))) - f(\tau, x_\psi(\tau,\xi), \psi(\tau, x_\psi(\tau,\xi)))\| \\
&\le c2^{q+1} C^q \delta^q e^{q\bar{a}(\tau-s)+qas-q\alpha s} \\
&\quad \times (\|x_\varphi(\tau,\xi)\| \cdot \|\varphi - \psi\| + 3\|x_\varphi(\tau,\xi) - x_\psi(\tau,\xi)\|).
\end{aligned} \tag{6.46}$$

We now apply Gronwall's lemma. Set

$$\bar{\rho}(t) = \|x_\varphi(t,\xi) - x_\psi(t,\xi)\| \quad \text{and} \quad \bar{\eta} = c2^{q+1} C^q \delta^q. \tag{6.47}$$

Note that $\bar{\eta} \le c2^{2q+1} C^q$, and that the last constant is independent of δ. Using the first inequality in (2.17), (6.40) in Lemma 6.12, and (6.46), it follows from (4.23) that

$$\begin{aligned}
\bar{\rho}(t) &\le \bar{\eta} \int_s^t \|U(t,\tau)\| e^{q\bar{a}(\tau-s)+qas-q\alpha s} \|x_\varphi(\tau,\xi)\| \cdot \|\varphi - \psi\| \, d\tau \\
&\quad + 3\bar{\eta} \int_s^t \|U(t,\tau)\| e^{q\bar{a}(\tau-s)+qas-q\alpha s} \|x_\varphi(\tau,\xi) - x_\psi(\tau,\xi)\| \, d\tau \\
&\le \bar{\eta} D_1 C \|\xi\| \cdot \|\varphi - \psi\| e^{((q+1)a-q\alpha)s} \int_s^t e^{\bar{a}(t-s)+((q-1)\bar{a}+a)(\tau-s)} \, d\tau \\
&\quad + 3\bar{\eta} D_1 e^{q(a-\alpha)s} \int_s^t e^{\bar{a}(t-\tau)+a\tau} e^{q\bar{a}(\tau-s)} \bar{\rho}(\tau) \, d\tau.
\end{aligned}$$

We conclude that

$$
e^{-\bar{a}(t-s)}\bar{\rho}(t) \leq \bar{\eta} D_1 C e^{((q+1)a-q\alpha)s} \int_s^\infty e^{T_1(\tau-s)} \, d\tau \|\xi\| \cdot \|\varphi - \psi\|
$$

$$
+ 3\bar{\eta} D_1 e^{((q+1)a-q\alpha)s} \int_s^t e^{T_1(\tau-s)} \bar{\rho}(\tau) \, d\tau,
$$

$$
\leq \bar{\eta} D_1 C \int_s^\infty e^{T_1(\tau-s)} \, d\tau \|\xi\| \cdot \|\varphi - \psi\|
$$

$$
+ 3\bar{\eta} D_1 \int_s^t e^{T_1(\tau-s)} \bar{\rho}(\tau) \, d\tau,
$$

with $T_1 < 0$ as in (6.41), using also the definition of α. We can now apply Gronwall's lemma to the function $e^{-\bar{a}(t-s)}\bar{\rho}(t)$ to obtain

$$
\bar{\rho}(t) \leq \frac{\bar{\eta} D_1 C}{|T_1|} e^{3\bar{\eta} D_1/|T_1|} e^{\bar{a}(t-s)} \|\xi\| \cdot \|\varphi - \psi\|.
$$

This completes the proof of the lemma.   □

6.3.7 Construction of the stable manifolds

In order to establish the existence of a function $\varphi \in Z_\alpha^*$ satisfying (4.24) when x is the function x_φ given by Lemma 6.12 with $x_\varphi(s, \xi) = \xi$, we first transform this problem into another one.

Lemma 6.14. *Given $\delta > 0$ sufficiently small and $\varphi \in Z_\alpha^*$, the following properties hold:*

1. *if (4.45) holds for every $(s, \xi) \in X_\alpha$ and $t \geq s$, then (4.46) holds for every $(s, \xi) \in X_\alpha$ (including the requirement that the integral is well-defined);*
2. *if (4.46) holds for every $(s, \xi) \in X_\alpha = X_\alpha(\delta)$, then (4.45) holds for every $(s, \xi) \in X_\beta = X_\beta(\delta/C)$ and $t \geq s$.*

The proof can be obtained by repeating arguments in the proof of Lemma 4.7 and thus will be omitted.

We now put together the information given by the former lemmas to establish the existence of a function $\varphi \in Z_\alpha^*$ satisfying (4.24) when $x = x_\varphi$ (with the function x_φ given by Lemma 6.12).

Lemma 6.15. *Given $\delta > 0$ sufficiently small, there exists a unique function $\varphi \in Z_\alpha^*$ such that (4.46) holds for every $(s, \xi) \in X_\alpha$.*

Proof. We look for a fixed point of the operator Φ defined for each $\varphi \in Z_\alpha^*$ by

$$
(\Phi\varphi)(s, \xi) = -\int_s^\infty V(\tau, s)^{-1} f(\tau, x_\varphi(\tau, \xi), \varphi(\tau, x_\varphi(\tau, \xi))) \, d\tau \tag{6.48}
$$

when $(s,\xi) \in X_\alpha$, where x_φ is the unique function given by Lemma 6.12 such that $x_\varphi(s,\xi) = \xi$, and by

$$(\Phi\varphi)(s,\xi) = (\Phi\varphi)(s, \delta e^{-\alpha s}\xi/\|\xi\|)$$

otherwise. In view of Proposition 6.5, it suffices to prove that Φ is a contraction with the norm $\|\cdot\|$ in (4.25) (or more precisely the norm $\varphi \mapsto \|\varphi|X_\alpha\|$).

We first show that $\Phi\varphi|X_\alpha$ is of class C^k in ξ for each $\varphi \in \mathcal{Z}_\alpha^*$. We consider again the function $f^*(t,\xi)$ in (6.25). It follows from the last inequality in (2.17) and Lemma 6.8 that for $j = 1, \ldots, k$, and $\bar{s} \geq s$,

$$\int_{\bar{s}}^\infty \|V(\tau,s)^{-1}\partial^j f^*(\tau,\xi)\| \, d\tau$$

$$\leq B\delta^{q+1-j}e^{-\alpha(q+1-j)s}e^{(q+1+j)as}D_2 \int_{\bar{s}}^\infty e^{-\underline{b}(\tau-s)+\overline{b}\tau}e^{(q+1)\bar{a}(\tau-s)} \, d\tau \quad (6.49)$$

$$= B\delta^{q+1-j}e^{-\alpha(q+1-j)s}e^{((q+1+j)a+b)s}D_2 \int_{\bar{s}}^\infty e^{(T_2+q\bar{a})(\tau-s)} \, d\tau < \infty,$$

where (see (6.6))

$$T_2 = \bar{a} - \underline{b} + b < 0. \quad (6.50)$$

In particular, the integral $\int_s^\infty V(\tau,s)^{-1}\partial^j f^*(\tau,\xi) \, d\tau$ is well-defined for $j = 1, \ldots, k$. Furthermore, by Lemma 6.8, for $j = 0, \ldots, k-1$, $(s,\xi), (s,\bar{\xi}) \in X_\alpha$, and $t \geq s$,

$$\|\partial^j f^*(t,\xi) - \partial^j f^*(t,\bar{\xi})\| \leq B\delta^{q-j}e^{-\alpha(q-j)s}e^{(q+1)\rho(t)+(j+1)as}\|\xi-\bar{\xi}\|. \quad (6.51)$$

In a similar manner, it follows from (6.51) that for $j = 0, \ldots, k-1$, (s,ξ), $(s,\xi+rv) \in X_\alpha$, $r > 0$, and $\bar{s} \geq s$,

$$\int_{\bar{s}}^\infty \|V(\tau,s)^{-1}\Delta_j(\tau,r)\| \, d\tau$$

$$\leq D_2 B\delta^{q-j}e^{-\alpha(q-j)s}e^{[b+(j+q+2)a]s}\|v\| \int_{\bar{s}}^\infty e^{(T_2+q\bar{a})(\tau-s)} \, d\tau, \quad (6.52)$$

where

$$\Delta_j(\tau,r) = \frac{\partial^j f^*(\tau,\xi+rv) - \partial^j f^*(\tau,\xi)}{r}.$$

Thus by (6.52) and (6.49), given $\rho > 0$ there exists $\bar{s} > 0$ such that for (s,ξ) and $(s,\xi+rv)$ as above,

$$\int_{\bar{s}}^\infty \|V(\tau,s)^{-1}\Delta_j(\tau,r)\| \, d\tau \leq \rho\|v\|,$$

and

$$\int_{\bar{s}}^\infty \|V(\tau,s)^{-1}\partial^{j+1} f^*(\tau,\xi)\| \, d\tau < \rho.$$

For this $\bar{s}$ and provided that r is sufficiently small, we also have

$$\left\| \int_s^{\bar{s}} \|V(\tau,s)^{-1}[\Delta_j(\tau,r) - \partial^{j+1} f^*(\tau,\xi)v]\, d\tau \right\| < \rho\|v\|.$$

Hence, by the former three inequalities

$$\left\| \int_s^{\infty} \|V(\tau,s)^{-1}[\Delta_j(\tau,r) - \partial^{j+1} f^*(\tau,\xi)v]\, d\tau \right\| < 3\rho\|v\|.$$

Since ρ is arbitrary, there exists the limit

$$\lim_{r \to 0} \int_s^{\infty} V(\tau,s)^{-1} \Delta_j(\tau,r)\, d\tau = \int_s^{\infty} V(\tau,s)^{-1} \partial^{j+1} f^*(\tau,\xi)v\, d\tau.$$

Proceeding by induction in j we find that $\Phi\varphi|X_\alpha$ is of class C^k in ξ, with derivatives given by

$$\partial^j (\Phi\varphi)(s,\xi) = - \int_s^{\infty} V(\tau,s)^{-1} \partial^j f^*(\tau,\xi)\, d\tau \qquad (6.53)$$

for $j = 1, \ldots, k$. Since $x_\varphi(t,0) = 0$ for every $\varphi \in Z_\alpha^*$ and $t \geq s$ (see Lemma 6.12), it follows from (6.48) that $(\Phi\varphi)(s,0) = 0$ for every $s \geq 0$. Furthermore, by (6.53),

$$\partial(\Phi\varphi)(s,0) = - \int_s^{\infty} V(\tau,s)^{-1} \partial f(\tau,0) \partial a_\varphi(\tau,0)\, d\tau,$$

where $a_\varphi(\tau,\xi) = (x_\varphi(\tau,\xi), \varphi(\tau, x_\varphi(\tau,\xi)))$. Since $\partial f(\tau,0) = 0$, we conclude that $\partial(\Phi\varphi)(s,0) = 0$ for every $s \geq 0$.

Using the second inequality in (2.17) together with the definition of α in (6.1), proceeding as in (6.49) we obtain

$$\|\partial^j (\Phi\varphi)(s,\xi)\| \leq \int_s^{\infty} \|V(\tau,s)^{-1}\| \cdot \|\partial^j f^*(\tau,\xi)\|\, d\tau$$

$$\leq B\delta^{q+1-j} e^{-\alpha(q+1-j)s} e^{((q+1+j)a+b)s} D_2 \int_s^{\infty} e^{(T_2+q\bar{a})(\tau-s)}\, d\tau$$

$$\leq B\delta^{q+1-j} D_2 \int_s^{\infty} e^{(T_2+q\bar{a})(\tau-s)}\, d\tau \leq \frac{B\delta^{q+1-j} D_2}{|T_2 + q\bar{a}|},$$

with $T_2 < 0$ as in (6.50). Taking δ sufficiently small so that

$$B\delta^{q+1-k} D_2 / |T_2 + q\bar{a}| < 1,$$

we have $\|\partial^j (\Phi\varphi)(s,\xi)\| \leq 1$ for every $s \geq 0$ and $\xi \in R_s(\delta e^{-\alpha s})$. Furthermore, by Lemma 6.11,

$$\|\partial^k f^*(\tau,\xi) - \partial^k f^*(\tau,\bar{\xi})\| \leq D'(\delta e^{-\alpha s})^{q-k} e^{q\rho(\tau)+(k+1)as} \|\xi - \bar{\xi}\|^{\varepsilon}.$$

Using again the second inequality in (2.17) together with the definition of α in (6.1) we conclude that

$$\|\partial^k(\Phi\varphi)(s,\xi) - \partial^k(\Phi\varphi)(s,\bar{\xi})\| \leq$$

$$\leq \int_s^\infty \|V(\tau,s)^{-1}\| \cdot \|\partial^k f^*(\tau,\xi) - \partial^k f^*(\tau,\bar{\xi})\| \, d\tau$$

$$\leq D'\delta^{q-k} e^{-\alpha(q-k)s} e^{(q+k+1)as} \|\xi - \bar{\xi}\|^\varepsilon D_2 \int_s^\infty e^{-\underline{b}(\tau-s)+b\tau+\bar{a}q(\tau-s)} \, d\tau$$

$$= D'\delta^{q-k} e^{((q+k+1)a+b-\alpha(q-k))s} \|\xi - \bar{\xi}\|^\varepsilon D_2 \int_s^\infty e^{(-\underline{b}+b+\bar{a}q)(\tau-s)} \, d\tau$$

$$= D'\delta^{q-k} D_2 \|\xi - \bar{\xi}\|^\varepsilon \int_s^\infty e^{(T_2+(q-1)\bar{a})(\tau-s)} \, d\tau = \frac{D'\delta^{q-k} D_2}{|T_2 + (q-1)\bar{a}|} \|\xi - \bar{\xi}\|^\varepsilon.$$

Taking δ sufficiently small so that the last fraction is at most 1, for every $s \geq 0$ and $\xi \in R_s(\delta e^{-\alpha s})$ we have

$$\|\partial^k(\Phi\varphi)(s,\xi) - \partial^k(\Phi\varphi)(s,\bar{\xi})\| \leq \|\xi - \bar{\xi}\|^\varepsilon.$$

This shows that $\Phi(Z_\alpha^*) \subset Z_\alpha^*$, and hence, $\Phi \colon Z_\alpha^* \to Z_\alpha^*$ is well-defined.

We now show that $\Phi \colon Z_\alpha^* \to Z_\alpha^*$ is a contraction. Given $\varphi, \psi \in Z_\alpha^*$, and $(s,\xi) \in X_\alpha$, let x_φ and x_ψ be the unique functions given by Lemma 6.12 such that $x_\varphi(s,\xi) = x_\psi(s,\xi) = \xi$. Proceeding in a similar manner to that in (6.44) and (6.45), with g replaced by h, we obtain

$$\begin{aligned}
b(\tau) :&= \|f(\tau, x_\varphi(\tau,\xi), \varphi(\tau, x_\varphi(\tau,\xi))) - f(\tau, x_\psi(\tau,\xi), \psi(\tau, x_\psi(\tau,\xi)))\| \\
&\leq c2^q \|(x_\varphi(\tau,\xi) - x_\psi(\tau,\xi), \varphi(\tau, x_\varphi(\tau,\xi)) - \psi(\tau, x_\psi(\tau,\xi)))\| \\
&\quad \times (\|x_\varphi(\tau,\xi)\|^q + \|x_\psi(\tau,\xi)\|^q) \\
&\leq c2^q (\|x_\varphi(\tau,\xi)\| \cdot \|\varphi - \psi\| + 2\|x_\varphi(\tau,\xi) - x_\psi(\tau,\xi)\|) \\
&\quad \times (\|x_\varphi(\tau,\xi)\|^q + \|x_\psi(\tau,\xi)\|^q).
\end{aligned}$$

It follows from Lemma 6.12, and (6.43) in Lemma 6.13 that

$$\begin{aligned}
b(\tau) &\leq \bar{\eta} e^{q\bar{a}(\tau-s)+q(a-\alpha)s} (\|x_\varphi(\tau,\xi)\| \cdot \|\varphi - \psi\| + 3\|x_\varphi(\tau,\xi) - x_\psi(\tau,\xi)\|) \\
&\leq \bar{\eta} e^{(q+1)\bar{a}(\tau-s)+(q+1)as-\alpha qs} (C + 3Ke^{-as}) \|\xi\| \cdot \|\varphi - \psi\|,
\end{aligned}$$

with $\bar{\eta}$ as in (6.47). Setting $G = \bar{\eta}(C + 3K)$ and using (2.17), we obtain

$$\|(\Phi\varphi)(s,\xi) - (\Phi\psi)(s,\xi)\| \leq \int_s^\infty \|V(\tau,s)^{-1}\| \cdot b(\tau) \, d\tau$$

$$\leq D_2 G \|\xi\| \cdot \|\varphi - \psi\| e^{((q+1)a+b-\alpha q)s} \int_s^\infty e^{(-\underline{b}+b+(q+1)\bar{a})(\tau-s)} \, d\tau$$

$$= D_2 G \|\xi\| \cdot \|\varphi - \psi\| \int_s^\infty e^{(T_2+q\bar{a})(\tau-s)} \, d\tau \leq \frac{D_2 G}{|T_2 + q\bar{a}|} \|\xi\| \cdot \|\varphi - \psi\|.$$

Taking $\delta > 0$ sufficiently small, we have

$$\bar{\theta} = \frac{D_2 G}{|T_2 + q\bar{a}|} = \frac{D_2 c 2^{q+1} C^q \delta^q (C + 3K)}{|T_2 + q\bar{a}|} < 1.$$

Therefore

$$\|\Phi\varphi_1 - \Phi\varphi_2\| \le \bar{\theta}\|\varphi_1 - \varphi_2\|,$$

and $\Phi\colon \mathcal{Z}_\alpha^* \to \mathcal{Z}_\alpha^*$ is a contraction in the complete metric space $\mathcal{Z}_\alpha^*$ (see Proposition 6.5). Hence, there exists a unique function $\varphi \in \mathcal{Z}_\alpha^*$ satisfying $\Phi\varphi = \varphi$. In particular, in view of (6.48), the identity (4.46) holds for every $(s, \xi) \in X_\alpha$. This completes the proof of the lemma. $\square$

We can now establish Theorem 6.1.

Proof of Theorem 6.1. As explained in the beginning of Section 6.3.1, in view of the required forward invariance property in (6.7), to show the existence of a stable manifold $\mathcal{V}$ is equivalent to find a function φ satisfying (4.23) and (4.24) in some appropriate domain. If follows from Lemma 6.12 that for each fixed $\varphi \in \mathcal{Z}_\alpha^*$ there exists a unique function $x = x_\varphi$ satisfying (4.23) and thus it remains to solve (4.24) setting $x = x_\varphi$ or, equivalently, to solve (4.45) in Lemma 6.14. This lemma indicates that the problem can be reduced to solve the equation in (4.46), that is, to find $\varphi \in \mathcal{Z}_\alpha^*$ such that (4.46) holds for every $(s, \xi) \in X_\alpha$. More precisely, it follows from the second property in Lemma 6.14 that if (4.46) holds for every $(s, \xi) \in X_\alpha = X_\alpha(\delta)$, then (4.45) holds for every $(s, \xi) \in X_\beta = X_\beta(\delta/C)$ and $t \ge s$. Finally, Lemma 6.15 shows that there exists a unique function $\varphi \in \mathcal{Z}_\alpha^*$ such that (4.46) holds for every $(s, \xi) \in X_\alpha$.

Furthermore, by (6.40), provided that δ is sufficiently small and $(s, \xi) \in X_\beta$ we have $(t, x_\varphi(t)) \in X_\alpha$ for every $t \ge s$. This ensures that we can replace the function φ in (4.23)–(4.24) by its restriction $\varphi|X_\alpha$. In other words, there exists a unique function $\varphi \in \mathcal{Z}_\alpha$ such that the corresponding set $\mathcal{V}$ in (6.5) is forward invariant under the semiflow Ψ_τ for the initial conditions $(s, \xi) \in X_\beta$, and thus (6.7) holds.

We now establish the remaining properties in the theorem. For each $s > 0$, we consider a constant $\varepsilon = \varepsilon(s) \in (0, s)$ and we consider the map

$$F_s\colon \{(t, \xi)\colon t \in (-\varepsilon, \varepsilon) \text{ and } \xi \in R_{s+t}(\delta e^{-\beta(s+t)})\} \to \mathbb{R}^+ \times X$$

defined by (5.54). Since A and f are of class C^k, the map $(t, s, \xi, \eta) \mapsto \Psi_t(s, \xi, \eta)$ is also of class C^k on $\mathbb{R}^+ \times \mathbb{R}^+ \times X$ (see for example [46]). Since $\varphi \in \mathcal{Z}_\alpha$, the map F_s is also of class C^k (for each fixed s). Furthermore, F_s is injective and is thus a parametrization of class C^k of an open subset of $\mathcal{V}$ containing $(s, 0)$ (see the proof of Theorem 4.1). Therefore, $\mathcal{V}$ is a smooth manifold of class C^k. The second property in Theorem 6.1 is an immediate consequence of the above discussion (or of the first property in Lemma 6.14). To prove the third property, we denote again by $x = x_\varphi$ the unique function given by Lemma 6.12 such that $x_\varphi(s, \xi) = \xi$. With the notation in (6.24) we have

$$\|\partial_\xi^j(\Psi_\tau(s,\xi,\varphi(s,\xi))) - \partial_\xi^j(\Psi_\tau(s,\bar{\xi},\varphi(s,\bar{\xi})))\|$$
$$= \|\partial_\xi^j(t,x(t,\xi),\varphi^*(t,\xi)) - \partial_\xi^j(t,x(t,\bar{\xi}),\varphi^*(t,\bar{\xi}))\| \qquad (6.54)$$
$$\leq \|\partial^j x(t,\xi) - \partial^j x(t,\bar{\xi})\| + \|\partial^j \varphi^*(t,\xi) - \partial^j \varphi^*(t,\bar{\xi})\|$$

for every $\tau \geq 0$ and $t = s + \tau$. When $\xi, \bar{\xi} \in R_s(\delta e^{-\beta s})$, in view of (6.40) we can replace the function φ in (6.54) by its restriction to X_α. Note that by Lemma 6.7, we have

$$\|\partial^j \varphi^*(t,\xi) - \partial^j \varphi^*(t,\bar{\xi})\| \leq A e^{\rho(t)+jas}\|\xi - \bar{\xi}\|$$

for $j = 0, \ldots, k - 1$. Thus, for these values of j the inequality in (6.8) follows readily from Lemmas 6.9 and 6.7, taking into account that $a \geq 0$. For $j = k$ the inequality in (6.9) follows from the fact that $x_\varphi \in \mathcal{B}$ (see Lemma 6.12) and Lemma 6.10, since $\|\xi - \bar{\xi}\| \leq \|\xi - \bar{\xi}\|^\varepsilon$ (recall that $\delta < 1$) and $a \geq 0$. This completes the proof of the theorem. $\qquad \square$

7

A nonautonomous Grobman–Hartman theorem

A fundamental problem in the study of the local behavior of a dynamical system is whether the linearization of the system along a given solution approximates well the solution itself in some open neighborhood. In other words, we look for an appropriate local change of variables, called a conjugacy, that can transform the system into a linear one. Moreover, as a means to distinguish the dynamics in a neighborhood of the solution further than in the topological category (such as, for example, to distinguish different types of nodes), the change of variables should be as regular as possible. The problem goes back to the pioneering work of Poincaré, that can be interpreted today as looking for an *analytic* change of variables which transforms the initial system into a linear one. The work of Sternberg [89, 90] showed that there are algebraic obstructions, expressed in terms of resonances between the eigenvalues of the linear approximation, that prevent the existence of conjugacies with a prescribed high regularity (see also [19, 20, 87, 61] for further related work). The main purpose of this chapter is to establish a nonautonomous and nonuniform version of the Grobman–Hartman theorem in Banach spaces. In addition, we show that the conjugacies are always *Hölder continuous*, with Hölder exponent expressed in terms of ratios of Lyapunov exponents. We follow closely [17, 10].

7.1 Conjugacies for flows

Let X be a Banach space. We continue to denote by $B(X)$ the set of bounded linear operators in X and we assume that:

E1. the function $A\colon \mathbb{R} \to B(X)$ is continuous and satisfies (2.2);
E2. the function $f\colon \mathbb{R} \times X \to X$ is continuous, and there exist $\bar{\delta} > 0$ and $\bar{\beta} \geq 0$ such that for every $t \in \mathbb{R}$ and $x, y \in X$,

$$\|f(t,x) - f(t,y)\| \leq \bar{\delta} e^{-\bar{\beta}|t|} \min\{1, \|x - y\|\}. \tag{7.1}$$

Under the assumptions E1–E2 both the linear equation $v' = A(t)v$ and the perturbed equation $v' = A(t)v + f(t, v)$ define evolution operators in the whole $\mathbb{R}$, that we denote respectively by $T(t, s)$ and $R(t, s)$, with $t, s \in \mathbb{R}$.

Theorem 7.1 ([17]). *Assume that E1–E2 hold and that the linear equation $v' = A(t)v$ admits a strong nonuniform exponential dichotomy in $\mathbb{R}$ with $\underline{b} > 0$ and $\bar{\beta} = 6\max\{a, b\}$. If $\bar{\delta}$ is sufficiently small, then there exist homeomorphisms $h_t \colon X \to X$ for $t \in \mathbb{R}$ such that*

$$T(t, s) \circ h_s = h_t \circ R(t, s), \quad t, s \in \mathbb{R}. \tag{7.2}$$

The proof of Theorem 7.1 is given in Section 7.4 as a consequence of the corresponding result for maps (see Section 7.2). We note that this is the only place in the book where we reduce the study of flows or semiflows to the study of the associated (time-1) maps.

We also show in Section 7.4 that the topological conjugacies h_t and their inverses are Hölder continuous. We assume that there exists a strong nonuniform exponential dichotomy and we set

$$\alpha_0 = \min\{\bar{a}/\underline{a}, \underline{b}/\bar{b}\}, \tag{7.3}$$

with the same constants as in (2.6). It follows from (2.6) that $\alpha_0 \in (0, 1]$.

Theorem 7.2 ([17]). *Assume that E1–E2 hold and that the linear equation $v' = A(t)v$ admits a strong nonuniform exponential dichotomy in $\mathbb{R}$ with $\underline{b} > 0$ and $\bar{\beta} = 6\max\{a, b\}$. For each $\alpha \in (0, \alpha_0)$, if $\bar{\delta}$ is sufficiently small (depending on α), then there exist homeomorphisms h_t as in Theorem 7.1, and a constant $\bar{K} > 0$ (depending on α and δ) such that*

$$\|h_t(x) - h_t(y)\| \le \bar{K} e^{2\max\{a,b\}(2+3\alpha)|t|} \|x - y\|^\alpha,$$

$$\|h_t^{-1}(x) - h_t^{-1}(y)\| \le \bar{K} e^{2\max\{a,b\}(2+3\alpha)|t|} \|x - y\|^\alpha$$

for every $t \in \mathbb{R}$ and $x, y \in X$ with $\|x - y\| \le e^{-3\max\{a,b\}|t|}$.

The proof of Theorem 7.2 is given in Section 7.4, also as a consequence of a corresponding result for maps.

We note that in the classical autonomous case of *uniform* exponential dichotomies, the Hölder regularity of the conjugacies seems to have been known by some experts, although, apparently, no published proof can be found in the literature (see the detailed discussion in Section 1.3).

7.2 Conjugacies for maps

We establish here a nonautonomous and nonuniform version of the Grobman–Hartman theorem in the case of discrete time. We show in Section 7.3 that the conjugacies are always Hölder continuous. These results will be used in Section 7.4 to prove Theorems 7.1 and 7.2.

7.2.1 Setup

We assume here that:

F1. there exist invertible linear operators $A_m \in B(X)$, $m \in \mathbb{Z}$ with inverse $A_m^{-1} \in B(X)$;

F2. there exist continuous maps $f_m \colon X \to X$, $m \in \mathbb{Z}$, and constants $\delta > 0$ and $\vartheta \geq 0$ such that for each $m \in \mathbb{Z}$ the map $A_m + f_m$ is a homeomorphism and

$$\|f_m\|_\infty := \sup\{\|f_m(x)\| : x \in X\} \leq \delta e^{-\vartheta|m|}; \qquad (7.4)$$

F3. there exists $\beta \geq 0$ such that for every $x, y \in X$ we have

$$\|f_m(x) - f_m(y)\| \leq \delta e^{-\beta|m|}\|x - y\|, \quad m \in \mathbb{Z}. \qquad (7.5)$$

The conditions F1 and F2 will be assumed throughout Section 7.2, while F3 will only be needed in some results (this will be made explicit in each statement). We note that when the conditions F1–F3 hold, provided that δ is sufficiently small the requirement in F2 that the map $G_m = A_m + f_m$ is a homeomorphism is not needed: in this case it is easy to verify, using the remaining conditions, that G_m is invertible, and it follows from Lemma 7.15 that the inverse is Lipschitz.

Set

$$\mathcal{A}(m,n) = \begin{cases} A_{m-1} \cdots A_n, & m > n \\ \mathrm{Id}, & m = n \\ A_m^{-1} \cdots A_{n-1}^{-1}, & m < n \end{cases}.$$

We now introduce the notion of nonuniform exponential dichotomy in the case of discrete time.

Definition 7.3. *We say that the sequence of linear operators $(A_m)_{m \in \mathbb{Z}}$ admits a nonuniform exponential dichotomy if there exist projections $P_n \in B(X)$ for $n \in \mathbb{Z}$, with*

$$P_m \mathcal{A}(m,n) = \mathcal{A}(m,n)P_n \text{ for every } m, n \in \mathbb{Z} \text{ with } m \geq n, \qquad (7.6)$$

and there exist constants

$$\bar{a} < 0 \leq \underline{b}, \quad a, b \geq 0, \quad and \quad \bar{D} \geq 1$$

such that for every $m, n \in \mathbb{Z}$ with $m \geq n$ we have

$$\|\mathcal{A}(m,n)P_n\| \leq \bar{D}e^{\bar{a}(m-n)+a|n|}, \quad \|\mathcal{A}(m,n)^{-1}Q_m\| \leq \bar{D}e^{-\underline{b}(m-n)+b|m|}, \quad (7.7)$$

where $Q_n = \mathrm{Id} - P_n$ are the complementary projections.

For a sequence $(A_m)_{m \in \mathbb{Z}}$ admitting a nonuniform exponential dichotomy, we consider the linear subspaces

$$E_m = P_m X \quad \text{and} \quad F_m = Q_m X$$

for each $m \in \mathbb{Z}$. We call E_m and F_m respectively the *stable* and *unstable* *subspaces* at time $m \in \mathbb{Z}$. Clearly $X = E_m \oplus F_m$ for every $m \in \mathbb{Z}$, and the dimensions $\dim E_m$ and $\dim F_m$ are independent of m. We define the operators

$$B_m := A_m | E_m \colon E_m \to E_{m+1} \quad \text{and} \quad C_m := A_m | F_m \colon F_m \to F_{m+1}$$

for each $m \in \mathbb{Z}$. Clearly, these are invertible continuous linear operators with continuous inverse. Furthermore, with respect to the decompositions $X = E_m \oplus F_m$, we have the block form

$$A_m = \begin{pmatrix} B_m & 0 \\ 0 & C_m \end{pmatrix}, \quad m \in \mathbb{Z}. \tag{7.8}$$

Each sequence $(z_m)_{m \in \mathbb{Z}} \subset X$ satisfying $z_{m+1} = A_m z_m$ for every $m \in \mathbb{Z}$ can be written in the form

$$z_m = \mathcal{A}(m,n) z_n = (\mathcal{B}(m,n) x_n, \mathcal{C}(m,n) y_n), \quad m, n \in \mathbb{Z},$$

where $z_n = (x_n, y_n) \in E_m \times F_m$, and

$$\mathcal{B}(m,n) = \begin{cases} B_{m-1} \cdots B_n, & m > n \\ \mathrm{Id}, & m = n \\ B_m^{-1} \cdots B_{n-1}^{-1}, & m < n \end{cases}, \quad \mathcal{C}(m,n) = \begin{cases} C_{m-1} \cdots C_n, & m > n \\ \mathrm{Id}, & m = n \\ C_m^{-1} \cdots C_{n-1}^{-1}, & m < n \end{cases}.$$

Furthermore, the inequalities in (7.7) can be written in the form

$$\|\mathcal{B}(m,n)\| \le \bar{D} e^{\bar{a}(m-n) + a|n|}, \quad \|\mathcal{C}(m,n)^{-1}\| \le \bar{D} e^{-\underline{b}(m-n) + b|m|}. \tag{7.9}$$

We also introduce appropriate Lyapunov norms now in the case of discrete time. Choose $\varrho > 0$ such that $\varrho < \min\{-\bar{a}, \underline{b}\}$. For each $m \in \mathbb{Z}$ we define

$$\begin{aligned}
\|x\|_m' &= \sum_{k \ge m} \|\mathcal{B}(k,m)x\| e^{(-\bar{a}-\varrho)(k-m)} \quad \text{for } x \in E_m, \\
\|y\|_m' &= \sum_{k \le m} \|\mathcal{C}(m,k)^{-1}y\| e^{(\underline{b}-\varrho)(m-k)} \quad \text{for } y \in F_m,
\end{aligned} \tag{7.10}$$

and we set

$$\|(x,y)\|_m' = \|x\|_m' + \|y\|_m' \quad \text{for each } (x,y) \in E_m \times F_m. \tag{7.11}$$

Using (7.9) it is straightforward to verify that each series in (7.10) is finite, and

$$\|x\| \le \|x\|_m' \le \frac{\bar{D} e^{a|m|}}{1 - e^{-\varrho}} \|x\| \quad \text{and} \quad \|y\| \le \|y\|_m' \le \frac{\bar{D} e^{b|m|}}{1 - e^{-\varrho}} \|y\|.$$

Lemma 7.4. *For each $z \in X$ and $m \in \mathbb{Z}$ we have*

$$\|z\| \le \|z\|_m' \le \frac{2\bar{D}^2}{1 - e^{-\varrho}} e^{2 \max\{a,b\}|m|} \|z\|. \tag{7.12}$$

Proof. Clearly,

$$\|(x,y)\| \le \|x\| + \|y\| \le \|x\|_m' + \|y\|_m' = \|(x,y)\|_m'.$$

Since $P_m(x,y) = x$ and $Q_m(x,y) = y$, it follows from (7.7) that

$$\|(x,y)\|_m' \le \frac{\bar{D}}{1 - e^{-\varrho}} e^{\max\{a,b\}|m|} (\|P_m\| + \|Q_m\|)\|(x,y)\|$$

$$\le \frac{2\bar{D}^2}{1 - e^{-\varrho}} e^{2 \max\{a,b\}|m|} \|(x,y)\|,$$

which gives the desired result.   □

Furthermore, whenever $m \ge n$,

$$\|\mathcal{B}(m,n)\|' := \sup_{x \in E \setminus \{0\}} \frac{\|\mathcal{B}(m,n)x\|_m'}{\|x\|_n'} \le e^{(\bar{a}+\varrho)(m-n)},$$

$$\|\mathcal{C}(m,n)^{-1}\|' := \sup_{y \in F \setminus \{0\}} \frac{\|\mathcal{C}(m,n)^{-1}y\|_n'}{\|y\|_m'} \le e^{(-\underline{b}+\varrho)(m-n)}.$$

7.2.2 Existence of topological conjugacies

We construct here *topological* conjugacies between the sequences formed respectively by the maps A_m and $A_m + f_m$. We proceed in three steps:

1. we show that there exist unique continuous functions $\hat{u}_m$ satisfying

$$A_m \circ \hat{u}_m = \hat{u}_{m+1} \circ (A_m + f_m) \tag{7.13}$$

   such that $\hat{u}_m - \text{Id}$ is bounded for each $m \in \mathbb{Z}$ (see Theorem 7.5);
2. we show that there exist unique continuous functions $\hat{v}_m$ satisfying

$$\hat{v}_{m+1} \circ A_m = (A_m + f_m) \circ \hat{v}_m \tag{7.14}$$

   such that $\hat{v}_m - \text{Id}$ is bounded for each $m \in \mathbb{Z}$ (see Theorem 7.6);
3. we verify that for each $m \in \mathbb{Z}$ these functions satisfy

$$\hat{u}_m \circ \hat{v}_m = \hat{v}_m \circ \hat{u}_m = \text{Id},$$

   and thus they are the desired topological conjugacies (see Corollary 7.7).

The Hölder regularity of the conjugacies will be obtained in Section 7.3.

We note that the problem in (7.14) is obtained from that in (7.13) by interchanging the order in the compositions in each side. We emphasize that Theorem 7.5 does not show that the unique maps $\widehat{u}_m$ are invertible, and thus, in order to show the existence of topological conjugacies, we must also consider the problem in (7.14). One could of course consider instead the general problem

$$(A_m + \bar{f}_m) \circ \widehat{w}_m = \widehat{u}_{m+1} \circ (A_m + f_m) \tag{7.15}$$

by showing the uniqueness of the continuous functions $\widehat{w}_m$ satisfying (7.15) such that $\widehat{w}_m - \mathrm{Id}$ for each $m \in \mathbb{Z}$, one would immediately conclude the existence and continuity of the inverse of each function $\widehat{u}_m$. Namely, one can simply take $\bar{f}_m = 0$ in (7.15), which corresponds to Theorem 7.5, and $f_m = 0$ in (7.15), which corresponds to Theorem 7.6. However, the difficulty involved in considering the general equation in (7.15) is essentially the same as that of considering the two separate problems in Theorems 7.5 and 7.6, that is, equations (7.13) and (7.14). This is caused by the fact that rewriting an equation of this type in terms of a fixed point problem, for a contraction operator obtained from a composition of maps such that the first one is nonlinear, increases the difficulty of the estimates required in the proofs. By considering separately the equations in (7.13) and (7.14), and thus two fixed points problems instead of only one, we avoid this difficulty (see (7.19)–(7.20) in the proof of Theorem 7.5, and (7.29)–(7.30) in the proof of Theorem 7.6).

We consider the space $\mathcal{X}$ of sequences $u = (u_m)_{m \in \mathbb{Z}}$ of continuous functions $u_m \colon X \to X$ such that

$$\|u\|'_\infty := \sup\{\|u_m\|'_m : m \in \mathbb{Z}\} < \infty, \tag{7.16}$$

where

$$\|u_m\|'_m := \sup\{\|u_m(x)\|'_m : x \in X\}.$$

One can easily verify that $\mathcal{X}$ is a complete metric space with this norm.

We now present the first main result.

Theorem 7.5. *Assume that F1–F2 hold. If the sequence $(A_m)_{m \in \mathbb{Z}}$ admits a nonuniform exponential dichotomy with $\underline{b} > 0$ and $\max\{a, b\} \leq \vartheta$, then there is a unique $(u_m)_{m \in \mathbb{Z}} \in \mathcal{X}$ such that for every $m \in \mathbb{Z}$ we have*

$$A_m \circ \widehat{u}_m = \widehat{u}_{m+1} \circ (A_m + f_m), \quad \text{where } \widehat{u}_m = \mathrm{Id} + u_m. \tag{7.17}$$

Proof. Setting $G_m = A_m + f_m$, the equation in (7.17) is equivalent to

$$A_m \circ u_m - u_{m+1} \circ G_m = f_m. \tag{7.18}$$

Writing $u_m = (b_m, c_m)$ and $f_m = (g_m, h_m)$, with values in $E_m \times F_m$, using F1 we find that (7.18) holds for every $m \in \mathbb{Z}$ if and only if $(b_m, c_m) = (\bar{b}_m, \bar{c}_m)$ for every $m \in \mathbb{Z}$, where

$$\bar{b}_m = (B_{m-1} \circ b_{m-1} - g_{m-1}) \circ G_{m-1}^{-1}, \tag{7.19}$$

$$\bar{c}_m = C_m^{-1} \circ (c_{m+1} \circ G_m + h_m). \tag{7.20}$$

Given $u = (u_m)_{m \in \mathbb{Z}} = (b_m, c_m)_{m \in \mathbb{Z}} \in \mathfrak{X}$, we define $S(u) = (\bar{b}_m, \bar{c}_m)_{m \in \mathbb{Z}}$. The statement in the theorem is thus equivalent to the existence of a unique fixed point of S in the space $\mathfrak{X}$. We will prove that $S(\mathfrak{X}) \subset \mathfrak{X}$ and that S is a contraction in the complete metric space $\mathfrak{X}$.

Since G_m is a homeomorphism, $(\bar{b}_m, \bar{c}_m)$ is continuous for every $m \in \mathbb{Z}$. Furthermore, using the Lyapunov norms in (7.10) for each $z \in X$ we obtain

$$\|\bar{b}_m(z)\|_m' \le \sum_{k \ge m} \|\mathcal{B}(k,m)B_{m-1}b_{m-1}(G_{m-1}^{-1}(z))\|e^{(-\bar{a}-\varrho)(k-m)}$$

$$+ \sum_{k \ge m} \|\mathcal{B}(k,m)g_{m-1}(G_{m-1}^{-1}(z))\|e^{(-\bar{a}-\varrho)(k-m)}$$

$$\le e^{\bar{a}+\varrho} \sum_{k \ge m-1} \|\mathcal{B}(k,m-1)b_{m-1}(G_{m-1}^{-1}(z))\|e^{(-\bar{a}-\varrho)(k-(m-1))}$$

$$+ \sum_{k \ge m} \|\mathcal{B}(k,m)\| \cdot \|g_{m-1}\|_\infty e^{(-\bar{a}-\varrho)(k-m)} \tag{7.21}$$

$$\le e^{\bar{a}+\varrho} \|b_{m-1}(G_{m-1}^{-1}(z))\|_{m-1}'$$

$$+ \bar{D}\delta \sum_{k \ge m} e^{\bar{a}(k-m)+\vartheta|m|}e^{-\vartheta|m-1|}e^{-(\bar{a}+\varrho)(k-m)}$$

$$\le e^{\bar{a}+\varrho}\|b_{m-1}(G_{m-1}^{-1}(z))\|_{m-1}' + \bar{D}\delta e^{\vartheta} \sum_{k \ge m} e^{\varrho(m-k)}.$$

Setting $\theta = \bar{D}\delta e^{\vartheta}/(1 - e^{-\varrho})$, for the sequences $b = (b_m)_{m \in \mathbb{Z}}$ and $\bar{b} = (\bar{b}_m)_{m \in \mathbb{Z}}$ we have

$$\|\bar{b}\|_\infty' = \sup\{\|\bar{b}_m\|_m' : m \in \mathbb{Z}\} \le e^{\bar{a}+\varrho}\|b\|_\infty' + \theta < \infty. \tag{7.22}$$

In an analogous manner, for each $z \in X$ we obtain

$$\|\bar{c}_m(z)\|_m' \le \sum_{k \le m} \|\mathcal{C}(m,k)^{-1}C_m^{-1}c_{m+1}(G_m(z))\|e^{(\underline{b}-\varrho)(m-k)}$$

$$+ \sum_{k \le m} \|\mathcal{C}(m,k)^{-1}C_m^{-1}h_m(z)\|e^{(\underline{b}-\varrho)(m-k)}$$

$$\le e^{-\underline{b}+\varrho} \sum_{k \le m+1} \|\mathcal{C}(m+1,k)^{-1}c_{m+1}(G_m(z))\|e^{(\underline{b}-\varrho)(m+1-k)}$$

$$+ \sum_{k \le m} \|\mathcal{C}(m+1,k)^{-1}\| \cdot \|h_m\|_\infty e^{(\underline{b}-\varrho)(m-k)} \tag{7.23}$$

$$\le e^{-\underline{b}+\varrho}\|c_{m+1}(G_m(z))\|_{m+1}'$$

$$+ \bar{D}\delta \sum_{k \le m} e^{-\underline{b}(m+1-k)+\vartheta|m+1|}e^{-\vartheta|m|}e^{(\underline{b}-\varrho)(m-k)}$$

$$\le e^{-\underline{b}+\varrho}\|c_{m+1}(G_m(z))\|_{m+1}' + \bar{D}\delta e^{\vartheta-\underline{b}} \sum_{k \le m} e^{\varrho(k-m)}.$$

For the sequences $c = (c_m)_{m \in \mathbb{Z}}$ and $\bar{c} = (\bar{c}_m)_{m \in \mathbb{Z}}$ we have

$$\|\bar{c}\|_\infty' \le e^{-\underline{b}+\varrho} \|c\|_\infty' + \theta < \infty. \tag{7.24}$$

By (7.22) and (7.24) we have $S(u) \in \mathcal{X}$, and thus $S \colon \mathcal{X} \to \mathcal{X}$ is well-defined.

We now prove that S is a contraction. Given $u_1 = (b_{1,m}, c_{1,m})_{m \in \mathbb{Z}}$ and $u_2 = (b_{2,m}, c_{2,m})_{m \in \mathbb{Z}}$ in $\mathcal{X}$, proceeding as in (7.21) for each $z \in X$ we obtain

$$\|\bar{b}_{1,m}(z) - \bar{b}_{2,m}(z)\|_m' \le e^{\bar{a}+\varrho} \|b_{1,m-1}(G_{m-1}^{-1}(z)) - b_{2,m-1}(G_{m-1}^{-1}(z))\|_{m-1}'$$
$$\le e^{\bar{a}+\varrho} \|b_{1,m-1} - b_{2,m-1}\|_{m-1}'.$$

Thus

$$\|\bar{b}_1 - \bar{b}_2\|_\infty' \le e^{\bar{a}+\varrho} \|b_1 - b_2\|_\infty'. \tag{7.25}$$

Analogously, proceeding as in (7.23) we obtain

$$\|\bar{c}_{1,m}(z) - \bar{c}_{2,m}(z)\|_m' \le e^{-\underline{b}+\varrho} \|c_{1,m+1}(G_m(z)) - c_{2,m+1}(G_m(z))\|_{m+1}'$$
$$\le e^{-\underline{b}+\varrho} \|c_{1,m+1} - c_{2,m+1}\|_{m+1}',$$

and

$$\|\bar{c}_1 - \bar{c}_2\|_\infty' \le e^{-\underline{b}+\varrho} \|c_1 - c_2\|_\infty'. \tag{7.26}$$

Since $\varrho < \min\{-\bar{a}, \underline{b}\}$, it follows from (7.25) and (7.26) that

$$\|S(v_1) - S(v_2)\|_\infty' \le \max\{e^{\bar{a}+\varrho}, e^{-\underline{b}+\varrho}\} \|v_1 - v_2\|_\infty',$$

and the operator S is a contraction. Thus, there exists a unique sequence $u \in \mathcal{X}$ such that $S(u) = u$. This completes the proof of the theorem.   □

We note that the following result is the first place where we use the condition F3.

Theorem 7.6. *Assume that F1–F3 hold with $\beta = \vartheta$. If the sequence $(A_m)_{m \in \mathbb{Z}}$ admits a nonuniform exponential dichotomy with $\underline{b} > 0$ and $\max\{a, b\} \le \vartheta$, and δ is sufficiently small, then there exists a unique $(v_m)_{m \in \mathbb{Z}} \in \mathcal{X}$ such that for every $m \in \mathbb{Z}$ we have*

$$\hat{v}_{m+1} \circ A_m = (A_m + f_m) \circ \hat{v}_m, \quad \text{where } \hat{v}_m = \mathrm{Id} + v_m. \tag{7.27}$$

Proof. The equation in (7.27) is equivalent to

$$v_{m+1} \circ A_m - A_m \circ v_m = f_m \circ \hat{v}_m. \tag{7.28}$$

Writing $v_m = (d_m, e_m)$ and $f_m = (g_m, h_m)$, again with values in $E_m \times F_m$, using F1 we find that (7.28) holds for every $m \in \mathbb{Z}$ if and only if $(d_m, e_m) = (\bar{d}_m, \bar{e}_m)$ for every $m \in \mathbb{Z}$, where

$$\bar{d}_m = (B_{m-1} \circ d_{m-1} + g_{m-1} \circ \hat{v}_{m-1}) \circ A_{m-1}^{-1}, \tag{7.29}$$

$$\bar{e}_m = C_m^{-1} \circ (e_{m+1} \circ A_m - h_m \circ \hat{v}_m). \tag{7.30}$$

Given $v = (v_m)_{m \in \mathbb{Z}} = (d_m, e_m)_{m \in \mathbb{Z}} \in \mathcal{X}$, we define $T(v) = (\bar{d}_m, \bar{e}_m)_{m \in \mathbb{Z}}$. To prove the theorem we must show that T has a unique fixed point in $\mathcal{X}$.

We first show that $T(\mathcal{X}) \subset \mathcal{X}$. By the condition F1 the map A_m^{-1} is continuous, and thus $(\bar{d}_m, \bar{e}_m)$ is continuous for every $m \in \mathbb{Z}$. We obtain

$$\|\bar{d}_m(z)\|'_m \leq \sum_{k \geq m} \|\mathcal{B}(k, m) B_{m-1} d_{m-1}(A_{m-1}^{-1} z)\| e^{(-\bar{a} - \varrho)(k-m)}$$

$$+ \sum_{k \geq m} \|\mathcal{B}(k, m) g_{m-1}(\hat{v}_{m-1}(A_{m-1}^{-1} z))\| e^{(-\bar{a} - \varrho)(k-m)}$$

$$\leq e^{\bar{a} + \varrho} \sum_{k \geq m-1} \|\mathcal{B}(k, m-1) d_{m-1}(A_{m-1}^{-1} z)\| e^{(-\bar{a} - \varrho)(k-(m-1))}$$

$$+ \sum_{k \geq m} \|\mathcal{B}(k, m)\| \cdot \|g_{m-1}\|_\infty e^{(-\bar{a} - \varrho)(k-m)}.$$

Proceeding as in (7.21) we conclude that $\|\bar{d}\|'_\infty < \infty$. In an analogous manner,

$$\|\bar{e}_m(z)\|'_m \leq \sum_{k \leq m} \|\mathcal{C}(m, k)^{-1} C_m^{-1} e_{m+1}(A_m z)\| e^{(\underline{b} - \varrho)(m-k)}$$

$$+ \sum_{k \leq m} \|\mathcal{C}(m, k)^{-1} h_m(\hat{v}_m(z))\| e^{(\underline{b} - \varrho)(m-k)}$$

$$\leq e^{-\underline{b} + \varrho} \sum_{k \leq m+1} \|\mathcal{C}(m+1, k)^{-1} e_{m+1}(A_m z)\| e^{(\underline{b} - \varrho)(m+1-k)}$$

$$+ \sum_{k \leq m} \|\mathcal{C}(m, k)^{-1}\| \cdot \|h_m\|_\infty e^{(\underline{b} - \varrho)(m-k)},$$

and proceeding as in (7.23) we conclude that $\|\bar{e}\|'_\infty < \infty$. This shows that $T(v) \in \mathcal{X}$, and thus $T: \mathcal{X} \to \mathcal{X}$ is well-defined.

We now prove that T is a contraction. Given $v_i = (d_{i,m}, e_{i,m})_{m \in \mathbb{Z}} \in \mathcal{X}$ for $i = 1, 2$, and setting

$$\hat{v}_{i,m} = \text{Id} + v_{i,m} \quad \text{and} \quad G_{i,m} = \hat{v}_{i,m} \circ A_{m-1}^{-1},$$

proceeding as in (7.21) for each $z \in X$ we obtain

$$\|\bar{d}_{1,m}(z) - \bar{d}_{2,m}(z)\|'_m$$

$$\leq e^{\bar{a} + \varrho} \sum_{k \geq m-1} \|\mathcal{B}(k, m-1)(d_{1,m-1} - d_{2,m-1})(A_{m-1}^{-1} z)\| e^{(-\bar{a} - \varrho)(k-m)}$$

$$+ \sum_{k \geq m} \|\mathcal{B}(k, m)[g_{m-1}(G_{1,m-1}(z)) - g_{m-1}(G_{2,m-1}(z))]\| e^{(-\bar{a} - \varrho)(k-m)}$$

$$\leq e^{\bar{a} + \varrho} \|d_{1,m-1}(A_{m-1}^{-1} z) - d_{2,m-1}(A_{m-1}^{-1} z)\|'_{m-1}$$

$$+ \theta \|\hat{v}_{1,m-1}(A_{m-1}^{-1} z) - \hat{v}_{2,m-1}(A_{m-1}^{-1} z)\|$$

$$\leq e^{\bar{a} + \varrho} \|d_{1,m-1} - d_{2,m-1}\|'_{m-1} + \theta \|v_{1,m-1} - v_{2,m-1}\|'_{m-1},$$

using (7.12). Thus

$$\|\bar{d}_{1,m} - \bar{d}_{2,m}\|'_m \le e^{\bar{a}+\varrho}\|d_{1,m-1} - d_{2,m-1}\|'_{m-1} + \theta\|v_{1,m-1} - v_{2,m-1}\|'_{m-1}. \quad (7.31)$$

Analogously, proceeding as in (7.23) we obtain

$$\|\bar{e}_{1,m}(z) - \bar{e}_{2,m}(z)\|'_m$$

$$\le e^{-\underline{b}+\varrho} \sum_{k \le m+1} \|\mathcal{C}(m+1,k)^{-1}(e_{1,m+1}(A_m z) - e_{2,m+1}(A_m z))\|e^{(\underline{b}-\varrho)(m-k)}$$

$$+ \sum_{k \ge m} \|\mathcal{C}(m+1,k)^{-1}[h_{m-1}(\widehat{v}_{1,m}(z)) - h_{m-1}(\widehat{v}_{2,m}(z))]\|e^{(\underline{b}-\varrho)(m-k)}$$

$$\le e^{-\underline{b}+\varrho}\|e_{1,m+1} - e_{2,m+1}\|'_{m+1} + \theta\|\widehat{v}_{1,m} - \widehat{v}_{2,m}\|'_m,$$

and thus,

$$\|\bar{e}_{1,m} - \bar{e}_{2,m}\|'_m \le e^{-\underline{b}+\varrho}\|e_{1,m+1} - e_{2,m+1}\|'_{m+1} + \theta\|v_{1,m} - v_{2,m}\|'_m. \quad (7.32)$$

By (7.31) and (7.32) we conclude that

$$\|T(v_1) - T(v_2)\|'_\infty \le (\max\{e^{\bar{a}+\varrho}, e^{-\underline{b}+\varrho}\} + 2\theta)\|v_1 - v_2\|'_\infty.$$

Since $\varrho < \min\{-\bar{a}, \underline{b}\}$, taking δ sufficiently small the operator T is a contraction. Thus, there exists a unique $v \in X$ such that $T(v) = v$. This completes the proof of the theorem. □

We now combine the information provided by Theorems 7.5 and 7.6 to obtain the topological conjugacies.

Corollary 7.7. *Assume that F1–F3 hold with $\beta = \vartheta$. If the sequence $(A_m)_{m \in \mathbb{Z}}$ admits a nonuniform exponential dichotomy with $\underline{b} > 0$ and $\max\{a, b\} \le \vartheta$, and δ in (7.4) and (7.5) is sufficiently small, then the maps $\widehat{u}_m = \mathrm{Id} + u_m$ and $\widehat{v}_m = \mathrm{Id} + v_m$, with u_m as in Theorem 7.5 and v_m as in Theorem 7.6, are homeomorphisms and satisfy*

$$\widehat{u}_m \circ \widehat{v}_m = \widehat{v}_m \circ \widehat{u}_m = \mathrm{Id}, \quad m \in \mathbb{Z}. \quad (7.33)$$

Proof. In view of the continuity of the functions u_m in Theorem 7.5 and v_m in Theorem 7.6, it is sufficient to show that the identities in (7.33) hold. We continue to set $G_m = A_m + f_m$. By (7.17) and (7.27) we have

$$\widehat{u}_{m+1} \circ \widehat{v}_{m+1} \circ A_m = \widehat{u}_{m+1} \circ G_m \circ \widehat{v}_m = A_m \circ \widehat{u}_m \circ \widehat{v}_m \quad (7.34)$$

for every $m \in \mathbb{Z}$. Since

$$\widehat{u}_m \circ \widehat{v}_m - \mathrm{Id} = u_m + v_m + u_m \circ v_m,$$

we have

$$\sup\{\|\widehat{u}_m \circ \widehat{v}_m - \mathrm{Id}\|'_m : m \in \mathbb{Z}\} < \infty,$$

and thus $(\widehat{u}_m \circ \widehat{v}_m)_{m \in \mathbb{Z}} \in X$. It follows from (7.34) and the uniqueness statements in Theorems 7.5 or 7.6 (for the perturbations $f_m = 0$) that $\widehat{u}_m \circ \widehat{v}_m = \mathrm{Id}$ for every $m \in \mathbb{Z}$. This shows that the maps $\widehat{u}_m$ and $\widehat{v}_m$ are homeomorphisms and (7.33) holds. □

7.3 Hölder regularity of the conjugacies

We show in this section that the topological conjugacies u_m and v_m in Corollary 7.7 are in fact Hölder continuous.

7.3.1 Main statement

We need a stronger version of dichotomy in this section.

Definition 7.8. *We say that the sequence of linear operators* $(A_m)_{m \in \mathbb{Z}}$ *admits a strong nonuniform exponential dichotomy if there exist projections* $P_n \in B(X)$ *for* $n \in \mathbb{Z}$ *satisfying* (7.6), *and there exist constants*

$$\underline{a} \leq \bar{a} < 0 \leq \underline{b} \leq \bar{b}, \quad a, b \geq 0, \quad and \quad \bar{D} \geq 1$$

such that for every $m, n \in \mathbb{Z}$ *with* $m \geq n$ *we have*

$$\|\mathcal{A}(m,n)P_n\| \leq \bar{D}e^{\bar{a}(m-n)+a|n|}, \quad \|\mathcal{A}(m,n)^{-1}Q_m\| \leq \bar{D}e^{-\underline{b}(m-n)+b|m|},$$

and for every $m, n \in \mathbb{Z}$ *with* $m \leq n$ *we have*

$$\|\mathcal{A}(m,n)P_n\| \leq \bar{D}e^{-\underline{a}(n-m)+a|n|}, \quad \|\mathcal{A}(m,n)^{-1}Q_m\| \leq \bar{D}e^{\bar{b}(n-m)+b|m|}.$$

Clearly, any sequence $(A_m)_{m \in \mathbb{Z}}$ admitting a strong nonuniform exponential dichotomy admits a nonuniform exponential dichotomy.

We now assume that there exists a strong nonuniform exponential dichotomy, and we recall the constant α_0 introduced in (7.3). The following is the main statement concerning the Hölder regularity of the conjugacies in Corollary 7.7.

Theorem 7.9. *Assume that F1–F3 hold with* $\beta = 4\vartheta$. *If the sequence of linear operators* $(A_m)_{m \in \mathbb{Z}}$ *admits a strong nonuniform exponential dichotomy with* $\underline{b} > 0$ *and* $\max\{a, b\} \leq \vartheta$, *then for each* $\alpha \in (0, \alpha_0)$, *provided that* δ *in* (7.4) *and* (7.5) *is sufficiently small (depending on* α), *for the unique sequences* $(u_m)_{m \in \mathbb{Z}} \in \mathfrak{X}$ *in Theorem 7.5 and* $(v_m)_{m \in \mathbb{Z}} \in \mathfrak{X}$ *in Theorem 7.6 there exists* $K > 0$ *(depending on* α *and* δ) *such that*

$$\|u_m(x) - u_m(y)\| \leq Ke^{2\max\{a,b\}\alpha|m|}\|x - y\|^{\alpha},$$
$$\|v_m(x) - v_m(y)\| \leq Ke^{2\max\{a,b\}\alpha|m|}\|x - y\|^{\alpha}$$

for every $m \in \mathbb{Z}$ *and* $x, y \in X$ *with* $\|x - y\| \leq e^{-2\max\{a,b\}|m|}$.

Theorem 7.9 is a simple consequence of slightly stronger statements in Theorems 7.12 and 7.13.

7.3.2 Lyapunov norms

We need to introduce new Lyapunov norms, which are adapted to the "two-sided" notion of strong nonuniform exponential dichotomy (instead of the "one-sided" notion of nonuniform exponential dichotomy). We continue to choose $\varrho > 0$ such that $\varrho < \min\{-\bar{a}, \underline{b}\}$. For each $m \in \mathbb{Z}$ we set

$$\|x\|_m^* = \sum_{k \geq m} \|\mathcal{B}(k,m)x\| e^{-(\bar{a}+\varrho)(k-m)} + \sum_{k<m} \|\mathcal{B}(k,m)x\| e^{(\underline{a}-\varrho)(m-k)} \quad (7.35)$$

for $x \in E_m$, and

$$\|y\|_m^* = \sum_{k \geq m} \|\mathcal{C}(m,k)^{-1}y\| e^{-(\bar{b}+\varrho)(k-m)} + \sum_{k<m} \|\mathcal{C}(m,k)^{-1}y\| e^{(\underline{b}-\varrho)(m-k)} \tag{7.36}$$

for $y \in F_m$. We also set

$$\|(x,y)\|_m^* = \|x\|_m^* + \|y\|_m^*.$$

It is straightforward to verify that each series in (7.35)–(7.36) converges, and setting

$$N = \bar{D}(1 + e^{-\varrho})/(1 - e^{-\varrho}), \tag{7.37}$$

for each $(x,y) \in E_m \times F_m$ we have

$$\|x\| \leq \|x\|_m^* \leq Ne^{a|m|}\|x\| \quad \text{and} \quad \|y\| \leq \|y\|_m^* \leq Ne^{b|m|}\|y\|.$$

Lemma 7.10. *For each $z \in X$ and $m \in \mathbb{Z}$ we have*

$$\|z\| \leq \|z\|_m^* \leq 2\bar{D}Ne^{2\max\{a,b\}|m|}\|z\|. \tag{7.38}$$

The proof is analogous to the one of Lemma 7.4, and thus it will be omitted.

We now obtain estimates for the norms of the linear operators with respect to the new Lyapunov norms.

Lemma 7.11. *For each $m \in \mathbb{Z}$ we have*

$$\|A_m\|^* := \sup_{z \in X \setminus \{0\}} \frac{\|A_m z\|_{m+1}^*}{\|z\|_m^*} \leq e^{\bar{b}+\varrho},$$

$$\|A_m^{-1}\|^* := \sup_{z \in X \setminus \{0\}} \frac{\|A_m^{-1}z\|_m^*}{\|z\|_{m+1}^*} \leq e^{-\underline{a}+\varrho}.$$

Proof. Setting $z = (x,y) \in E_m \times F_m$ we have

$$\|A_m z\|_{m+1}^* = \|B_m x\|_{m+1}^* + \|C_m y\|_{m+1}^*.$$

Furthermore

$$\|B_m x\|_{m+1}^* = e^{\bar{a}+\varrho} \sum_{k \geq m+1} \|\mathcal{B}(k,m)x\| e^{-(\bar{a}+\varrho)(k-m)}$$

$$+ e^{\underline{a}-\varrho} \sum_{k \leq m} \|\mathcal{B}(k,m)x\| e^{(\underline{a}-\varrho)(m-k)} \leq e^{\bar{a}+\varrho} \|x\|_m^*,$$

and

$$\|C_m y\|_{m+1}^* = e^{\bar{b}+\varrho} \sum_{k \geq m+1} \|\mathcal{C}(m,k)^{-1}y\| e^{-(\bar{b}+\varrho)(k-m)}$$

$$+ e^{\underline{b}-\varrho} \sum_{k \leq m} \|\mathcal{C}(m,k)^{-1}y\| e^{(\underline{b}-\varrho)(m-k)} \leq e^{\bar{b}+\varrho} \|y\|_m^*.$$

Since $\bar{b} + \varrho > 0 > \bar{a} + \varrho$, we obtain $\|A_m z\|_{m+1}^* \leq e^{\bar{b}+\varrho} \|z\|_m^*$. Similarly, we have

$$\|A_{m-1}^{-1} v\|_{m-1}^* = \|B_{m-1}^{-1} x\|_{m-1}^* + \|C_{m-1}^{-1} y\|_{m-1}^*,$$

with

$$\|B_{m-1}^{-1} x\|_{m-1}^* = e^{-(\bar{a}+\varrho)} \sum_{k \geq m-1} \|\mathcal{B}(k,m)x\| e^{-(\bar{a}+\varrho)(k-m)}$$

$$+ e^{-\underline{a}+\varrho} \sum_{k < m-1} \|\mathcal{B}(k,m)x\| e^{(\underline{a}-\varrho)(m-k)} \leq e^{-\underline{a}+\varrho} \|x\|_m^*,$$

and

$$\|C_{m-1}^{-1} y\|_{m-1}^* = e^{-\bar{b}+\varrho} \sum_{k \geq m-1} \|\mathcal{C}(m,k)^{-1}y\| e^{(-\bar{b}+\varrho)(k-m)}$$

$$+ e^{-\underline{b}+\varrho} \sum_{k < m-1} \|\mathcal{C}(m,k)^{-1}y\| e^{(\underline{b}-\varrho)(m-k)} \leq e^{-\underline{b}+\varrho} \|y\|_m^*.$$

Since $-\underline{a} + \varrho > 0 > -\underline{b} + \varrho$, we obtain $\|A_{m-1}^{-1} z\|_{m-1}^* \leq e^{-\underline{a}+\varrho} \|z\|_m^*$.   □

7.3.3 Proof of the Hölder regularity

We establish in this section the Hölder regularity of the conjugacies and of their inverses. For convenience of the proofs, we first consider the maps v_m (and then the maps u_m). We continue to consider the constant α_0 in (7.3). Let $(v_m)_{m \in \mathbb{Z}}$ be the unique sequence given by Theorem 7.6, and write $v_m = (d_m, e_m)$ with values in $E_m \times F_m$.

Theorem 7.12. *Assume that F1–F3 hold with $\beta = \vartheta$. If the sequence of linear operators $(A_m)_{m \in \mathbb{Z}}$ admits a strong nonuniform exponential dichotomy with $\underline{b} > 0$ and $\max\{a, b\} \leq \vartheta$, then for each $\alpha \in (0, \alpha_0)$, provided that δ is sufficiently small (depending on α) there exists $K > 0$ (depending on α and δ) such that for every $m \in \mathbb{Z}$ and $x, y \in X$ with $\|x - y\|_m^* < 1$ we have*

$$\|v_m(x) - v_m(y)\|_m^* \leq K(\|x - y\|_m^*)^{\alpha}.$$

Proof. Given constants $K > 0$ and $\alpha \in (0, 1)$, we consider the subset $\mathfrak{X}_\alpha \subset \mathfrak{X}$ composed of the sequences $(v_m)_{m \in \mathbb{Z}}$ satisfying

$$\max\{\|d_m(x) - d_m(y)\|_m^*, \|e_m(x) - e_m(y)\|_m^*\} \leq K(\|x - y\|_m^*)^\alpha$$

for every $m \in \mathbb{Z}$ and $x, y \in X$ with $\|x - y\|_m^* < 1$. One can easily verify that $\mathfrak{X}_\alpha$ is closed with respect to the norm $\|\cdot\|_\infty$ in (7.16). Hence, the statement in the theorem will follow readily after showing that the contraction map $T: \mathfrak{X} \to \mathfrak{X}$ in the proof of Theorem 7.6 satisfies $T(\mathfrak{X}_\alpha) \subset \mathfrak{X}_\alpha$.

We assume that the constant $\varrho > 0$ in the construction of the Lyapunov norms in (7.35)–(7.36) is chosen so small such that, in addition, it satisfies

$$\alpha < \min\left\{\frac{-\bar{a} - \varrho}{-\underline{a} + \varrho}, \frac{\underline{b} - \varrho}{\bar{b} + \varrho}\right\}. \tag{7.39}$$

Let now $v = (v_m)_{m \in \mathbb{Z}} = (d_m, e_m)_{m \in \mathbb{Z}}$ be a sequence in $\mathfrak{X}_\alpha$. We must show that the sequence $T(v) = (\bar{d}_m, \bar{e}_m)_{m \in \mathbb{Z}}$, with $\bar{d}_m$ and $\bar{e}_m$ as in the proof of Theorem 7.6 (see (7.29) and (7.30)) is in $\mathfrak{X}_\alpha$. Take $x, y \in X$. By (7.29) we have

$$\|\bar{d}_m(x) - \bar{d}_m(y)\|_m^* \leq \|B_{m-1}\delta_{m-1}(x) - B_{m-1}\delta_{m-1}(y)\|_m^* \\ + \|g_{m-1}(\bar{v}_{m-1}(x)) - g_{m-1}(\bar{v}_{m-1}(y))\|_m^* := T_1 + T_2,$$

where

$$\delta_m = d_m \circ A_m^{-1} \quad \text{and} \quad \bar{v}_m = \hat{v}_m \circ A_m^{-1}, \quad \text{with} \quad \hat{v}_m = \mathrm{Id} + v_m.$$

Since $\bar{a} + \varrho > (\underline{a} - \varrho)$, we obtain

$$T_1 = \sum_{k \geq m} \|\mathcal{B}(k, m-1)(\delta_{m-1}(x) - \delta_{m-1}(y))\| e^{-(\bar{a}+\varrho)(k-m)}$$
$$+ \sum_{k < m} \|\mathcal{B}(k, m-1)(\delta_{m-1}(x) - \delta_{m-1}(y))\| e^{(\underline{a}-\varrho)(m-k)}$$
$$= e^{\bar{a}+\varrho} \sum_{k \geq m} \|\mathcal{B}(k, m-1)(\delta_{m-1}(x) - \delta_{m-1}(y))\| e^{-(\bar{a}+\varrho)(k-m+1)}$$
$$+ e^{(\underline{a}-\varrho)}\|\delta_{m-1}(x) - \delta_{m-1}(y)\|$$
$$+ e^{(\underline{a}-\varrho)} \sum_{k < m-1} \|\mathcal{B}(k, m-1)(\delta_{m-1}(x) - \delta_{m-1}(y))\| e^{(\underline{a}-\varrho)(m-1-k)}$$
$$\leq e^{\bar{a}+\varrho} \sum_{k \geq m-1} \|\mathcal{B}(k, m-1)(\delta_{m-1}(x) - \delta_{m-1}(y))\| e^{-(\bar{a}+\varrho)(k-m+1)}$$
$$+ e^{\bar{a}+\varrho} \sum_{k < m-1} \|\mathcal{B}(k, m-1)(\delta_{m-1}(x) - \delta_{m-1}(y))\| e^{(-\underline{a}+\varrho)(m-1-k)}$$
$$\leq e^{\bar{a}+\varrho}\|\delta_{m-1}(x) - \delta_{m-1}(y)\|_{m-1}^*$$
$$\leq e^{\bar{a}+\varrho}K(\|A_{m-1}^{-1}(x - y)\|_{m-1}^*)^\alpha.$$

Furthermore, using F3 with $\beta = \vartheta$ and (7.38) we obtain

$$
\begin{aligned}
T_2 &\leq \delta \sum_{k \geq m} \|\mathcal{B}(k,m)\| e^{-\beta|m-1|} \|\bar{v}_{m-1}(x) - \bar{v}_{m-1}(y)\| e^{-(\bar{a}+\varrho)(k-m)} \\
&\quad + \delta \sum_{k < m} \|\mathcal{B}(k,m)\| e^{-\beta|m-1|} \|\bar{v}_{m-1}(x) - \bar{v}_{m-1}(y)\| e^{(\underline{a}-\varrho)(m-k)} \\
&\leq N\delta e^{\beta} \|\bar{v}_{m-1}(x) - \bar{v}_{m-1}(y)\|_{m-1}^* \\
&\leq N\delta e^{\beta} L + NK\delta e^{\beta} L^{\alpha},
\end{aligned}
\tag{7.40}
$$

where $L = \|A_{m-1}^{-1}(x-y)\|_{m-1}^*$ and with N as in (7.37). By Lemma 7.11, for $x \neq y$ with $\|x - y\|_m^* < 1$ we obtain

$$
\frac{\|\bar{d}_m(x) - \bar{d}_m(y)\|_m^*}{(\|x-y\|_m^*)^{\alpha}} \leq K(e^{\bar{a}+\varrho} + N\delta e^{\beta})e^{\alpha(-\underline{a}+\varrho)} + N\delta e^{\beta} e^{-\underline{a}+\varrho}.
\tag{7.41}
$$

We now consider the second component $\bar{e}_m$. By (7.30), we have

$$
\begin{aligned}
\|\bar{e}_m(x) - \bar{e}_m(y)\|_m^* &\leq \|C_m^{-1}\tilde{e}_{m+1}(x) - C_m^{-1}\tilde{e}_{m+1}(y)\|_m^* \\
&\quad + \|C_m^{-1}(h_m \circ \hat{v}_m)(x) - C_m^{-1}(h_m \circ \hat{v}_m)(y)\|_m^* := T_3 + T_4,
\end{aligned}
$$

where $\tilde{e}_{m+1} = e_{m+1} \circ A_m$. We obtain

$$
\begin{aligned}
T_3 &= e^{-(\bar{b}+\varrho)} \sum_{k \geq m} \|\mathcal{C}(m+1,k)^{-1}(\tilde{e}_{m+1}(x) - \tilde{e}_{m+1}(y))\| e^{-(\bar{b}+\varrho)(k-m-1)} \\
&\quad + e^{-\underline{b}+\varrho} \sum_{k < m} \|\mathcal{C}(m+1,k)^{-1}(\tilde{e}_{m+1}(x) - \tilde{e}_{m+1}(y))\| e^{(\underline{b}-\varrho)(m+1-k)} \\
&\leq e^{-\underline{b}+\varrho} \sum_{k \geq m+1} \|\mathcal{C}(m+1,k)^{-1}(\tilde{e}_{m+1}(x) - \tilde{e}_{m+1}(y))\| e^{-(\bar{b}+\varrho)(k-m-1)} \\
&\quad + e^{-\underline{b}+\varrho} \sum_{k < m+1} \|\mathcal{C}(m+1,k)^{-1}(\tilde{e}_{m+1}(x) - \tilde{e}_{m+1}(y))\| e^{(\underline{b}-\varrho)(m+1-k)} \\
&= e^{-\underline{b}+\varrho} \|\tilde{e}_{m+1}(x) - \tilde{e}_{m+1}(y)\|_{m+1}^* \leq e^{-\underline{b}+\varrho} K(\|A_m(x-y)\|_{m+1}^*)^{\alpha},
\end{aligned}
\tag{7.42}
$$

and proceeding in a similar manner,

$$
\begin{aligned}
T_4 &= \delta \sum_{k > m} \|\mathcal{C}(m+1,k)^{-1}\| e^{-\beta|m|} \|\hat{v}_m(x) - \hat{v}_m(y)\| e^{-(\bar{b}+\varrho)(k-m)} \\
&\quad + \delta \sum_{k \leq m} \|\mathcal{C}(m+1,k)^{-1}\| e^{-\beta|m|} \|\hat{v}_m(x) - \hat{v}_m(y)\| e^{(\underline{b}-\varrho)(m-k)} \\
&\leq \delta \bar{D} \left(\frac{e^{-\bar{b}+\vartheta-\varrho}}{1 - e^{-\varrho}} + \frac{e^{-\underline{b}+\vartheta}}{1 - e^{-\varrho}} \right) \|\hat{v}_m(x) - \hat{v}_m(y)\|_m^* \\
&\leq \delta \bar{D} \frac{e^{-\bar{b}+\vartheta-\varrho} + e^{-\underline{b}+\vartheta}}{1 - e^{-\varrho}} [\|x-y\|_m^* + K(\|x-y\|_m^*)^{\alpha}].
\end{aligned}
\tag{7.43}
$$

By Lemma 7.11, for $x \neq y$ with $\|x - y\|_m^* < 1$ we obtain

$$\frac{\|\bar{e}_m(x) - \bar{e}_m(y)\|_m^*}{(\|x - y\|_m^*)^\alpha} \leq K e^{-\underline{b}+\varrho} e^{\alpha(\bar{b}+\varrho)} + (1 + K)\delta\bar{D}\frac{e^{-\bar{b}+\vartheta-\varrho} + e^{-\underline{b}+\vartheta}}{1 - e^{-\varrho}}. \quad (7.44)$$

It follows readily from (7.39) that

$$e^{\bar{a}+\varrho} e^{\alpha(-\underline{a}+\varrho)} < 1 \quad \text{and} \quad e^{-\underline{b}+\varrho} e^{\alpha(\bar{b}+\varrho)} < 1. \quad (7.45)$$

Hence, for each sufficiently small δ it follows from (7.41) and (7.44) that there exists $K > 0$ (independent of f) such that

$$\max\{\|\bar{d}_m(x) - \bar{d}_m(y)\|_m^*, \|\bar{e}_m(x) - \bar{e}_m(y)\|_m^*\} \leq K(\|x - y\|_m^*)^\alpha$$

for every $m \in \mathbb{Z}$ and $x, y \in X$ with $\|x-y\|_m^* < 1$, whenever $(d_m, e_m)_{m\in\mathbb{Z}} \in \mathfrak{X}_\alpha$. This completes the proof of the theorem. $\square$

Let now $(u_m)_{m\in\mathbb{Z}}$ be the unique sequence given by Theorem 7.5, and write $u_m = (b_m, c_m)$ with values in $E_m \times F_m$.

Theorem 7.13. *Assume that F1–F3 hold with $\beta = 4\vartheta$. If the sequence of linear operators $(A_m)_{m\in\mathbb{Z}}$ admits a strong nonuniform exponential dichotomy with $\underline{b} > 0$ and $\max\{a, b\} \leq \vartheta$, then for each $\alpha \in (0, \alpha_0)$, provided that δ is sufficiently small (depending on α) there exists $K > 0$ (depending on α and δ) such that for every $m \in \mathbb{Z}$ and $x, y \in X$ with $\|x - y\|_m^* < 1$ we have*

$$\|u_m(x) - u_m(y)\|_m^* \leq K(\|x - y\|_m^*)^\alpha.$$

Proof. Set

$$C_1 = e^{\bar{b}+\varrho} + 2\bar{D}N\delta e^{\vartheta} \quad \text{and} \quad C_2 = (1 - 2\bar{D}Ne^{-\underline{a}+\varrho+2\vartheta})^{-1}.$$

Lemma 7.14. *For each $x, y \in X$ we have*

$$\|G_m(x) - G_m(y)\|_{m+1}^* \leq C_1\|x - y\|_m^*.$$

Proof of the lemma. In view of Lemmas 7.10 and 7.11, we have

$$\begin{aligned}
\|G_m(x) - G_m(y)\|_{m+1}^* &\leq \|A_m(x - y)\|_{m+1}^* + \|f_m(x) - f_m(y)\|_{m+1}^* \\
&\leq e^{\bar{b}+\varrho}\|x - y\|_m^* + 2\bar{D}Ne^{2\vartheta|m+1|}\|f_m(x) - f_m(y)\| \\
&\leq e^{\bar{b}+\varrho}\|x - y\|_m^* + 2\bar{D}N\delta e^{-2\vartheta(|m|-1)}\|x - y\| \\
&\leq C_1\|x - y\|_m^*,
\end{aligned}$$

which gives the desired inequality. $\square$

Lemma 7.15. *For each δ sufficiently small and each $x, y \in X$ we have*

$$\|G_m^{-1}(x) - G_m^{-1}(y)\|_m^* \leq C_2 e^{-\underline{a}+\varrho}\|x - y\|_{m+1}^*.$$

Proof of the lemma. Again by Lemmas 7.10 and 7.11,

$$\begin{aligned}
\|G_m(x) - G_m(y)\|_{m+1}^* &\geq \|A_m(x-y)\|_{m+1}^* - \|f_m(x) - f_m(y)\|_{m+1}^* \\
&\geq (e^{\underline{a}-\varrho} - 2\bar{D}Ne^{2\vartheta|m+1|}\delta e^{-4\vartheta|m|})\|x-y\|_m^* \\
&\geq (e^{\underline{a}-\varrho} - 2\bar{D}N\delta e^{2\vartheta(1-|m|)})\|x-y\|_m^* \\
&\geq C_2^{-1}e^{\underline{a}-\varrho}\|x-y\|_m^*,
\end{aligned}$$

thus giving the desired inequality. $\square$

We now proceed with the proof of the theorem. We use the same notation as in the proof of Theorem 7.12, and we proceed in a similar manner. Namely, we let ϱ and α be as in (7.39), and we show that if $u = (u_m)_{m\in\mathbb{Z}} = (b_m, c_m)_{m\in\mathbb{Z}}$ is in $\mathfrak{X}_\alpha$ (see the proof of Theorem 7.12 for the definition), then the sequence $S(u) = (\bar{b}_m, \bar{c}_m)_{m\in\mathbb{Z}}$, with $\bar{b}_m$ and $\bar{c}_m$ as in the proof of Theorem 7.5 (see (7.19) and (7.20)), is also in $\mathfrak{X}_\alpha$.

Take $x, y \in X$ with $\|x - y\|_m^* < 1$. By (7.19), proceeding as in (7.40) we obtain

$$\begin{aligned}
\|\bar{b}_m(x) - \bar{b}_m(y)\|_m^* &\leq e^{\bar{a}+\varrho}K(\|G_{m-1}^{-1}(x) - G_{m-1}^{-1}(y)\|_{m-1}^*)^\alpha \\
&\quad + N\delta e^\beta \|G_{m-1}^{-1}(x) - G_{m-1}^{-1}(y)\|_{m-1}^*.
\end{aligned}$$

It follows from Lemma 7.15 that

$$\begin{aligned}
&\|\bar{b}_m(x) - \bar{b}_m(y)\|_m^* \\
&\leq e^{\bar{a}+\varrho}KC_2^\alpha e^{\alpha(-\underline{a}+\varrho)}(\|x-y\|_m^*)^\alpha + N\delta C_2 e^{-\underline{a}+\varrho+\beta}\|x-y\|_m^* \qquad (7.46)\\
&\leq \left(Ke^{\bar{a}+\varrho}e^{\alpha(-\underline{a}+\varrho)}C_2^\alpha + N\delta C_2 e^{-\underline{a}+\varrho+\beta}\right)(\|x-y\|_m^*)^\alpha
\end{aligned}$$

(since $\|x - y\|_m^* < 1$). Furthermore, setting

$$C_3 = (e^{-\bar{b}+\vartheta-\varrho} + e^{-\underline{b}+\vartheta})/(1 - e^{-\varrho}),$$

and proceeding as in (7.43) we obtain

$$\|C_m^{-1}(h_m(x) - h_m(y))\|_m^* \leq \delta\bar{D}C_3\|x-y\|_m^*.$$

Thus, proceeding as in (7.42) and using Lemma 7.14 yields

$$\begin{aligned}
&\|\bar{c}_m(x) - \bar{c}_m(y)\|_m^* \\
&\leq e^{-\underline{b}+\varrho}K(\|G_m(x) - G_m(y)\|_{m+1}^*)^\alpha + \delta\bar{D}C_3\|x-y\|_m^* \qquad (7.47)\\
&\leq \left(Ke^{-\underline{b}+\varrho}(e^{\bar{b}+\varrho} + 2\bar{D}N\delta e^\vartheta)^\alpha + \delta\bar{D}C_3\right)(\|x-y\|_m^*)^\alpha,
\end{aligned}$$

again since $\|x - y\|_m^* < 1$. We now proceed as in the proof of Theorem 7.12. Namely, it follows readily from (7.39) that the inequalities in (7.45) hold for every $\alpha \in (0, \alpha_0)$. Thus, by (7.46) and (7.47), for each sufficiently small δ there exists $K > 0$ such that

$$\max\{\|\bar{b}_m(x) - \bar{b}_m(y)\|_m^*, \|\bar{c}_m(x) - \bar{c}_m(y)\|_m^*\} \le K(\|x - y\|_m^*)^\alpha,$$

for each $m \in \mathbb{Z}$ and $x, y \in X$ with $\|x - y\|_m^* < 1$. This completes the proof of the theorem. $\qquad\square$

Theorem 7.9 follows now readily from Theorems 7.12 and 7.13 together with the inequalities in (7.38).

7.4 Proofs of the results for flows

7.4.1 Reduction to discrete time

The first step in the proofs of Theorems 7.1 and 7.2 is the reduction of the problem to discrete time. Fix $r \in [-1, 1]$. For each $m \in \mathbb{Z}$ we define invertible linear operators

$$A_m = A_{m,r} = T(m + r, m + r - 1), \tag{7.48}$$

and maps

$$f_m(u) = \int_{m+r-1}^{m+r} T(m + r, \tau) f(\tau, v(\tau, u)) \, d\tau, \tag{7.49}$$

where $v(t, u)$ is the solution of the differential equation $v' = A(t)v + f(t, v)$ with $v(m + r - 1) = u$. For every $t \in \mathbb{R}$ we have

$$v(t, u) = T(t, m + r - 1)u + \int_{m+r-1}^{t} T(t, \tau) f(\tau, v(\tau, u)) \, d\tau, \tag{7.50}$$

and thus in particular $v(m + r, u) = A_m u + f_m(u)$.

Lemma 7.16. *Assume that E1–E2 hold with $\bar{\beta} = 6 \max\{a, b\}$. For each $\delta > 0$, if $\bar{\delta}$ is sufficiently small, then:*

1. *the maps A_m and f_m satisfy the conditions F1–F3 with $\beta = 4\vartheta$ and the given δ;*
2. *$(A_m)_{m \in \mathbb{Z}}$ admits a strong nonuniform exponential dichotomy if $v' = A(t)v$ admits a strong nonuniform exponential dichotomy in $\mathbb{R}$.*

Proof. The second statement is clear from the definitions. In particular, condition F1 holds. Set now $\vartheta = \max\{a, b\}$. By (2.17), for any $t \ge s$ we have

$$\|T(t, s)\| \le \|T(t, s)P(s)\| + \|T(t, s)Q(s)\|$$
$$= \|U(t, s)\| + \|V(t, s)\| \le (D_1 + D_2)e^{\bar{b}(t-s)+\vartheta|s|},$$

and hence

$$\|T(t, s)\| \le De^{\bar{b}(t-s)+\vartheta|s|}, \tag{7.51}$$

where $D = D_1 + D_2$. By condition E2, we obtain

$$\|f_m\|_\infty = \sup_{u \in X} \left\| \int_{m+r-1}^{m+r} T(m+r,\tau) f(\tau, v(\tau, u))\, d\tau \right\|$$

$$\le \int_{m+r-1}^{m+r} \|T(m+r,\tau)\| \cdot \sup_{u \in X} \|f(\tau, v(\tau, u))\|\, d\tau$$

$$\le D\bar\delta \int_{m+r-1}^{m+r} e^{\bar b(m+r-\tau)} e^{-5\vartheta|\tau|}\, d\tau$$

$$\le D\bar\delta e^{10\vartheta} e^{-5\vartheta|m|} \int_{m+r-1}^{m+r} e^{\bar b(m+r-\tau)}\, d\tau$$

$$\le \frac{D\bar\delta e^{10\vartheta} e^{-5\vartheta|m|}(e^{\bar b}-1)}{\bar b}.$$

Thus, provided that $\bar\delta$ is sufficiently small (see also the proof of Lemma 7.15 and the discussion after condition F3 in Section 7.2.1), we obtain condition F2.

It remains to establish condition F3. We first prove that there exists $D' > 0$ such that for any $m + r - 1 \le t \le m + r$ and $x, y \in X$,

$$\|v(t,x) - v(t,y)\| \le D' e^{\vartheta|m|} \|x - y\|. \tag{7.52}$$

By (7.50) and condition E2,

$$\|v(t,x) - v(t,y)\| \le \|T(t, m+r-1)\| \cdot \|x - y\|$$
$$+ \bar\delta \int_{m+r-1}^{t} \|T(t,\tau)\| e^{-6\vartheta|\tau|} \|v(\tau, x) - v(\tau, y)\|\, d\tau.$$

Using (7.51), since $t \le m + r$ we have

$$\|v(t,x) - v(t,y)\| \le D e^{\bar b + \vartheta|m+r-1|} \|x - y\|$$
$$+ \bar\delta D e^{\bar b} \int_{m+r-1}^{t} \|v(\tau, x) - v(\tau, y)\|\, d\tau.$$

Applying Gronwall's lemma we obtain

$$\|v(t,x) - v(t,y)\| \le D e^{\bar b + 2\vartheta + \vartheta|m|} e^{\bar\delta D e^{\bar b}} \|x - y\|$$

for any $m + r - 1 \le t \le m + 1$, which proves (7.52). By (7.51), condition E2, and (7.52), for any $x, y \in X$ and $m \in \mathbb{Z}$,

$$\|f_m(x) - f_m(y)\| \le \bar\delta D \int_{m+r-1}^{m+r} e^{\bar b(m+r-\tau)} e^{-5\vartheta|\tau|} \|v(\tau, x) - v(\tau, y)\|\, d\tau$$

$$\le \bar\delta D D' e^{\vartheta|m|} \|x - y\| \int_{m+r-1}^{m+r} e^{\bar b(m+r-\tau)} e^{-5\vartheta|\tau|}\, d\tau$$

$$\le \bar\delta D D' e^{5\vartheta} \frac{e^{\bar b} - 1}{\bar b} e^{-4\vartheta|m|} \|x - y\|.$$

Provided that $\bar\delta$ is sufficiently small this yields F3. $\qquad\square$

7.4.2 Proofs

We recall that $T(t,s)$ and $R(t,s)$ are respectively the evolution operators generated by the equations $v' = A(t)v$ and $v' = A(t) + f(t,v)$.

Proof of Theorem 7.1. Take $r \in [-1,1]$. In view of Lemma 7.16, by Corollary 7.7 there exist homeomorphisms $G_{m,r}\colon X \to X$, $m \in \mathbb{Z}$ such that

$$T(m+r, m+r-1)G_{m,r} = G_{m+1,r}R(m+r, m+r-1), \quad m \in \mathbb{Z}. \quad (7.53)$$

The uniqueness statements in Theorems 7.5 and 7.6 can be used to show the following.

Lemma 7.17. *We have* $G_{m,r} = G_{\bar{m},\bar{r}}$ *whenever* $\bar{m} + \bar{r} = m + r$.

Proof of the lemma. Let $\|\cdot\|'_{m,r}$ be the family of norms defined as in (7.11) when we replace the operators A_n in (7.8) by those in (7.48) (which now depend on r). One can easily verify that

$$\|x\|'_{m,r} = \|x\|'_{\bar{m},\bar{r}} \text{ for every } x \in X$$

whenever $m + r = \bar{m} + \bar{r}$: it follows from (7.48) that $A_{m,r} = A_{\bar{m},\bar{r}}$ and thus that

$$\mathcal{A}_r(m+k, m+l) = \mathcal{A}_{\bar{r}}(\bar{m}+k, \bar{m}+l), \quad k, l \in \mathbb{Z},$$

where $\mathcal{A}_r$ and $\mathcal{A}_{\bar{r}}$ are obtained as in (7.8) respectively using the sequences of operators $A_{n,r}$ and $A_{n,\bar{r}}$. In particular, this shows that

$$\sup_{m\in\mathbb{Z}} \sup_{x\in X} \|G_{m,r}(x) - x\|'_{m+1,r-1} = \sup_{m\in\mathbb{Z}} \sup_{x\in X} \|G_{m,r}(x) - x\|'_{m,r} < \infty \quad (7.54)$$

for every $r \in [0,1]$. Furthermore, by (7.53) we have that

$$T(m+\bar{r}, m+\bar{r}-1)G_{m-1,r} = G_{m,r}U(m+\bar{r}, m+\bar{r}-1), \quad m \in \mathbb{Z} \quad (7.55)$$

for any $\bar{r} = r - 1$ with $r \in [0,1]$. It follows from (7.54) and the uniqueness statement in Theorem 7.5 that the sequence $(G_{m-1,r})_{m\in\mathbb{Z}}$ coincides with the unique sequence of homeomorphisms in Theorem 7.5 satisfying (7.55), that is, the sequence $(G_{m,\bar{r}})_{m\in\mathbb{Z}}$. In other words, when $r \in [0,1]$ we have

$$G_{m,\bar{r}} = G_{m-1,r} \text{ for every } m \in \mathbb{Z}.$$

The case when $r \in [-1,0]$ can be treated in a similar manner. This establishes the desired statement. $\square$

By Lemma 7.17 we can define homeomorphisms

$$J_t = G_{m,r} \text{ with } m = [t] \text{ and } r = t - [t], \quad (7.56)$$

where $[t]$ denotes the integer part of t. For each $s \in \mathbb{R}$ we define the map $h_s\colon X \to X$ by

$$h_s(x) = \int_0^1 T(s, \tau + s) J_{\tau+s} R(\tau + s, s)(x) \, d\tau, \tag{7.57}$$

where for simplicity composition is denoted as multiplication. To verify that the integral is well-defined, we will show that for each $s \in \mathbb{R}$ and $x \in X$ the integrand in (7.57) is bounded in $\tau \in [-1, 1]$. We will use Theorem 7.9. We first note that for the functions A_m and f_m in (7.48)–(7.49), the constants K and α in Theorem 7.9 can be chosen independently of $r \in [-1, 1]$ (this follows immediately from the definition of α_0 in (7.3), and from the form of the constants in the right-hand sides of (7.41), (7.44), (7.46) and (7.47)).

In a similar way to that in (7.51), using (2.17) we have that for any $t \geq s$,

$$\|T(s, t)\| \leq De^{-\underline{a}(t-s)+\vartheta|t|}, \tag{7.58}$$

where $D = D_1 + D_2$ and $\vartheta = \max\{a, b\}$. By (7.58) and Theorem 7.9, when $M := \|R(\tau + s, s)(x)\| < e^{-2\vartheta(1+|s|)}$ we have

$$
\begin{aligned}
N : &= \|T(s, \tau + s) J_{\tau+s} R(\tau + s, s)(x)\| \\
&\leq De^{-\underline{a}\tau+\vartheta(1+|s|)} \|J_{\tau+s} R(\tau + s, s)(x)\| \\
&\leq De^{-\underline{a}\tau+\vartheta(1+|s|)} \left[e^{-2\vartheta(1+|s|)} + \|(J_{\tau+s} - \mathrm{Id}) R(\tau + s, s)(x)\| \right] \\
&\leq De^{-\underline{a}} (e^{-\vartheta(1+|s|)} + Ke^{\vartheta(1+|s|)}),
\end{aligned} \tag{7.59}
$$

and when $M \geq e^{-2\vartheta(1+|s|)}$ we have

$$N \leq De^{-\underline{a}+\vartheta(1+|s|)} [M + K(1 + e^{2\vartheta(1+|s|)} M)]. \tag{7.60}$$

We now estimate M. In view of (7.51) and (7.1), since $0 \leq \tau \leq 1$ we have

$$
\begin{aligned}
M &\leq \|T(\tau + s, s)x\| + \int_s^{\tau+s} \|T(\tau + s, r) f(r, R(r, s)(x))\| \, dr \\
&\leq De^{\overline{b}\tau+\vartheta|s|} \|x\| + D\overline{\delta} \int_s^{\tau+s} e^{\overline{b}(\tau+s-r)+\vartheta|r|} e^{-6\vartheta|r|} \, dr \\
&\leq De^{\overline{b}+\vartheta|s|} \|x\| + \frac{D\overline{\delta}e^{\overline{b}}}{\overline{b}}.
\end{aligned}
$$

Together with (7.59)–(7.60), this shows that for each s and x the integrand in (7.57) is bounded in $\tau \in [0, 1]$ and thus the integral is well-defined.

Since the maps $R(\tau + s, s)$ and $J_{\tau+s}$ are continuous, each map h_s is also continuous. Furthermore, the map h_s is invertible with inverse given by

$$h_s^{-1}(x) = \int_0^1 R(s, \tau + s) J_{\tau+s}^{-1} T(\tau + s, s)(x) \, d\tau.$$

Clearly, each map h_s^{-1} is also continuous.

We now establish the identity in (7.2). Note that for each $s, t \in \mathbb{R}$ we have

$$T(t,s)h_s = \int_0^1 T(t,\tau+s)J_{\tau+s}R(\tau+s,s)\,d\tau$$

$$= \int_0^1 T(t,\tau+s)J_{\tau+s}R(\tau+s,t)\,d\tau \circ R(t,s).$$
(7.61)

We will show that

$$P := \int_0^1 T(t,\tau+s)J_{\tau+s}R(\tau+s,t)\,d\tau = h_t.$$
(7.62)

Making the change of variables $w = \tau - t$, since $R(t,w+t)^{-1} = R(w+t,t)$ we obtain

$$P = \int_{s-t}^0 T(t,w+t)J_{w+t}R(w+t,t)\,dw$$

$$+ \int_0^{1+s-t} T(t,w+t)J_{w+t}R(w+t,t)\,dw.$$
(7.63)

But from (7.53) (and the uniqueness observation after this identity) we have

$$T(w+t+1,w+t)J_{w+t} = J_{w+t+1}R(w+t+1,w+t),$$

and the first integral in (7.63) can be written as

$$\int_{s-t}^0 T(t,w+t+1)T(w+t+1,w+t)J_{w+t}R(w+t,t)\,dw$$

$$= \int_{s-t}^0 T(t,w+t+1)J_{w+t+1}R(w+t+1,t)\,dw$$

$$= \int_{1+s-t}^1 T(t,t+\tau)J_{t+\tau}R(t+\tau,t)\,d\tau.$$

It follows from (7.63) that

$$P = \int_0^1 T(t,t+\tau)J_{t+\tau}R(t+\tau,t)\,d\tau = h_t,$$

and this establishes (7.62). It follows from (7.61) that (7.2) holds. This completes the proof. $\square$

Proof of Theorem 7.2. We continue to consider the homeomorphisms J_t and h_t constructed in the proof of Theorem 7.1 (see (7.56) and (7.57)). It follows from (7.57) and (7.58) that

$$\|h_s(x) - h_s(y)\| \le \int_0^1 De^{-\underline{a}\tau+\vartheta|s+\tau|}a(\tau)\,d\tau,$$
(7.64)

where

$$a(\tau) = \|J_{\tau+s}R(\tau+s,s)(x) - J_{\tau+s}R(\tau+s,s)(y)\|.$$

Assume now that $\|x - y\| \leq e^{-3\vartheta|s|}$. We claim that

$$\|R(\tau+s,s)(x) - R(\tau+s,s)(y)\| \leq Ke^{-2\vartheta(1+|s|)} \tag{7.65}$$

for some constant $K > 0$ (independent of s). Indeed,

$$R(t,s)(x) - R(t,s)(y)$$
$$= T(t,s)(x-y) + \int_s^t T(t,u)[f(u,R(u,s)(x)) - f(u,R(u,s)(y))]\,du,$$

and letting $b(t) = \|R(t,s)(x) - R(t,s)(y)\|$, for $t \geq s$ we obtain

$$b(t) \leq De^{\bar{b}(t-s)+\vartheta|s|}\|x-y\| + \int_s^t De^{\bar{b}(t-u)}\delta e^{-6\vartheta|u|}b(u)\,du,$$

using (7.51) and condition F3. Setting $\Phi(t) = e^{-\bar{b}(t-s)}b(t)$, for each $t \geq s$ we have

$$\Phi(t) \leq De^{\vartheta|s|}\|x-y\| + \int_s^t D\delta\Phi(u)\,du,$$

and by Gronwall's lemma,

$$\Phi(t) \leq De^{\vartheta|s|}e^{\delta D(t-s)}\|x-y\|.$$

Hence, for any $\tau \in [0,1]$ we have

$$b(\tau+s) \leq De^{\bar{b}\tau+\vartheta|s|}e^{\delta D\tau}\|x-y\|$$
$$\leq De^{\bar{b}+\delta D+\vartheta|s|}e^{-3\vartheta|s|}. \tag{7.66}$$

This establishes (7.65). We can thus apply Theorem 7.9 to obtain

$$a(\tau) \leq b(\tau+s) + K'e^{2\vartheta\alpha(1+|s|)}b(\tau+s)^\alpha,$$

since $J_{\tau+s} = \mathrm{Id} + (J_{\tau+s} - \mathrm{Id})$, for some constant $K' > 0$ (independent of s). It follows from the first inequality in (7.66) that since $\|x-y\| \leq 1$ we have

$$a(\tau) \leq De^{\bar{b}+\delta D+\vartheta|s|}\|x-y\| + K'e^{2\vartheta\alpha(1+|s|)}\left(De^{\bar{b}+\delta D+\vartheta|s|}\|x-y\|\right)^\alpha$$
$$\leq Le^{\vartheta(1+3\alpha)|s|}\|x-y\|^\alpha,$$

for some constant $L > 0$ (independent of s). By (7.64) we obtain

$$\|h_s(x) - h_s(y)\| \leq De^{-\underline{a}+\vartheta}Le^{\vartheta(2+3\alpha)|s|}\|x-y\|^\alpha.$$

This completes the proof of the theorem.   □

Part III

Center manifolds, symmetry and reversibility

Center manifolds in Banach spaces

Center manifold theorems are powerful tools in the analysis of the behavior of dynamical systems. For example, when the equation $v' = A(t)v$ has a (uniformly) partially hyperbolic behavior with no unstable directions, then under some mild additional assumptions all solutions of $v' = A(t)v + f(t,v)$ converge exponentially to the center manifold. Hence, the stability of the system is completely determined by the behavior on the center manifold. Therefore, one often considers a reduction to the center manifold. This has also the advantage of reducing the dimension of the system. Furthermore, since one often needs to approximate the center manifolds to sufficiently high order, it is also important to discuss their regularity. Our main goal is to establish the existence of smooth invariant center manifolds in the presence of *nonuniformly* partially hyperbolic behavior. The method of proof is inspired in the arguments of Chapter 6. In particular, the smoothness of the center manifolds is obtained with a single fixed point problem, instead of one for each additional derivative. We follow closely [15], now with arbitrary stable and unstable subspaces.

8.1 Standing assumptions

Let X be a Banach space and let $A \colon \mathbb{R} \to B(X)$ be a continuous function, where $B(X)$ continues to denote the set of bounded linear operators on X. Consider the initial value problem

$$v' = A(t)v, \quad v(s) = v_s, \tag{8.1}$$

with $s \in \mathbb{R}$ and $v_s \in X$. We assume that all solutions of (8.1) are global.

We write the unique solution of the initial value problem in (8.1) in the form $v(t) = T(t,s)v(s)$, where $T(t,s)$ is the associated evolution operator. Consider constants

$$0 \le a < b, \quad 0 \le c < d, \tag{8.2}$$

$$a', b', c', d' \geq 0. \tag{8.3}$$

Definition 8.1. *We say that the linear equation* $v' = A(t)v$ *admits a nonuniform exponential trichotomy if there exist functions* $P, Q_1, Q_2 \colon \mathbb{R} \to B(X)$ *such that* $P(t)$, $Q_1(t)$, *and* $Q_2(t)$ *are projections with*

$$P(t) + Q_1(t) + Q_2(t) = \mathrm{Id},$$

$$P(t)T(t, s) = T(t, s)P(s), \quad Q_i(t)T(t, s) = T(t, s)Q_i(s), \quad i = 1, 2$$

for every $t, s \in \mathbb{R}$, *and there exist constants as in* (8.2)–(8.3) *and* $D_i \geq 1$, $1 \leq i \leq 4$ *such that:*

1. for every $s, t \in \mathbb{R}$ *with* $t \geq s$,

$$\|T(t, s)P(s)\| \leq D_1 e^{a(t-s)+a'|s|}, \quad \|T(t, s)^{-1}Q_2(t)\| \leq D_3 e^{-b(t-s)+b'|t|}; \tag{8.4}$$

2. for every $s, t \in \mathbb{R}$ *with* $t \leq s$,

$$\|T(t, s)P(s)\| \leq D_2 e^{c(s-t)+c'|s|}, \quad \|T(t, s)^{-1}Q_1(t)\| \leq D_4 e^{-d(s-t)+d'|t|}. \tag{8.5}$$

The constants in (8.2) can be thought of as Lyapunov exponents, while the nonuniformity of the exponential behavior is controlled by the constants in (8.3). When the three components of the solutions respectively correspond to genuine center, stable, and unstable components of $A(t)$ we can take $a = c = 0$ (and thus $b > 0$ and $d > 0$). In a certain sense, the existence of a nonuniform exponential trichotomy is the weakest hypothesis under which one is able to establish the existence of center manifolds, or more precisely of "intermediate" manifolds.

We now present the standing assumptions on the vector field. Set

$$\beta = \max\{(k + 1)a' + b', (k + 1)c' + d'\}. \tag{8.6}$$

We denote by ∂ the partial derivative with respect to the second variable and we assume that there exists an integer $k \geq 1$ such that:

G1. $A \colon \mathbb{R} \to B(X)$ is of class C^k and satisfies (2.2);
G2. $f \colon \mathbb{R} \times X \to X$ is of class C^k and satisfies:
  1. $f(t, 0) = 0$ and $\partial f(t, 0) = 0$ for every $t \in \mathbb{R}$;
  2. there exist $\delta > 0$ and $c_j > 0$ for $j = 1, \ldots, k + 1$ such that for every $t \in \mathbb{R}$ and $u, v \in X$ we have

$$\|\partial^j f(t, u)\| \leq c_j \delta e^{-\beta|t|} \text{ for } j = 1, \ldots, k, \tag{8.7}$$

$$\|\partial^k f(t, u) - \partial^k f(t, v)\| \leq c_{k+1} \delta e^{-\beta|t|} \|u - v\|. \tag{8.8}$$

Note that for every $j = 0, \ldots, k-1$, $t \in \mathbb{R}$, and $u, v \in X$ we have

$$\|\partial^j f(t, u) - \partial^j f(t, v)\| \leq c_{j+1} \delta e^{-\beta|t|} \|u - v\|. \tag{8.9}$$

In the presence of a nonuniform exponential trichotomy we consider the subspaces

$$E(t) = P(t)X, \quad F_1(t) = Q_1(t)X, \quad F_2(t) = Q_2(t)X. \tag{8.10}$$

The unique solution of $v' = A(t)v$ can then be written in the form

$$v(t) = (U(t, s)\xi, V_1(t, s)\eta_1, V_2(t, s)\eta_2) \text{ for } t \in \mathbb{R}, \tag{8.11}$$

with $v_s = (\xi, \eta_1, \eta_2) \in E(s) \times F_1(s) \times F_2(s)$, where

$$U(t, s) := P(t)T(t, s)P(s), \quad V_i(t, s) := Q_i(t)T(t, s)Q_i(s), \quad i = 1, 2.$$

Given $s \in \mathbb{R}$ and an initial condition $v_s = (\xi, \eta_1, \eta_2) \in E(s) \times F_1(s) \times F_2(s)$, we denote by $(x(\cdot, s, v_s), y_1(\cdot, s, v_s), y_2(\cdot, s, v_s))$ the unique solution of the problem (4.4) or, equivalently, of the problem

$$x(t) = U(t, s)\xi + \int_s^t U(t, r)f(r, x(r), y_1(r), y_2(r))\, dr,$$

$$y_i(t) = V_i(t, s)\eta_i + \int_s^t V_i(t, r)f(r, x(r), y_1(r), y_2(r))\, dr, \quad i = 1, 2 \tag{8.12}$$

for $t \in \mathbb{R}$. For each $\tau \in \mathbb{R}$, we write

$$\Psi_\tau(s, v_s) = (s + \tau, x(s + \tau, s, v_s), y_1(s + \tau, s, v_s), y_2(s + \tau, s, v_s)).$$

This is the flow generated by the equation in (4.4).

8.2 Existence of center manifolds

We present in this section the center manifold theorem for the origin in the equation $v' = A(t)v + f(t, v)$. As an application, we also establish the existence of center manifolds for nonuniformly partially hyperbolic solutions of differential equations in Banach spaces.

The center manifolds will be obtained as graphs. We continue to denote by ∂ the partial derivative with respect to the second variable. Let $\mathfrak{X}$ be the space of continuous functions $\varphi = (\varphi_1, \varphi_2) \colon \{(s, \xi) \in \mathbb{R} \times X : \xi \in E(s)\} \to X$ of class C^k in ξ such that for every $s \in \mathbb{R}$ and $x, y \in E(s)$ we have:

1. $\varphi(s, E(s)) \subset F_1(s) \oplus F_2(s)$;
2. $\varphi(s, 0) = 0$ and $\partial \varphi(s, 0) = 0$;
3. $\|\partial^j \varphi(s, x)\| \leq 1$ for $j = 1, \ldots, k$, and

$$\|\partial^k \varphi(s, x) - \partial^k \varphi(s, y)\| \leq \|x - y\|. \tag{8.13}$$

We note that by the mean value theorem, for $j = 0, \ldots, k - 1$ we have

$$\|\partial^j \varphi(s, x) - \partial^j \varphi(s, y)\| \le \|x - y\| \tag{8.14}$$

for every $s \in \mathbb{R}$ and $x, y \in E(s)$. Given a function $\varphi \in \mathfrak{X}$ we consider its graph

$$\mathcal{V} = \{(s, \xi, \varphi(s, \xi)) : (s, \xi) \in \mathbb{R} \times E(s)\} \subset \mathbb{R} \times X. \tag{8.15}$$

We set

$$\alpha_i = 4c_1 D_i \delta \text{ for } i = 1, 2, \tag{8.16}$$

and we consider the conditions

$$\begin{aligned} T_1 &:= (k+1)a - b + \max\{(k+1)a', b'\} < 0, \\ T_2 &:= (k+1)c - d + \max\{(k+1)c', d'\} < 0. \end{aligned} \tag{8.17}$$

These can be thought of as spectral gap conditions.

We now present the center manifold theorem. We use again the notation $p_{s,\xi} = (s, \xi, \varphi(s, \xi))$.

Theorem 8.2 ([15]). *Assume that G1–G2 hold. If the equation $v' = A(t)v$ in the Banach space X admits a nonuniform exponential trichotomy, and the conditions in (8.17) hold, then provided that δ in (8.7)–(8.8) is sufficiently small there is a unique function $\varphi \in \mathfrak{X}$ such the set $\mathcal{V}$ in (8.15) is invariant under the semiflow Ψ_τ, that is,*

$$\text{if } (s, \xi) \in \mathbb{R} \times E(s) \text{ then } \Psi_\tau(p_{s,\xi}) \in \mathcal{V} \text{ for every } \tau \in \mathbb{R}. \tag{8.18}$$

Furthermore:

1. *$\mathcal{V}$ is a smooth manifold of class C^k containing the line $\mathbb{R} \times \{0\}$ and satisfying $T_{(s,0)}\mathcal{V} = \mathbb{R} \times E(s)$ for every $s \in \mathbb{R}$;*
2. *for every $(s, \xi) \in \mathbb{R} \times E(s)$ we have*

$$\varphi_1(s, \xi) = -\int_{-\infty}^{s} V_1(\tau, s)^{-1} f(\Psi_{\tau-s}(p_{s,\xi})) \, d\tau,$$
$$\varphi_2(s, \xi) = \int_{s}^{+\infty} V_2(\tau, s)^{-1} f(\Psi_{\tau-s}(p_{s,\xi})) \, d\tau;$$

3. *there exists $D > 0$ such that for each $s \in \mathbb{R}$, $\xi, \bar{\xi} \in E(s)$, $\tau \in \mathbb{R}$, and $j = 0, \ldots, k$, if $\tau \ge 0$ then*

$$\|\partial_\xi^j(\Psi_\tau(p_{s,\xi})) - \partial_\xi^j(\Psi_\tau(p_{s,\bar{\xi}}))\| \le De^{(j+1)[(a+\alpha_1)\tau + a'|s|]}\|\xi - \bar{\xi}\|, \tag{8.19}$$

and if $\tau \le 0$ then

$$\|\partial_\xi^j(\Psi_\tau(p_{s,\xi})) - \partial_\xi^j(\Psi_\tau(p_{s,\bar{\xi}}))\| \le De^{(j+1)[(c+\alpha_2)|\tau| + c'|s|]}\|\xi - \bar{\xi}\|. \tag{8.20}$$

The proof of Theorem 8.2 is given in Section 8.3.

We call the manifold $\mathcal{V}$ in (8.15) a *center manifold* for the origin in the equation (4.4). We observe that $\mathcal{V}$ is in fact the *unique* center manifold. Note that the constants α_1 and α_2 in (8.19)–(8.20) can be made arbitrarily small by taking δ sufficiently small. A version of Theorem 8.2 in the case of discrete time is established in [4].

We now explain how Theorem 8.2 can be used to establish the existence of center manifolds for nonuniformly partially hyperbolic solutions of a given differential equation. Consider a function $F \colon \mathbb{R} \times X \to X$ of class C^k (for some $k \in \mathbb{N}$), and the equation (4.18). We say that a solution $v_0(t)$ of (4.18) is *nonuniformly partially hyperbolic* if the linear equation defined by $A(t) = \partial F(t, v_0(t))$ admits a nonuniform exponential trichotomy (see (8.4)–(8.5)).

Theorem 8.3. *Assume that F is of class C^k (for some $k \in \mathbb{N}$), and let $v_0(t)$ be a nonuniformly partially hyperbolic solution of (4.18) such that for every $t \in \mathbb{R}$ and $u, v \in X$ we have*

$$\|F(t, u) - F(t, v) - A(t)(u - v)\| \le \delta e^{-\beta|t|}\|u - v\|, \qquad (8.21)$$

$$\|\partial^j F(t, u) - \partial^j F(t, v)\| \le \delta e^{-\beta|t|}\|u - v\| \text{ for } j = 1, \ldots, k. \qquad (8.22)$$

If the conditions in (8.17) hold and δ is sufficiently small, then there exists a unique function $\varphi \in X$ such that the set

$$\mathcal{V} = \{(s, \xi, \varphi(s, \xi)) + (0, v_0(s)) : (s, \xi) \in \mathbb{R} \times E(s)\}$$

is a smooth manifold of class C^k with the following properties:

1. *$(s, v_0(s)) \in \mathcal{V}$ and $T_{(s,v_0(s))}\mathcal{V} = \mathbb{R} \times E(s)$ for every $s \in \mathbb{R}$;*
2. *$\mathcal{V}$ is invariant under solutions of the equation*

$$t' = 1, \quad v' = F(t, v),$$

   that is, if $(s, v_s) \in \mathcal{V}$ then $(t, v(t)) \in \mathcal{V}$ for every $t \in \mathbb{R}$, where $v(t) = v(t, v_s)$ is the unique solution of (4.18) for $t \in \mathbb{R}$ with $v(s) = v_s$;
3. *given $\varepsilon > 0$, provided that δ is sufficiently small there exists $D > 0$ such that for every $s \in \mathbb{R}$ and (s, v_s), $(s, \bar{v}_s) \in \mathcal{V}$ we have*

$$\|v(t, v_s) - v(t, \bar{v}_s)\| \le D e^{(a+\varepsilon)(t-s)+a'|s|}\|v_s - \bar{v}_s\| \text{ for } t \ge s,$$

$$\|v(t, v_s) - v(t, \bar{v}_s)\| \le D e^{(c+\varepsilon)(s-t)+c'|s|}\|v_s - \bar{v}_s\| \text{ for } t \le s.$$

Proof. As in the proof of Theorem 4.2, setting $y(t) = v(t) - v_0(t)$, where $v(t)$ is a solution of (4.18), we obtain

$$y'(t) = A(t)y(t) + G(t, y(t)),$$

where $G(t, y)$ is given by (4.22). By hypothesis $A(t)$ satisfies the assumption G1. Furthermore, it follows from (4.22) that G is of class C^k ($k \ge 1$). Furthermore, also by (4.22) we have $G(t, 0) = 0$, and since $A(t) = \partial F(t, v_0(t))$,

$$\partial G(t, 0) = \partial F(t, v_0(t)) - A(t) = 0.$$

Moreover, from (4.22) and (8.21), for each $(t, y, u) \in \mathbb{R} \times \mathbb{R}^n \times \mathbb{R}^n$ we have

$$\|\partial F(t, y + v_0(t))u - A(t)u\|$$
$$= \left\| \lim_{h \to 0} \frac{F(t, y + v_0(t) + hu) - F(t, y + v_0(t))}{h} - A(t)u \right\|$$
$$= \lim_{h \to 0} \frac{1}{|h|} \|F(t, y + v_0(t) + hu) - F(t, y + v_0(t)) - A(t)hu\|$$
$$\leq \lim_{h \to 0} \frac{1}{|h|} \delta e^{-\beta|t|} \|hu\| \leq \delta e^{-\beta|t|} \|u\|,$$

and thus,

$$\|\partial G(t, y)\| = \|\partial F(t, y + v_0(t)) - A(t)\| \leq \delta e^{-\beta|t|}.$$

For $j = 2, \ldots, k$ it follows from (8.22) that

$$\|\partial^j F(t, y + v_0(t))u\|$$
$$= \left\| \lim_{h \to 0} \frac{\partial^{j-1} F(t, y + v_0(t) + hu) - \partial^{j-1} F(t, y + v_0(t))}{h} \right\|$$
$$\leq \lim_{h \to 0} \frac{1}{|h|} \delta e^{-\beta|t|} \|hu\| \leq \delta e^{-\beta|t|} \|u\|,$$

and hence,

$$\|\partial^j G(t, y)\| = \|\partial^j F(t, y + v_0(t))\| \leq \delta e^{-\beta|t|}.$$

It also follows from (8.22) that

$$\|\partial^k G(t, x) - \partial^k G(t, y)\|$$
$$= \|\partial^k F(t, x + v_0(t)) - \partial^k F(t, y + v_0(t))\| \leq \delta e^{-\beta|t|} \|x - y\|.$$

Thus, the function G satisfies the assumption G2. We can now apply Theorem 8.2 to obtain the desired statement. $\square$

8.3 Proof of the existence of center manifolds

8.3.1 Functional spaces

In view of the desired invariance of the manifold $\mathcal{V}$ under the flow Ψ_τ (see (8.18)), any solution with initial condition in $\mathcal{V}$ at a time $s \in \mathbb{R}$ must remain in $\mathcal{V}$ for every $t \in \mathbb{R}$ and thus must be of the form $(t, x(t), \varphi(t, x(t)))$ for each $t \in \mathbb{R}$. In particular, the equations in (8.12) can be written in the form

$$x(t) = U(t,s)\xi + \int_s^t U(t,\tau)f(\tau,x(\tau),\varphi(\tau,x(\tau)))\,d\tau,$$

$$\varphi_i(t,x(t)) = V_i(t,s)\varphi_i(s,\xi) \tag{8.23}$$

$$+ \int_s^t V_i(t,\tau)f(\tau,x(\tau),\varphi(\tau,x(\tau)))\,d\tau, \quad i = 1,2$$

for $t \in \mathbb{R}$, where $\varphi = (\varphi_1, \varphi_2)$. We equip the space $\mathfrak{X}$ (see Section 8.2 for the definition) with the norm

$$\|\varphi\| = \sup\{\|\varphi(t,x)\|/\|x\| : t \in \mathbb{R} \text{ and } x \in E(t) \setminus \{0\}\}. \tag{8.24}$$

It follows from (8.14) that $\|\varphi\| \le 1$ for every $\varphi \in \mathfrak{X}$. We want to show that $\mathfrak{X}$ is a complete metric space with the norm in (8.24).

Proposition 8.4. *With the norm in (8.24), $\mathfrak{X}$ is a complete metric space.*

Proof. Given a function $\varphi \in \mathfrak{X}$ we set $\bar{\varphi} = \varphi|(\{t\} \times B_R)$ for each fixed $t \in \mathbb{R}$ and $R > 0$, where $B_R \subset E(t)$ is the ball of radius R centered at 0. Then

$$\|\bar{\varphi}(x)\| = \|\varphi(t,x)\| \le \|\varphi\| \cdot \|x\| \le R$$

for each $x \in B_R$. Thus, if $(\varphi_n)_n \subset \mathfrak{X}$ is a Cauchy sequence with respect to the norm in (8.24), then $(\bar{\varphi}_n)_n \subset C_R^{k,\alpha}(B_R, X)$ is a Cauchy sequence in the supremum norm. Hence, there exists a function $\bar{\varphi} \colon B_R \to X$ such that $\|\bar{\varphi}_n - \bar{\varphi}\|_\infty \to 0$ as $n \to \infty$. By Proposition 6.3, we have $\bar{\varphi} \in C_R^{k,\alpha}(B_R, X)$. Furthermore, by the uniqueness of the limit $\bar{\varphi}$, we can uniquely define a continuous function $\varphi \colon \{(t,x) \in \mathbb{R} \times X : x \in E(t)\} \to X$ by $\varphi|(\{t\} \times B_R) = \bar{\varphi}$. Taking into account the pointwise convergence of the k-th derivative in Proposition 6.3 (and hence of the lower-order derivatives) of each sequence $(\bar{\varphi}_n)_n$, and thus of the sequence $(\varphi_n)_n$, one can easily verify that $\varphi \in \mathfrak{X}$. It remains to show that $\|\varphi_n - \varphi\| \to 0$ as $n \to \infty$. Since $(\varphi_n)_n$ is a Cauchy sequence, for each $\varepsilon > 0$, there exists $N = N(\varepsilon)$ such that for every $t \in \mathbb{R}$, $x \in E(t)$, and $n,m \ge N(\varepsilon)$ we have

$$\|\varphi_n(t,x) - \varphi_m(t,x)\| \le \varepsilon\|x\|. \tag{8.25}$$

Letting $n \to \infty$ in (8.25), we obtain $\|\varphi - \varphi_m\| \le \varepsilon$ whenever $m \ge N(\varepsilon)$. This completes the proof. $\qquad\square$

Let now α_1 be as in (8.16) and consider constants $C_j > 0$ for $j = 0, \ldots, k+1$. As in Section 8.2 we denote by ∂ the partial derivative with respect to the second variable. For a fixed $s \in \mathbb{R}$, set

$$\rho(t) = (a + \alpha_1)(t - s) + a'|s| \tag{8.26}$$

and let $\mathcal{B}_+$ be the space of continuous functions $x \colon [s, +\infty) \times E(s) \to X$ of class C^k ($k \ge 1$) in ξ such that $x(s,\xi) = \xi$ for every $\xi \in E(s)$, $x(t,\xi) \in E(t)$ for each $t \ge s$ and $\xi \in E(s)$, and

$$\|x\|' := \sup\left\{\frac{\|x(t,\xi)\|}{\|\xi\|}e^{-\rho(t)} : t \geq s, \xi \in E(s) \setminus \{0\}\right\} \leq C_0, \tag{8.27}$$

$$\|x\|_j := \sup\left\{\|\partial^j x(t,\xi)\|e^{-j\rho(t)} : t \geq s, \xi \in E(s)\right\} \leq C_j \text{ for } j = 1,\ldots,k,$$

$$L_k(x) := \sup\left\{\frac{\|\partial^k x(t,\xi) - \partial^k x(t,\bar{\xi})\|}{\|\xi - \bar{\xi}\|}e^{-(k+1)\rho(t)}\right\} \leq C_{k+1},$$

with the last supremum taken over all $t \geq s$ and $\xi, \bar{\xi} \in E(s)$ with $\xi \neq \bar{\xi}$. Note that given $x \in \mathcal{B}_+$, for every $t \geq s$ and $\xi, \bar{\xi} \in E(s)$ with $\xi \neq \bar{\xi}$ we have

$$\|x(t,\xi)\| \leq \|x\|'\|\xi\|e^{\rho(t)} \leq C_0\|\xi\|e^{\rho(t)}, \tag{8.28}$$

$$\|\partial^j x(t,\xi)\| \leq \|x\|_j e^{j\rho(t)} \leq C_j e^{j\rho(t)} \text{ for } j = 1,\ldots,k, \tag{8.29}$$

$$\frac{\|\partial^k x(t,\xi) - \partial^k x(t,\bar{\xi})\|}{\|\xi - \bar{\xi}\|} \leq L_k(x)e^{(k+1)\rho(t)} \leq C_k e^{(k+1)\rho(t)}. \tag{8.30}$$

Proposition 8.5. *With the norm in (8.27), $\mathcal{B}_+$ is a complete metric space.*

Proof. Given a function $x \in \mathcal{B}_+$ we set $\bar{x} = x|(\{t\} \times B_R)$ for each fixed $t \geq s$ and $R > 0$, where $B_R \subset E(t)$ is the ball of radius R centered at 0. Then

$$\|\bar{x}(\xi)\| = \|x(t,\xi)\| \leq \|x\|'\|\xi\|e^{\rho(t)} \leq D$$

for each $\xi \in B_R$, where $D = C_0 R e^{\rho(t)}$. Thus, if $(x_n)_n \subset \mathcal{B}_+$ is a Cauchy sequence with respect to the norm in (8.27), then $(\bar{x}_n)_n \subset C_D^{k,\alpha}(B_R, E(t))$ is a Cauchy sequence in the supremum norm. Hence, there exists a function $\bar{x}: B_R \to E(t)$ such that $\|\bar{x}_n - \bar{x}\|_\infty \to 0$ as $n \to \infty$. We can now proceed as in the proof of Proposition 8.4 to obtain the desired statement. □

We now consider the past. With α_2 as in (8.16), we set

$$\sigma(t) = (c + \alpha_2)(s - t) + c'|s|,$$

and we introduce the space $\mathcal{B}_-$ of continuous functions $x: (-\infty, s] \times E(s) \to X$ of class C^k ($k \geq 1$) in ξ such that $x(s,\xi) = \xi$ for every $\xi \in E(s)$, $x(t,\xi) \in E(t)$ for each $t \leq s$ and $\xi \in E(s)$, and

$$\|x\|' := \sup\left\{\frac{\|x(t,\xi)\|}{\|\xi\|}e^{-\sigma(t)} : t \leq s, \xi \in E(s) \setminus \{0\}\right\} \leq C_0, \tag{8.31}$$

$$\|x\|_j := \sup\left\{\|\partial^j x(t,\xi)\|e^{-j\sigma(t)} : t \leq s, \xi \in E(s)\right\} \leq C_j \text{ for } j = 1,\ldots,k,$$

$$L_k(x) := \sup\left\{\frac{\|\partial^k x(t,\xi) - \partial^k x(t,\bar{\xi})\|}{\|\xi - \bar{\xi}\|}e^{-(k+1)\sigma(t)}\right\} \leq C_{k+1}, \tag{8.32}$$

with the last supremum taken over all $t \leq s$ and $\xi, \bar{\xi} \in E(s)$ with $\xi \neq \bar{\xi}$.

Proposition 8.6. *With the norm in (8.31), $\mathcal{B}_-$ is a complete metric space.*

The proof of Proposition 8.6 is analogous to that of Proposition 8.5.

8.3.2 Lipschitz property of the derivatives

We now use the inequalities in Section 6.3.2 to obtain several bounds for the norms of the derivatives of solutions and of the vector field along solutions. Given $\varphi \in X$ and $x \in \mathcal{B}_+ \cup \mathcal{B}_-$ we write

$$\varphi^*(t, \xi) = \varphi(t, x(t, \xi)). \tag{8.33}$$

Lemma 8.7. *For each* $j = 1, \ldots, k$ *there exist constants* A_j *and* B_j *such that given* $\varphi \in X$ *and* $(s, \xi) \in \mathbb{R} \times E(s)$ *we have*

$$\|\partial^j \varphi^*(t, \xi)\| \leq \begin{cases} A_j e^{j\rho(t)}, & t \geq s \text{ and } x \in \mathcal{B}_+ \\ B_j e^{j\sigma(t)}, & t \leq s \text{ and } x \in \mathcal{B}_- \end{cases}.$$

Proof. We will only consider the case when $t \geq s$ and $x \in \mathcal{B}_+$, since the other case can be treated in an analogous manner. Using (6.16) for the derivative $\partial^j \varphi^*$ we obtain

$$\|\partial^j \varphi^*(t, \xi)\| \leq c \sum_{m=1}^{j} \|\partial^m \varphi(t, x(t, \xi))\| \sum_{p(j,m)} \prod_{l=1}^{j} \|\partial^l x(t, \xi)\|^{k_l},$$

with $p(j, m)$ given by (6.17). Since $\varphi \in X$, using the identity $\sum_{l=1}^{j} lk_l = j$ in (6.17) together with (8.29), we have

$$\|\partial^j \varphi^*(t, \xi)\| \leq c \sum_{m=1}^{j} \sum_{p(j,m)} \prod_{l=1}^{j} (C_l e^{l\rho(t)})^{k_l} \leq A_j e^{j\rho(t)}$$

for some constant $A_j > 0$. $\qquad \square$

In the following lemmas, as in Lemma 8.7, we will continue to give the proofs only when $t \geq s$ and $x \in \mathcal{B}_+$. The other case is analogous. Given $\varphi \in X$ and $x \in \mathcal{B}_+ \cup \mathcal{B}_-$ we write

$$f^*(t, \xi) = f(t, x(t, \xi), \varphi(t, x(t, \xi))) = f(t, x(t, \xi), \varphi^*(t, \xi)), \tag{8.34}$$

with $\varphi^*(t, \xi)$ as in (8.33).

Lemma 8.8. *For each* $j = 1, \ldots, k$ *there exist constants* $\bar{A}_j$ *and* $\bar{B}_j$ *such that given* $\varphi \in X$ *and* $(s, \xi) \in \mathbb{R} \times E(s)$ *we have*

$$\|\partial^j f^*(t, \xi)\| \leq \delta e^{-\beta|t|} \begin{cases} \bar{A}_j e^{j\rho(t)}, & t \geq s \text{ and } x \in \mathcal{B}_+ \\ \bar{B}_j e^{j\sigma(t)}, & t \leq s \text{ and } x \in \mathcal{B}_- \end{cases}.$$

Proof. Using (6.19) for the derivative $\partial^j f^*(t, \xi)$ we obtain

$$\|\partial^j f^*(t,\xi)\| \le c \sum_{q(j)} \|\partial^{\lambda_1,\lambda_2}_{x(t,\xi),\varphi(t,x(t,\xi))} f(t,\cdot)\|$$

$$\times \sum_{s=1}^{j} \sum_{p_s(j,\lambda)} \prod_{m=1}^{s} \|\partial^{l_m} x(t,\xi)\|^{k_{m1}} \|\partial^{l_m} \varphi^*(t,\xi)\|^{k_{m2}},$$

with $p_s(j,\lambda)$ as in (6.20). Since $\varphi \in \mathfrak{X}$, using the identity $\sum_{m=1}^{s} l_m(k_{m1} + k_{m2}) = j$ in (6.20), together with (8.7) and (8.29), it follows from Lemma 8.7 that

$$\|\partial^j f^*(t,\xi)\| \le c\delta e^{-\beta|t|} \sum_{q(j)} c_j \sum_{s=1}^{j} \sum_{p_s(j,\lambda)} \prod_{m=1}^{s} (C_{l_m} e^{l_m \rho(t)})^{k_{m1}} (A_{l_m} e^{l_m \rho(t)})^{k_{m2}}$$

$$\le \bar{A}_j \delta e^{-\beta|t|} e^{j\rho(t)}$$

for some constant $\bar{A}_j > 0$. This establishes the desired statement. □

Lemma 8.9. *For every $j = 0, \ldots, k$ and $(s,\xi), (s,\bar{\xi}) \in \mathbb{R} \times E(s)$ we have*

$$\|\partial^j x(t,\xi) - \partial^j x(t,\bar{\xi})\| \le C_{j+1} \|\xi - \bar{\xi}\| \begin{cases} e^{(j+1)\rho(t)}, & t \ge s \text{ and } x \in \mathcal{B}_+ \\ e^{(j+1)\sigma(t)}, & t \le s \text{ and } x \in \mathcal{B}_- \end{cases}.$$

Proof. By the definition of the spaces $\mathcal{B}_+$ and $\mathcal{B}_-$, the statement is automatically true for $j = k$ (see (8.30) and (8.32)). For $j < k$, it suffices to observe that by (8.29) we have

$$\|\partial^j x(t,\xi) - \partial^j x(t,\bar{\xi})\| \le \sup_{r\in[0,1]} \|\partial^{j+1} x(t,\xi + r(\bar{\xi} - \xi))\| \cdot \|\xi - \bar{\xi}\|$$

$$\le C_{j+1} e^{(j+1)\rho(t)} \|\xi - \bar{\xi}\|,$$

with an application of the mean value theorem. □

Lemma 8.10. *For each $j = 0, \ldots, k$ there exist constants $\widetilde{A}_j$ and $\widetilde{B}_j$ such that given $\varphi \in \mathfrak{X}$ and $(s,\xi), (s,\bar{\xi}) \in \mathbb{R} \times E(s)$ we have*

$$\|\partial^j \varphi^*(t,\xi) - \partial^j \varphi^*(t,\bar{\xi})\| \le \|\xi - \bar{\xi}\| \begin{cases} \widetilde{A}_j e^{(j+1)\rho(t)}, & t \ge s \text{ and } x \in \mathcal{B}_+ \\ \widetilde{B}_j e^{(j+1)\sigma(t)}, & t \le s \text{ and } x \in \mathcal{B}_- \end{cases}.$$

Proof. For $j < k$ the result follows immediately from Lemma 8.7. However, the proof does not simplify by considering only the case $j = k$. By (6.18) we have

$$\|\partial^j \varphi^*(t,\xi) - \partial^j \varphi^*(t,\bar{\xi})\|$$

$$\le c \sum_{m=1}^{j} \|\partial^m \varphi(t,x(t,\xi)) - \partial^m \varphi(t,x(t,\bar{\xi}))\| \sum_{p(j,m)} \prod_{l=1}^{j} \|\partial^l x(t,\xi)\|^{k_l} \tag{8.35}$$

$$+ c' \sum_{m=1}^{j} \|\partial^m \varphi(t,x(t,\bar{\xi}))\| S_j,$$

with $p(j, m)$ as in (6.17) and with

$$S_j := \sum_{p(j,m)} \sum_{l=1}^{j} T_l \prod_{i=1}^{l-1} \|\partial^i x(t, \bar{\xi})\|^{k_i} \prod_{i=l+1}^{j} \|\partial^i x(t, \xi)\|^{k_i}, \qquad (8.36)$$

where

$$T_l := \|\partial^l x(t, \xi) - \partial^l x(t, \bar{\xi})\| \sum_{k=0}^{k_l-1} \|\partial^l x(t, \xi)\|^{k_l-1-k} \|\partial^l x(t, \bar{\xi})\|^k.$$

Since $\varphi \in \mathcal{X}$, by (8.13) and (8.14) for $m = 1, \ldots, k$ we have

$$\|\partial^m \varphi(t, x(t, \xi)) - \partial^m \varphi(t, x(t, \bar{\xi}))\| \leq \|x(t, \xi) - x(t, \bar{\xi})\|.$$

Using Lemma 8.9 with $j = 0$ we obtain

$$\|\partial^m \varphi(t, x(t, \xi)) - \partial^m \varphi(t, x(t, \bar{\xi}))\| \leq C_1 e^{\rho(t)} \|\xi - \bar{\xi}\|. \qquad (8.37)$$

Furthermore, by Lemma 8.9 and (8.29) with $j = l$,

$$\begin{aligned} T_l &\leq C_{l+1} e^{(l+1)\rho(t)} \|\xi - \bar{\xi}\| \sum_{k=0}^{k_l-1} (C_l e^{l\rho(t)})^{k_l-1} \\ &\leq C_{l+1} C_l^{k_l-1} k_l e^{(lk_l+1)\rho(t)} \|\xi - \bar{\xi}\|. \end{aligned} \qquad (8.38)$$

By (8.38) and (8.29) it follows from (8.36) that

$$\begin{aligned} S_j &\leq \sum_{p(j,m)} \sum_{l=1}^{j} T_l \prod_{i=1, i \neq l}^{j} (C_i e^{i\rho(t)})^{k_i} \\ &\leq \widehat{C}_j \sum_{p(j,m)} \sum_{l=1}^{j} e^{(lk_l+1)\rho(t)} \|\xi - \bar{\xi}\| \prod_{i=1, i \neq l}^{j} e^{ik_i\rho(t)} \\ &= \widehat{C}_j e^{(j+1)\rho(t)} \|\xi - \bar{\xi}\|, \end{aligned} \qquad (8.39)$$

where $\widehat{C}_j$ is a positive constant, using the identity $\sum_{m=1}^{j} mk_m = j$ (see (6.17)). Thus, by (8.35), (8.37), (8.39), (8.29), and the fact that $\varphi \in \mathcal{X}$, we obtain

$$\begin{aligned} \|\partial^j \varphi^*(t, \xi) - \partial^j \varphi^*(t, \bar{\xi})\| &\leq cC_1 e^{\rho(t)} \|\xi - \bar{\xi}\| \sum_{m=1}^{j} \sum_{p(j,m)} \prod_{l=1}^{j} (C_l e^{l\rho(t)})^{k_l} \\ &\quad + c'j\widehat{C}_j e^{(j+1)\rho(t)} \|\xi - \bar{\xi}\| \\ &= \widetilde{A}_j e^{(j+1)\rho(t)} \|\xi - \bar{\xi}\|, \end{aligned}$$

since $\sum_{l=1}^{j} lk_l = j$ (see (6.17)), for some constant $\widetilde{A}_j > 0$. We have thus obtained the desired statement. □

Lemma 8.11. *There exist constants $\widehat{A}_k$ and $\widehat{B}_k$ such that given $\varphi \in \mathfrak{X}$ and $(s, \xi), (s, \bar{\xi}) \in \mathbb{R} \times E(s)$ we have*

$$\|\partial^k f^*(t, \xi) - \partial^k f^*(t, \bar{\xi})\| \leq \delta e^{-\beta|t|} \|\xi - \bar{\xi}\| \begin{cases} \widehat{A}_k e^{(k+1)\rho(t)}, & t \geq s \text{ and } x \in \mathcal{B}_+ \\ \widehat{B}_k e^{(k+1)\sigma(t)}, & t \leq s \text{ and } x \in \mathcal{B}_- \end{cases}.$$

Proof. By (6.21) we have

$$\|\partial^k f^*(t, \xi) - \partial^k f^*(t, \bar{\xi})\|$$

$$\leq c \sum_{q(k)} G_{\lambda_1, \lambda_2} \sum_{s=1}^{k} \sum_{p_s(k, \lambda)} \prod_{m=1}^{s} \|\partial^{l_m} x(t, \xi)\|^{k_{m1}} \|\partial^{l_m} \varphi^*(t, \xi)\|^{k_{m2}} \tag{8.40}$$

$$+ c' \sum_{q(k)} \|\partial_{x(t, \bar{\xi}), \varphi^*(t, \bar{\xi})}^{\lambda_1, \lambda_2} f(t, \cdot)\| \sum_{s=1}^{k} \widetilde{S}_s,$$

where

$$G_{\lambda_1, \lambda_2} := \|\partial_{x(t, \xi), \varphi^*(t, \xi)}^{\lambda_1, \lambda_2} f(t, \cdot) - \partial_{x(t, \bar{\xi}), \varphi^*(t, \bar{\xi})}^{\lambda_1, \lambda_2} f(t, \cdot)\|,$$

and where

$$\widetilde{S}_s := \sum_{p_s(k, \lambda)} \sum_{m=1}^{s} \widetilde{T}_{k_{m1}, k_{m2}, l_m} \left(\prod_{i=1}^{l_m - 1} \|\partial^{l_i} x(t, \bar{\xi})\|^{k_{i1}} \|\partial^{l_i} \varphi^*(t, \bar{\xi})\|^{k_{i2}} \right.$$

$$\left. \times \prod_{i=l_m + 1}^{s} \|\partial^{l_i} x(t, \xi)\|^{k_{i1}} \|\partial^{l_i} \varphi^*(t, \xi)\|^{k_{i2}} \right), \tag{8.41}$$

with

$$\widetilde{T}_{k_{m1}, k_{m2}, l_m} := \|\partial^{l_m} \varphi^*(t, \xi)\|^{k_{m2}} \|\partial^{l_m} x(t, \xi) - \partial^{l_m} x(t, \bar{\xi})\|$$

$$\times \sum_{k=0}^{k_{m1} - 1} \|\partial^{l_m} x(t, \xi)\|^{k_{m1} - 1 - k} \|\partial^{l_m} x(t, \bar{\xi})\|^{k}$$

$$+ \|\partial^{l_m} x(t, \bar{\xi})\|^{k_{m1}} \|\partial^{l_m} \varphi^*(t, \xi) - \partial^{l_m} \varphi^*(t, \bar{\xi})\|$$

$$\times \sum_{k=0}^{k_{m2} - 1} \|\partial^{l_m} \varphi^*(t, \xi)\|^{k_{m2} - 1 - k} \|\partial^{l_m} \varphi^*(t, \bar{\xi})\|^{k}.$$

By the mean value theorem, (8.7), and Lemmas 8.9 and 8.10 with $j = 0$, for $\lambda_1 + \lambda_2 = 1, \ldots, k - 1$ we have

$$G_{\lambda_1, \lambda_2} \leq \sup_{r \in [0,1]} \|\partial_{a(r)}^{\lambda_1 + 1, \lambda_2} f(t, \cdot)\| \cdot \|x(t, \xi) - x(t, \bar{\xi})\|$$

$$+ \sup_{r \in [0,1]} \|\partial_{b(r)}^{\lambda_1, \lambda_2 + 1} f(t, \cdot)\| \cdot \|\varphi^*(t, \xi) - \varphi^*(t, \bar{\xi})\|$$

$$\leq c_{\lambda_1 + \lambda_2 + 1} \delta e^{-\beta|t|} e^{\rho(t)} \|\xi - \bar{\xi}\| (C_1 + \widetilde{A}_0),$$

where

$$a(r) = (x(t,\xi) + r(x(t,\bar\xi) - x(t,\xi)), \varphi^*(t,\xi)),$$
$$b(r) = (x(t,\xi), \varphi^*(t,\xi) + r(\varphi^*(t,\bar\xi) - \varphi^*(t,\xi))).$$

Furthermore, when $\lambda_1 + \lambda_2 = k$ it follows from (8.8) and Lemmas 8.9 and 8.10 that

$$G_{\lambda_1,\lambda_2} = \|\partial^{\lambda_1,\lambda_2}_{x(t,\xi),\varphi^*(t,\xi)} f(t,\cdot) - \partial^{\lambda_1,\lambda_2}_{x(t,\bar\xi),\varphi^*(t,\bar\xi)} f(t,\cdot)\|$$
$$\le c_{k+1}\delta e^{-\beta|t|}\|(x(t,\xi),\varphi^*(t,\xi)) - (x(t,\bar\xi),\varphi^*(t,\bar\xi))\|$$
$$\le c_{k+1}\delta e^{-\beta|t|}e^{\rho(t)}\|\xi - \bar\xi\|(C_1 + \tilde A_0).$$

Thus, for each $(\lambda_1,\lambda_2) \in q(k)$ we have

$$G_{\lambda_1,\lambda_2} \le c_{\lambda_1+\lambda_2+1}\delta e^{-\beta|t|}e^{\rho(t)}\|\xi - \bar\xi\|(C_1 + \tilde A_0).$$

By (8.29), Lemma 8.7, and since $\sum_{m=1}^{s} l_m(k_{m1} + k_{m2}) = k$ (see (6.20)), the first summand in (8.40) is bounded by

$$G_k\delta e^{-\beta|t|}e^{(k+1)\rho(t)}\|\xi - \bar\xi\|, \tag{8.42}$$

for some constant $G_k > 0$. It follows from Lemmas 8.7, 8.9, and 8.10 that

$$\tilde T_{k_{m1},k_{m2},l_m} \le (A_{l_m}e^{l_m\rho(t)})^{k_{m2}}\|\xi - \bar\xi\|C_{l_m+1}e^{(l_m+1)\rho(t)}\sum_{k=0}^{k_{m1}-1}(C_{l_m}e^{l_m\rho(t)})^{k_{m1}-1}$$
$$+ (C_{l_m}e^{l_m\rho(t)})^{k_{m1}}\|\xi - \bar\xi\|\tilde A_{l_m}e^{(l_m+1)\rho(t)}\sum_{k=0}^{k_{m2}-1}(A_{l_m}e^{l_m\rho(t)})^{k_{m2}-1}$$
$$\le E_m e^{(1+l_m(k_{m1}+k_{m2}))\rho(t)}\|\xi - \bar\xi\|,$$

for some constant $E_m > 0$. By (8.41), (8.29), and Lemma 8.7 we obtain

$$\tilde S_s \le \sum_{p_s(k,\lambda)}\sum_{m=1}^{s} E_m e^{(1+l_m(k_{m1}+k_{m2}))\rho(t)}\|\xi - \bar\xi\|$$
$$\times e^{\sum_{i=1,i\ne m}^{s} l_i(k_{i1}+k_{i2})\rho(t)}\prod_{i=1,i\ne m}^{s} C_{l_i}^{k_{i1}}A_{l_i}^{k_{i2}}$$
$$\le F_s e^{(k+1)\rho(t)}\|\xi - \bar\xi\|,$$

for some constant $F_s > 0$, using the identity $\sum_{i=1}^{s} l_i(k_{i1}+k_{i2}) = k$. Therefore, the second summand in (8.40) is bounded by

$$H_k\delta e^{-\beta|t|}e^{(k+1)\rho(t)}\|\xi - \bar\xi\|, \tag{8.43}$$

for some constant $H_k > 0$. It follows from (8.40), (8.42), and (8.43) that

$$\|\partial^k f^*(t,\xi) - \partial^k f^*(t,\bar\xi)\| \le \delta e^{-\beta|t|}e^{(k+1)\rho(t)}\|\xi - \bar\xi\|(G_k + H_k).$$

Thus, the statement in the lemma follows setting $\hat A_k = G_k + H_k$. $\qquad\square$

8.3.3 Solution on the central direction

The proof of Theorem 8.2 will be obtained in several steps. We first establish the existence of a unique function $x = x_\varphi$ satisfying the first equation in (8.23) for each given $\varphi \in \mathfrak{X}$.

Lemma 8.12. *Provided that δ is sufficiently small, for each $\varphi \in \mathfrak{X}$ the following properties hold:*

1. *given $s \in \mathbb{R}$ there exists a unique function $x = x_\varphi \colon \mathbb{R} \times E(s) \to X$ with $x_\varphi(s, \xi) = \xi$ satisfying the first equation in (8.23) and $x_\varphi(t, \xi) \in E(t)$ for every $t \in \mathbb{R}$ and $\xi \in E(s)$;*
2. *the function x_φ satisfies*

$$x_\varphi|[s, +\infty) \times E(s) \in \mathcal{B}_+, \quad x_\varphi|(-\infty, s] \times E(s) \in \mathcal{B}_-,$$

and

$$\|x_\varphi(t, \xi)\| \leq \begin{cases} 2D_1 e^{\rho(t)} \|\xi\|, & t \geq s \\ 2D_2 e^{\rho(t)} \|\xi\|, & t \leq s \end{cases}. \tag{8.44}$$

Proof. We start with the case when $t \geq s$. Given $s \in \mathbb{R}$, $\varphi \in \mathfrak{X}$, and $\xi \in E(s)$, we define the operator

$$(Jx)(t, \xi) = U(t, s)\xi + \int_s^t U(t, \tau) f(\tau, x(\tau, \xi), \varphi(\tau, x(\tau, \xi))) \, d\tau$$

for each $x \in \mathcal{B}_+$ and $t \geq s$. Clearly Jx is a continuous function of class C^k in ξ. The fact that $(Jx)(s, \xi) = \xi$ is a consequence of the identity $U(s, s)\xi = \xi$. Furthermore, using (8.9) and (8.28) we obtain

$$\|f(\tau, x(\tau, \xi), \varphi(\tau, x(\tau, \xi)))\|$$
$$\leq c_1 \delta e^{-\beta|\tau|} \|(x(\tau, \xi), \varphi(\tau, x(\tau, \xi)))\| \leq 2c_1 \delta e^{-\beta|\tau|} \|x(\tau, \xi)\|$$
$$\leq 2c_1 C_0 \delta e^{\rho(\tau)} e^{-\beta|\tau|} \|\xi\|.$$

Thus, using the first inequality in (8.4) and the definition of β in (8.6),

$$\|(Jx)(t, \xi) - U(t, s)\xi\| \leq \int_s^t \|U(t, \tau)\| \cdot \|f(\tau, x(\tau, \xi), \varphi(\tau, x(\tau, \xi)))\| \, d\tau$$

$$\leq 2c_1 C_0 \delta D_1 \|\xi\| e^{a'|s|} \int_s^t e^{(a+\alpha_1)(\tau-s)} e^{a(t-\tau)} e^{(a'-\beta)|\tau|} \, d\tau$$

$$\leq 2c_1 C_0 \delta D_1 \|\xi\| e^{\rho(t)} \int_s^t e^{-\alpha_1(t-\tau)} e^{(a'-\beta)|\tau|} \, d\tau \leq \theta \|\xi\| e^{\rho(t)},$$

where

$$\theta = 2c_1 C_0 D_1 \delta / \alpha_1.$$

Furthermore, by (8.4) and (8.26), we have $\|U(t,s)\xi\| \le D_1 e^{\rho(t)}\|\xi\|$. Thus, choosing a constant $C_0 > D_1$ and taking δ sufficiently small, we obtain $\|Jx\|' \le D_1 + \theta < C_0$.

We now consider the derivatives $\partial^j(Jx)$. By Lemma 8.8 applied to the function f^* in (8.34), for $j = 1, \ldots, k$, we have

$$\|\partial^j f^*(\tau, \xi)\| \le \bar{A}_j \delta e^{-\beta|\tau|} e^{j\rho(\tau)}.$$

Thus, by the first inequality in (8.4), for $j = 2, \ldots, k$,

$$\|\partial^j(Jx)(t, \xi)\| \le \int_s^t \|U(t, \tau)\| \cdot \|\partial^j f^*(\tau, \xi)\| \, d\tau$$

$$\le \bar{A}_j \delta D_1 e^{ja'|s|} \int_s^t e^{j(a+\alpha_1)(\tau-s)} e^{a(t-\tau)} e^{(a'-\beta)|\tau|} \, d\tau$$

$$\le \bar{A}_j \delta D_1 e^{j\rho(t)} \int_s^t e^{-((j-1)a+\alpha_1 j)(t-\tau)} e^{(a'-\beta)|\tau|} \, d\tau$$

$$\le \frac{\bar{A}_j D_1 \delta}{(j-1)a + \alpha_1 j} e^{j\rho(t)}.$$

Therefore, taking δ sufficiently small, for $j = 2, \ldots, k$ we have

$$\|\partial^j(Jx)\|_j \le \frac{\bar{A}_j D_1 \delta}{(j-1)a + \alpha_1 j} \le C_j.$$

When $j = 1$ the term $U(t, s)$ is also present in the derivative, and thus

$$\|\partial(Jx)(t, \xi)\| \le \|U(t, s)\| + \bar{A}_1 D_1 \delta/\alpha_1.$$

Choosing a constant $C_1 > D_1$ and taking δ sufficiently small we obtain

$$\|\partial(Jx)\|_1 \le D_1 + \frac{\bar{A}_1 D_1 \delta}{\alpha_1} < C_1.$$

Finally, by Lemma 8.11, and the first inequality in (8.4), for each $t \ge s$ and $\xi, \bar{\xi} \in E(s)$ with $\xi \ne \bar{\xi}$ we have

$$\|\partial^k(Jx)(t, \xi) - \partial^k(Jx)(t, \bar{\xi})\|$$

$$\le \int_s^t \|U(t, \tau)\| \cdot \|\partial^k f^*(\tau, \xi) - \partial^k f^*(\tau, \bar{\xi})\| \, d\tau$$

$$\le \hat{A}_k D_1 \delta e^{(k+1)a'|s|} \|\xi - \bar{\xi}\| \int_s^t e^{(k+1)(a+\alpha_1)(\tau-s)} e^{a(t-\tau)} e^{(a'-\beta)|\tau|} \, d\tau$$

$$\le \hat{A}_k D_1 \delta \|\xi - \bar{\xi}\| e^{(k+1)\rho(t)} \int_s^t e^{-(ka+(k+1)\alpha_1)(t-\tau)} e^{(a'-\beta)|\tau|} \, d\tau$$

$$\le \frac{\hat{A}_k D_1 \delta}{ka + (k+1)\alpha_1} \|\xi - \bar{\xi}\| e^{(k+1)\rho(t)}.$$

By taking δ sufficiently small, we obtain

$$L_k(Jx) \le \frac{\widehat{A}_k D_1 \delta}{ka + (k+1)\alpha_1} \le C_{k+1}.$$

Hence, $Jx \in \mathcal{B}_+$ and $J \colon \mathcal{B}_+ \to \mathcal{B}_+$ is a well-defined operator.

We now prove that J is a contraction with the norm $\|\cdot\|'$ in (8.27). Given $x, y \in \mathcal{B}_+$ and $\tau \ge s$, it follows from (8.9) and the definition of α_1 that

$$\begin{aligned}
&\|f(\tau, x(\tau, \xi), \varphi(\tau, x(\tau, \xi))) - f(\tau, y(\tau, \xi), \varphi(\tau, y(\tau, \xi)))\| \\
&\le \delta e^{-\beta|\tau|} \|(x(\tau, \xi), \varphi(\tau, x(\tau, \xi))) - (y(\tau, \xi), \varphi(\tau, y(\tau, \xi)))\| \\
&\le 2c_1 \delta e^{-\beta|\tau|} \|x(\tau, \xi) - y(\tau, \xi)\| \\
&\le \frac{\alpha_1}{2D_1} e^{\rho(\tau)} e^{-\beta|\tau|} \|\xi\| \cdot \|x - y\|'.
\end{aligned} \tag{8.45}$$

By the first inequality in (8.4) and (8.45) we obtain

$$\begin{aligned}
&\|(Jx)(t, \xi) - (Jy)(t, \xi)\| \\
&\le \int_s^t \|U(t, \tau)\| \cdot \|f(\tau, x(\tau, \xi), \varphi(\tau, x(\tau, \xi))) - f(\tau, y(\tau, \xi), \varphi(\tau, y(\tau, \xi)))\| \, d\tau \\
&\le \frac{\alpha_1}{2} \|\xi\| \cdot \|x - y\|' e^{(a+\alpha_1)(t-s)+a'|s|} \int_s^t e^{-\alpha_1(t-\tau)} e^{(a'-\beta)|\tau|} \, d\tau \\
&\le \frac{\|\xi\|}{2} \cdot \|x - y\|' e^{\rho(t)}
\end{aligned}$$

for each $t \ge s$, using the fact that $\beta \ge a'$. Therefore

$$\|Jx - Jy\|' \le \frac{1}{2} \|x - y\|', \tag{8.46}$$

and J is a contraction. Thus, by Proposition 8.5, there exists a unique function $x = x_\varphi \in \mathcal{B}_+$ such that $Jx = x$. Set

$$z(t, \xi) = (J0)(t, \xi) = U(t, s)\xi.$$

The function x can be obtained by

$$x(t, \xi) = \lim_{n \to +\infty} (J^n 0)(t, \xi) = \sum_{k=0}^{+\infty} [(J^{k+1} 0)(t, \xi) - (J^k 0)(t, \xi)]$$

for each $t \ge s$. It follows from (8.46) that

$$\|x\|' \le \sum_{k=0}^{+\infty} \|(J^k z)(t, \xi) - (J^k 0)(t, \xi)\|' \le \sum_{k=0}^{+\infty} \frac{1}{2^k} \|z\|' = 2\|z\|' \le 2D_1,$$

which together with (8.27) yields the desired results for $t \ge s$.

The case when $t \le s$ can be treated in a similar manner, considering now the space $\mathcal{B}_-$ with the norm (8.31), using the first inequality in (8.5) as well as the fact that $\beta \ge c'$, together with Proposition 8.6. $\square$

By Lemma 8.12 we have

$$x_\varphi|[s, +\infty) \times E(s) \in \mathcal{B}_+ \quad \text{and} \quad x_\varphi|(-\infty, s] \times E(s) \in \mathcal{B}_-.$$

Thus, if we denote by x_φ and $\bar{x}_\varphi$ the unique functions given by Lemma 8.12 such that $x_\varphi(s, \xi) = \xi$ and $\bar{x}_\varphi(s, \bar{\xi}) = \bar{\xi}$, it follows from Lemma 8.9 that

$$\|x_\varphi(t, \xi) - \bar{x}_\varphi(t, \bar{\xi})\| \le C_1 \|\xi - \bar{\xi}\| \begin{cases} e^{\rho(t)}, & t \ge s \\ e^{\sigma(t)}, & t \le s \end{cases}.$$

8.3.4 Reduction to an equivalent problem

In order to establish the existence of a function $\varphi \in X$ satisfying the second identity in (8.23) when $x = x_\varphi$, where x_φ is the continuous function given by Lemma 8.12 with $x_\varphi(s, \xi) = \xi$, we first transform this problem into an equivalent problem.

Lemma 8.13. *Provided that δ is sufficiently small, given $\varphi \in X$ the following properties are equivalent:*

1.

$$\varphi_1(t, x_\varphi(t, \xi)) = V_1(t, s)\varphi_1(s, \xi)$$
$$+ \int_s^t V_1(t, \tau) f(\tau, x_\varphi(\tau, \xi), \varphi(\tau, x_\varphi(\tau, \xi))) \, d\tau \quad (8.47)$$

for every $(s, \xi) \in \mathbb{R} \times E(s)$ and $t \le s$, and

$$\varphi_2(t, x_\varphi(t, \xi)) = V_2(t, s) f(s, \xi)$$
$$+ \int_s^t V_2(t, \tau) f(\tau, x_\varphi(\tau, \xi), \varphi(\tau, x_\varphi(\tau, \xi))) \, d\tau \quad (8.48)$$

for every $(s, \xi) \in \mathbb{R} \times E(s)$ and $t \ge s$;

2.

$$\varphi_1(s, \xi) = \int_{-\infty}^s V_1(\tau, s)^{-1} f(\tau, x_\varphi(\tau, \xi), \varphi(\tau, x_\varphi(\tau, \xi))) \, d\tau,$$
$$\varphi_2(s, \xi) = - \int_s^{+\infty} V_2(\tau, s)^{-1} f(\tau, x_\varphi(\tau, \xi), \varphi(\tau, x_\varphi(\tau, \xi))) \, d\tau \quad (8.49)$$

for every $(s, \xi) \in \mathbb{R} \times E(s)$ (including the requirement that the integrals are well-defined).

Proof. We start by showing that the integrals in (8.49) are well-defined for each $(s, \xi) \in \mathbb{R} \times E(s)$. By the second inequality in (8.44) in Lemma 8.12, and (8.9), for each $\tau \le s$ we have

$$\|f(\tau, x_\varphi(\tau, \xi), \varphi(\tau, x_\varphi(\tau, \xi)))\|$$
$$\leq 2c_1\delta e^{-\beta|\tau|}\|x_\varphi(\tau, \xi)\| \leq 4c_1\delta e^{-\beta|\tau|}D_2 e^{\sigma(\tau)}\|\xi\|. \tag{8.50}$$

Proceeding in a similar manner, using now the first inequality in (8.44), for each $\tau \geq s$ we have

$$\|f(\tau, x_\varphi(\tau, \xi), \varphi(\tau, x_\varphi(\tau, \xi)))\| \leq 4c_1\delta e^{-\beta|\tau|}D_1 e^{\rho(\tau)}\|\xi\|. \tag{8.51}$$

It follows from the second inequality in (8.5), and (8.50), using the inequality $|\tau| \leq |\tau - s| + |s|$, that

$$\int_{-\infty}^{s} \|V_1(\tau, s)^{-1}f(\tau, x_\varphi(\tau, \xi), \varphi(\tau, x_\varphi(\tau, \xi)))\|\, d\tau$$

$$\leq 4c_1\delta D_4 D_2 e^{c'|s|}\|\xi\| \int_{-\infty}^{s} e^{(c-d+\alpha_2)(s-\tau)}e^{-(\beta-d')|\tau|}\, d\tau \tag{8.52}$$

$$\leq 4c_1\delta D_4 D_2 e^{c'|s|}\|\xi\| \int_{-\infty}^{s} e^{(T_2+\alpha_2)(s-\tau)}\, d\tau,$$

since $T_2 \geq c-d$ (because $c \geq 0$) and where we have used that $\beta \geq d'$. By (8.17) we have $T_2 < 0$ and choosing δ sufficiently small we can make α_2 sufficiently small so that $T_2 + \alpha_2 < 0$. This shows that the first integral in (8.49) is well-defined. In a similar manner, using the second inequality in (8.4) and (8.51) we obtain

$$\int_{s}^{+\infty} \|V_2(\tau, s)^{-1}f(\tau, x_\varphi(\tau, \xi), \varphi(\tau, x_\varphi(\tau, \xi)))\|\, d\tau$$

$$\leq 4c_1\delta D_3 D_1 e^{a'|s|}\|\xi\| \int_{s}^{+\infty} e^{(T_1+\alpha_1)(\tau-s)}\, d\tau, \tag{8.53}$$

since $T_1 \geq a - b$ (because $a \geq 0$) and where we have used that $\beta \geq b'$. By (8.17) we have $T_1 < 0$ and choosing δ sufficiently small we have $T_1 + \alpha_1 < 0$. Thus, the second integral in (8.49) is also well-defined.

We now assume that the identities (8.47)–(8.48) hold, and we rewrite them in the equivalent form

$$\varphi_i(s, \xi) = V_i(t, s)^{-1}\varphi_i(t, x_\varphi(t, \xi))$$
$$- \int_{s}^{t} V_i(\tau, s)^{-1}f(\tau, x_\varphi(\tau, \xi), \varphi(\tau, x_\varphi(\tau, \xi)))\, d\tau \tag{8.54}$$

for $t \leq s$ when $i = 1$, and $t \geq s$ when $i = 2$. By the second inequality in (8.44) and the second inequality in (8.5), for every $t \leq s$ we have

$$\|V_1(t, s)^{-1}\varphi_1(t, x_\varphi(t, \xi))\| \leq D_4 e^{-d(s-t)+d'|t|}\|x_\varphi(t, \xi)\|$$
$$\leq 2D_4 D_2\|\xi\|e^{(c-d+d'+\alpha_2)(s-t)+(c'+d')|s|}$$
$$\leq 2D_4 D_2\|\xi\|e^{(T_2+\alpha_2)(s-t)+(c'+d')|s|}.$$

Thus, letting $t \to -\infty$ in (8.54) when $i = 1$, we obtain the first identity in (8.49). To establish the second identity we proceed in a similar manner using the first inequality in (8.44) and the second inequality in (8.4) to obtain

$$\|V_2(t,s)^{-1}\varphi_2(t,x_\varphi(t,\xi))\| \le 2D_3 D_1 \|\xi\| e^{(\overline{T}_1+\alpha_1)(t-s)+(a'+b')|s|}$$

for every $t \ge s$, and thus, letting $t \to +\infty$ in (8.54) when $i = 2$, we obtain the second identity in (8.49).

We now assume that the identities in (8.49) hold for every $(s,\xi) \in \mathbb{R} \times E(s)$. Since

$$V_i(t,s)V_i(\tau,s)^{-1} = V_i(t,\tau) \quad \text{for } i = 1,2,$$

we obtain

$$V_1(t,s)\varphi_1(s,\xi) + \int_s^t V_1(t,\tau)f(\tau,x_\varphi(\tau,\xi),\varphi(\tau,x_\varphi(\tau,\xi)))\,d\tau$$
$$= \int_{-\infty}^t V_1(\tau,t)^{-1}f(\tau,x_\varphi(\tau,\xi),\varphi(\tau,x_\varphi(\tau,\xi)))\,d\tau \tag{8.55}$$

for each $t \le s$, and

$$V_2(t,s)\varphi_2(s,\xi) + \int_s^t V_2(t,\tau)f(\tau,x_\varphi(\tau,\xi),\varphi(\tau,x_\varphi(\tau,\xi)))\,d\tau$$
$$= -\int_t^{+\infty} V_2(\tau,t)^{-1}f(\tau,x_\varphi(\tau,\xi),\varphi(\tau,x_\varphi(\tau,\xi)))\,d\tau \tag{8.56}$$

for each $t \ge s$. We want to show that the right-hand sides of (8.55) and (8.56) are respectively $\varphi_1(t,x_\varphi(t,\xi))$ and $\varphi_2(t,x_\varphi(t,\xi))$. We first define a flow F_τ for each $\tau \in \mathbb{R}$ and $(s,\xi) \in \mathbb{R} \times E(s)$ by

$$F_\tau(s,\xi) = (s+\tau, x_\varphi(s+\tau,\xi)).$$

In view of (8.49), we have

$$\varphi_1(s,\xi) = \int_{-\infty}^s V_1(\tau,s)^{-1}f(F_{\tau-s}(s,\xi),\varphi(F_{\tau-s}(s,\xi)))\,d\tau,$$
$$\varphi_2(s,\xi) = -\int_s^{+\infty} V_2(\tau,s)^{-1}f(F_{\tau-s}(s,\xi),\varphi(F_{\tau-s}(s,\xi)))\,d\tau. \tag{8.57}$$

Furthermore,

$$F_{\tau-t}(t,x_\varphi(t,\xi)) = F_{\tau-t}(F_{t-s}(s,\xi)) = F_{\tau-s}(s,\xi) = (\tau,x_\varphi(\tau,\xi)),$$

and thus, by (8.57) with (s,ξ) replaced by $(t,x_\varphi(t,\xi))$,

$$\varphi_1(t, x_\varphi(t, \xi)) = \int_{-\infty}^t V_1(\tau, t)^{-1} f(F_{\tau - t}(t, x_\varphi(t, \xi)), \varphi(F_{\tau - t}(t, x_\varphi(t, \xi)))) \, d\tau$$

$$= \int_{-\infty}^t V_1(\tau, t)^{-1} f(\tau, x_\varphi(\tau, \xi), \varphi(\tau, x_\varphi(\tau, \xi))) \, d\tau,$$

$$\varphi_2(t, x_\varphi(t, \xi)) = -\int_t^{+\infty} V_2(\tau, t)^{-1} f(F_{\tau - t}(t, x_\varphi(t, \xi)), \varphi(F_{\tau - t}(t, x_\varphi(t, \xi)))) \, d\tau$$

$$= -\int_t^{+\infty} V_2(\tau, t)^{-1} f(\tau, x_\varphi(\tau, \xi), \varphi(\tau, x_\varphi(\tau, \xi))) \, d\tau$$

$$(8.58)$$

for every $t \in \mathbb{R}$. Combining (8.55)–(8.56) and (8.58), we conclude that (8.47) and (8.48) hold on the respective domains. This completes the proof of the lemma. $\qquad\square$

We also need to have some information on how the function x_φ varies with φ. Given φ, $\psi \in X$ and $(s, \xi) \in \mathbb{R} \times E(s)$, we denote by x_φ and x_ψ the continuous functions given by Lemma 8.12 such that $x_\varphi(s, \xi) = x_\psi(s, \xi) = \xi$.

Lemma 8.14. *Provided that δ is sufficiently small, for every φ, $\psi \in X$ and $(s, \xi) \in \mathbb{R} \times E(s)$ we have*

$$\|x_\varphi(t, \xi) - x_\psi(t, \xi)\| \leq \begin{cases} D_1 e^{\rho(t)} \|\xi\| \cdot \|\varphi - \psi\|, & t \geq s \\ D_2 e^{\sigma(t)} \|\xi\| \cdot \|\varphi - \psi\|, & t \leq s \end{cases}.$$

Proof. Take $\tau \geq s$. Proceeding in a similar manner to that in (8.45), we obtain

$$\|f(\tau, x_\varphi(\tau, \xi), \varphi(\tau, x_\varphi(\tau, \xi))) - f(\tau, x_\psi(\tau, \xi), \psi(\tau, x_\psi(\tau, \xi)))\|$$
$$\leq c_1 \delta e^{-\beta |\tau|} \|(x_\varphi(\tau, \xi) - x_\psi(\tau, \xi), \varphi(\tau, x_\varphi(\tau, \xi)) - \psi(\tau, x_\psi(\tau, \xi)))\|.$$

Furthermore,

$$\|\varphi(\tau, x_\varphi(\tau, \xi)) - \psi(\tau, x_\psi(\tau, \xi))\|$$
$$\leq \|\varphi(\tau, x_\varphi(\tau, \xi)) - \psi(\tau, x_\varphi(\tau, \xi))\| + \|\psi(\tau, x_\varphi(\tau, \xi)) - \psi(\tau, x_\psi(\tau, \xi))\|$$
$$\leq \|x_\varphi(\tau, \xi)\| \cdot \|\varphi - \psi\| + \|x_\varphi(\tau, \xi) - x_\psi(\tau, \xi)\|,$$

and hence,

$$\|f(\tau, x_\varphi(\tau, \xi), \varphi(\tau, x_\varphi(\tau, \xi))) - f(\tau, x_\psi(\tau, \xi), \psi(\tau, x_\psi(\tau, \xi)))\|$$
$$\leq c_1 \delta e^{-\beta |\tau|} (\|x_\varphi(\tau, \xi)\| \cdot \|\varphi - \psi\| + 2\|x_\varphi(\tau, \xi) - x_\psi(\tau, \xi)\|). \tag{8.59}$$

Set now $\bar{\rho}(t) = \|x_\varphi(t, \xi) - x_\psi(t, \xi)\|$. Using the first inequality in (8.4), the first inequality in (8.44) in Lemma 8.12, and (8.59), it follows from (8.23) and the definition of α_1 that

$$\bar{\rho}(t) \leq c_1 \delta \int_s^t \|U(t,\tau)\| \cdot \|x_\varphi(\tau,\xi)\| \cdot \|\varphi - \psi\| e^{-\beta|\tau|} \, d\tau$$

$$+ c_1 \delta \int_s^t \|U(t,\tau)\| \cdot 2\|x_\varphi(\tau,\xi) - x_\psi(\tau,\xi)\| e^{-\beta|\tau|} \, d\tau$$

$$\leq 2c_1 \delta D_1^2 \|\xi\| \cdot \|\varphi - \psi\| e^{a(t-s)+a'|s|} \int_s^t e^{\alpha_1(\tau-s)} \, d\tau$$

$$+ \frac{\alpha_1}{2} \int_s^t e^{a(t-\tau)} \bar{\rho}(\tau) \, d\tau$$

for each $t \geq s$, where we have used that $\beta \geq a'$. Therefore,

$$e^{a(s-t)} \bar{\rho}(t) \leq 2c_1 \delta D_1^2 \|\xi\| \cdot \|\varphi - \psi\| e^{a'|s|} \int_s^t e^{\alpha_1(\tau-s)} \, d\tau$$

$$+ \frac{\alpha_1}{2} \int_s^t e^{a(s-\tau)} \bar{\rho}(\tau) \, d\tau.$$

We now use the following version of Gronwall's lemma (see for example [29, page 37]): given continuous functions $u, v, w \colon [p,q] \to \mathbb{R}_0^+$ with v differentiable, if

$$u(t) \leq v(t) + \int_p^t w(\tau)u(\tau) \, d\tau$$

for every $t \in [p,q]$, then

$$u(t) \leq v(p) \exp\left(\int_p^t w(\tau)\,d\tau\right) + \int_p^t v'(\tau) \exp\left(\int_\tau^t w(r)\,dr\right) d\tau$$

for every $t \in [p,q]$. Applying this result to the function $u(t) = e^{a(s-t)}\bar{\rho}(t)$ with $p = s$ we readily obtain

$$\bar{\rho}(t) \leq 2c_1 \delta D_1^2 e^{a(t-s)+a'|s|} \int_s^t e^{\alpha_1(\tau-s)+(\alpha_1/2)(t-\tau)} \, d\tau \|\xi\| \cdot \|\varphi - \psi\|$$

$$\leq D_1 e^{(a+\alpha_1)(t-s)+a'|s|} \|\xi\| \cdot \|\varphi - \psi\|$$

for each $t \geq s$. This completes the proof of the lemma when $t \geq s$. The case when $t \leq s$ can be treated in an analogous manner, using the first inequality in (8.5) and the second inequality in (8.44). □

8.3.5 Construction of the center manifolds

We now use the former lemmas to establish the existence of a function $\varphi \in \mathfrak{X}$ satisfying the second equation in (8.23) when $x = x_\varphi$, via the equivalence in Lemma 8.13.

Lemma 8.15. *Provided that δ is sufficiently small, there exists a unique function $\varphi \in \mathfrak{X}$ such that (8.49) holds for every $(s,\xi) \in \mathbb{R} \times E(s)$.*

Proof. We look for a fixed point of the operator Φ defined for each $\varphi \in \mathcal{X}$ by

$$
(\Phi\varphi)(s,\xi) = \left(\int_{-\infty}^{s} V_1(\tau,s)^{-1} f(\tau, x_\varphi(\tau,\xi), \varphi(\tau, x_\varphi(\tau,\xi)))\, d\tau, \right.
$$
$$
\left. - \int_{s}^{+\infty} V_2(\tau,s)^{-1} f(\tau, x_\varphi(\tau,\xi), \varphi(\tau, x_\varphi(\tau,\xi)))\, d\tau \right)
\tag{8.60}
$$

for $(s,\xi) \in \mathbb{R} \times E(s)$, where x_φ is the unique function given by Lemma 8.12 such that $x_\varphi(s,\xi) = \xi$. In view of Proposition 8.4, it is sufficient to prove that Φ is a contraction with the norm in (8.24).

Proceeding in a similar manner to that in the proof of Lemma 6.15 (with the help of Lemma 8.8), we can show that the continuous function $\Phi\varphi$ is of class C^k in ξ for each $\varphi \in \mathcal{X}$. Since $x_\varphi(t,0) = 0$ for every $\varphi \in \mathcal{X}$ and $t \in \mathbb{R}$ (see (8.44)), it follows from (8.60) that $(\Phi\varphi)(s,0) = 0$ for every $s \in \mathbb{R}$. Furthermore, also by (8.60),

$$
\partial(\Phi\varphi)(s,0) = \left(\int_{-\infty}^{s} V_1(\tau,s)^{-1} \partial f(\tau,0) \partial a_\varphi(\tau,0)\, d\tau, \right.
$$
$$
\left. - \int_{s}^{+\infty} V_2(\tau,s)^{-1} \partial f(\tau,0) \partial a_\varphi(\tau,0)\, d\tau \right),
$$

where

$$
a_\varphi(\tau,\xi) = (x_\varphi(\tau,\xi), \varphi(\tau, x_\varphi(\tau,\xi))).
$$

Since $\partial f(\tau,0) = 0$ we have $\partial(\Phi\varphi)(s,0) = 0$ for every $s \in \mathbb{R}$.

Using the second inequalities in (8.5) and in (8.4), together with Lemma 8.8 and the definition of β in (8.6), we conclude that

$$
\|\partial^j(\Phi\varphi)(s,\xi)\| \le \int_{-\infty}^{s} \|V_1(\tau,s)^{-1}\| \cdot \|\partial^j f^*(\tau,\xi)\|\, d\tau
$$
$$
+ \int_{s}^{+\infty} \|V_2(\tau,s)^{-1}\| \cdot \|\partial^j f^*(\tau,\xi)\|\, d\tau
$$
$$
\le \delta \bar{B}_j D_4 \int_{-\infty}^{s} e^{(j(c+\alpha_2)-d)(s-\tau)-(\beta-d')|\tau|+jc'|s|}\, d\tau
$$
$$
+ \delta \bar{A}_j D_3 \int_{s}^{+\infty} e^{(j(a+\alpha_1)-b)(\tau-s)-(\beta-b')|\tau|+ja'|s|}\, d\tau.
$$

Since $c'|s| \le c'(s-\tau) + c'|\tau|$ for $\tau \le s$, and $a'|s| \le a'(\tau-s) + a'|\tau|$ for $\tau \ge s$, together with the fact that $a \ge 0$ and $c \ge 0$, we obtain

$$
\|\partial^j(\Phi\varphi)(s,\xi)\| \le \delta \bar{B}_j D_4 \int_{-\infty}^{s} e^{(T_2+j\alpha_2)(s-\tau)}\, d\tau
$$
$$
+ \delta \bar{A}_j D_3 \int_{s}^{+\infty} e^{(T_1+j\alpha_1)(\tau-s)}\, d\tau
\tag{8.61}
$$

with T_1, $T_2 < 0$ as in (8.17) and where we have used again (8.6). Choosing δ sufficiently small we can make α_1 and α_2 sufficiently small so that

$$T_1 + k\alpha_1 < 0 \quad \text{and} \quad T_2 + k\alpha_2 < 0,$$

and, for $j = 1, \ldots, k$,

$$\delta \left(\frac{\bar{B}_j D_4}{|T_2 + j\alpha_2|} + \frac{\bar{A}_j D_3}{|T_1 + j\alpha_1|} \right) < 1. \tag{8.62}$$

With these choices, we have $\|\partial^j(\Phi\varphi)(s,\xi)\| \le 1$ for every $s \in \mathbb{R}$ and $\xi \in E(s)$. Set

$$b_m^k(\tau) := \|\partial^k f^*(\tau, \xi) - \partial^k f^*(\tau, \bar{\xi})\|.$$

Using again the second inequalities in (8.5) and in (8.4), together with Lemma 8.11 and the definition of β, we conclude that

$$
\begin{aligned}
&\|\partial^k(\Phi\varphi)(s,\xi) - \partial^k(\Phi\varphi)(s,\bar{\xi})\| \\
&\le \int_{-\infty}^s \|V_1(\tau,s)^{-1}\| b_1^k(\tau)\, d\tau + \int_s^{+\infty} \|V_2(\tau,s)^{-1}\| b_2^k(\tau)\, d\tau \\
&\le \delta \widehat{B}_k D_4 \|\xi - \bar{\xi}\| \int_{-\infty}^s e^{((k+1)(c+\alpha_2)-d)(s-\tau)-(\beta-d')|\tau|+(k+1)c'|s|}\, d\tau \\
&\quad + \delta \widehat{A}_k D_3 \|\xi - \bar{\xi}\| \int_s^{+\infty} e^{((k+1)(a+\alpha_1)-b)(\tau-s)-(\beta-b')|\tau|+(k+1)a'|s|}\, d\tau \\
&\le \delta \widehat{B}_k D_4 \|\xi - \bar{\xi}\| \int_{-\infty}^s e^{(T_2+(k+1)\alpha_2)(s-\tau)}\, d\tau \\
&\quad + \delta \widehat{A}_k D_3 \|\xi - \bar{\xi}\| \int_s^{+\infty} e^{(T_1+(k+1)\alpha_1)(\tau-s)}\, d\tau.
\end{aligned}
\tag{8.63}
$$

Eventually choosing again δ sufficiently small we can make α_1 and α_2 sufficiently small so that

$$T_1 + (k+1)\alpha_1 < 0 \quad \text{and} \quad T_2 + (k+1)\alpha_2 < 0, \tag{8.64}$$

and

$$\delta \left(\frac{\widehat{B}_k D_4}{|T_2 + (k+1)\alpha_2|} + \frac{\widehat{A}_k D_3}{|T_1 + (k+1)\alpha_1|} \right) < 1. \tag{8.65}$$

With these choices, for every $s \in \mathbb{R}$ and $\xi \in E(s)$ we have

$$\|\partial^k(\Phi\varphi)(s,\xi) - \partial^k(\Phi\varphi)(s,\bar{\xi})\| \le 1.$$

This shows that $\Phi(\mathcal{X}) \subset \mathcal{X}$, and hence, $\Phi \colon \mathcal{X} \to \mathcal{X}$ is well-defined.

We now show that $\Phi \colon \mathcal{X} \to \mathcal{X}$ is a contraction with the norm in (8.24). Given $\varphi, \psi \in \mathcal{X}$, and $(s,\xi) \in \mathbb{R} \times E(s)$, let x_φ and x_ψ be the unique continuous functions given by Lemma 8.12 such that $x_\varphi(s, \xi) = x_\psi(s, \xi) = \xi$. Using (8.9) and Lemmas 8.12 and 8.14 we obtain

$$
\begin{aligned}
b_j(\tau) :&= \|h_j(\tau, x_\varphi(\tau, \xi), \varphi(\tau, x_\varphi(\tau, \xi))) - h_j(\tau, x_\psi(\tau, \xi), \psi(\tau, x_\psi(\tau, \xi)))\| \\
&\le c_1 \delta e^{-\beta|\tau|} \|(x_\varphi(\tau, \xi) - x_\psi(\tau, \xi), \varphi(\tau, x_\varphi(\tau, \xi)) - \psi(\tau, x_\psi(\tau, \xi)))\| \\
&\le c_1 \delta e^{-\beta|\tau|}(\|x_\varphi(\tau, \xi)\| \cdot \|\varphi - \psi\| + 2\|x_\varphi(\tau, \xi) - x_\psi(\tau, \xi)\|) \\
&\le 4c_1 \delta e^{-\beta|\tau|} \|\xi\| \cdot \|\varphi - \psi\|
\begin{cases}
D_1 e^{\rho(\tau)}, & \tau \ge s \\
D_2 e^{\sigma(\tau)}, & \tau \le s
\end{cases}
\end{aligned}
$$

for $j = 1, 2$. Using the second inequalities in (8.5) and in (8.4), together with the definition of β in (8.6), we conclude that

$$
\begin{aligned}
&\|(\Phi\varphi)(s, \xi) - (\Phi\psi)(s, \xi)\| \\
&\le \int_{-\infty}^{s} \|V_1(\tau, s)^{-1}\| b_1(\tau) \, d\tau + \int_{s}^{+\infty} \|V_2(\tau, s)^{-1}\| b_2(\tau) \, d\tau \\
&\le 4c_1 \delta \|\xi\| \cdot \|\varphi - \psi\| D_4 D_2 \int_{-\infty}^{s} e^{(c+\alpha_2-d)(s-\tau)-(\beta-d')|\tau|+c'|s|} \, d\tau \\
&\quad + 4c_1 \delta \|\xi\| \cdot \|\varphi - \psi\| D_3 D_1 \int_{s}^{+\infty} e^{(a+\alpha_1-b)(\tau-s)-(\beta-b')|\tau|+a'|s|} \, d\tau.
\end{aligned}
$$

Since $c'|s| \le c'(s - \tau) + c'|\tau|$ for $\tau \le s$, and $a'|s| \le a'(\tau - s) + a'|\tau|$ for $\tau \ge s$, we obtain

$$
\begin{aligned}
&\|(\Phi\varphi)(s, \xi) - (\Phi\psi)(s, \xi)\| \\
&\le 4c_1 \delta \|\xi\| \cdot \|\varphi - \psi\| D_4 D_2 \int_{-\infty}^{s} e^{(T_2+\alpha_2)(s-\tau)} \, d\tau \\
&\quad + 4c_1 \delta \|\xi\| \cdot \|\varphi - \psi\| D_3 D_1 \int_{s}^{+\infty} e^{(T_1+\alpha_1)(\tau-s)} \, d\tau \\
&\le 4c_1 \delta \left(\frac{D_2 D_4}{|T_2 + \alpha_2|} + \frac{D_1 D_3}{|T_1 + \alpha_1|} \right) \|\xi\| \cdot \|\varphi - \psi\|,
\end{aligned}
$$

where we have used again (8.6) (recall that $T_1 + \alpha_1 < 0$ and $T_2 + \alpha_2 < 0$). Furthermore, eventually choosing again $\delta > 0$ sufficiently small we also have

$$
\theta = 4c_1 \delta \left(\frac{D_2 D_4}{|T_2 + \alpha_2|} + \frac{D_1 D_3}{|T_1 + \alpha_1|} \right) < 1. \tag{8.66}
$$

Therefore

$$
\|\Phi\varphi_1 - \Phi\varphi_2\| \le \theta \|\varphi_1 - \varphi_2\|,
$$

and $\Phi: \mathcal{X} \to \mathcal{X}$ is a contraction in the complete metric space $\mathcal{X}$ (see Proposition 8.4). Hence, there exists a unique function $\varphi \in \mathcal{X}$ satisfying $\Phi\varphi = \varphi$. This completes the proof of the lemma.   □

We can now establish Theorem 8.2.

Proof of Theorem 8.2. As explained in Section 8.3.1, in view of the required invariance property in (8.18), to show the existence of a center manifold $\mathcal{V}$ is equivalent to find a function $\varphi \in \mathcal{X}$ satisfying (8.23). It follows from Lemma 8.12 that for each fixed $\varphi \in \mathcal{X}$ there exists a unique continuous function $x = x_\varphi$ satisfying the first equation in (8.23). Furthermore, this function is of class C^k in ξ. Thus, it is sufficient to solve the second equation in (8.23) setting $x = x_\varphi$ or, equivalently, to solve (8.47)–(8.48) in Lemma 8.13. This lemma indicates that to solve (8.47)–(8.48) is in its turn equivalent to solve the equations in (8.49), that is, to find $\varphi \in \mathcal{X}$ such that the identities in (8.49) hold for every $(s, \xi) \in \mathbb{R} \times E(s)$. Finally, Lemma 8.15 shows that there exists a unique function $\varphi \in \mathcal{X}$ such that (8.49) holds for every $(s, \xi) \in \mathbb{R} \times E(s)$.

We now define a map $K \colon \mathbb{R} \times E(s) \to \mathbb{R} \times X$ by

$$K(t, \xi) = \Psi_t(0, \xi, \varphi(0, \xi)). \tag{8.67}$$

Since $\varphi \in \mathcal{X}$, the map $\xi \mapsto \varphi(0, \xi)$ is of class C^k. Furthermore, by the assumptions G1 and G2 the map

$$(t, s, \xi, \eta) \mapsto \Psi_t(s, \xi, \eta)$$

is of class C^k (see for example [46]). Therefore, K is also of class C^k. In addition, the map K is injective: if $K(t, \xi) = K(t', \xi')$ then the first component of K gives $t = t'$; applying Ψ_{-t} to both sides of the identity $K(t, \xi) = K(t, \xi')$ yields $\zeta - \zeta'$. This shows that K is a parametrization of class C^k on $\mathbb{R} \times E(s)$ of the set $\mathcal{V}$. Therefore, $\mathcal{V}$ is a smooth manifold of class C^k.

It remains to establish the additional properties in the theorem. The first two properties are an immediate consequence of the above discussion and of Lemma 8.13. To prove the third property, we denote by x_φ the unique function given by Lemma 8.12 such that $x_\varphi(s, \xi) = \xi$. With the notation in (8.33) we have

$$
\begin{aligned}
&\|\partial_\xi^j(\Psi_\tau(s, \xi, \varphi(s, \xi))) - \partial_\xi^j(\Psi_\tau(s, \bar{\xi}, \varphi(s, \bar{\xi})))\| \\
&= \|\partial_\xi^j(t, x(t, \xi), \varphi^*(t, \xi)) - \partial_\xi^j(t, x(t, \bar{\xi}), \varphi^*(t, \bar{\xi}))\| \\
&\le \|\partial^j x(t, \xi) - \partial^j x(t, \bar{\xi})\| + \|\partial^j \varphi^*(t, \xi) - \partial^j \varphi^*(t, \bar{\xi})\|
\end{aligned}
\tag{8.68}
$$

for every $\tau \in \mathbb{R}$ and $t = s + \tau$. Note that by Lemma 8.10, we have

$$
\|\partial^j \varphi^*(t, \xi) - \partial^j \varphi^*(t, \bar{\xi})\| \le \|\xi - \bar{\xi}\| \begin{cases} \widetilde{A}_j e^{(j+1)\rho(t)}, & t \ge s \\ \widetilde{B}_j e^{(j+1)\sigma(t)}, & t \le s \end{cases}
\tag{8.69}
$$

for $j = 0, \ldots, k$. The desired inequalities in (8.19)–(8.20) follow readily from Lemma 8.9, (8.69), and (8.68). This completes the proof of the theorem.   □

It follows from the proof of Theorem 8.2 that δ can be any positive number satisfying (8.62), (8.64), (8.65), and (8.66) in the proof of Lemma 8.15. We note that the extra step in the proof of Theorem 8.2 involving the map K

in (8.67) has only the purpose of showing that V is also of class C^k in the time direction (we observe that the space X where we look for the function φ only requires the C^k differentiability in the second component). An alternative proof could be obtained observing that by (8.60) in the proof of Lemma 8.15 the function $\varphi \in X$ constructed satisfies the identity

$$
\begin{aligned}
\varphi(s, \xi) = \bigg(& \int_{-\infty}^{s} V_1(\tau, s)^{-1} f(\tau, x_\varphi(\tau, \xi), \varphi(\tau, x_\varphi(\tau, \xi))) \, d\tau, \\
& - \int_{s}^{+\infty} V_2(\tau, s)^{-1} f(\tau, x_\varphi(\tau, \xi), \varphi(\tau, x_\varphi(\tau, \xi))) \, d\tau \bigg)
\end{aligned}
\tag{8.70}
$$

for every $(s, \xi) \in \mathbb{R} \times E(s)$. The desired C^k differentiability of φ can then be obtained directly from (8.70).

When $a < 0$ or $c < 0$ (see (8.2)) we have the following generalization of Theorem 8.2.

Theorem 8.16. *Assume that G1–G2 hold. If the equation $v' = A(t)v$ in the Banach space X satisfies (8.4)–(8.5), and for $j = 1, \ldots, k+1$ we have*

$$
ja - b + \max\{ja', b'\} < 0 \quad and \quad jc - d + \max\{jc', d'\} < 0,
\tag{8.71}
$$

then the statement in Theorem 8.2 holds.

Proof. The only places where we use the conditions $a \geq 0$ and $c \geq 0$ are in the inequalities (8.52), (8.53), (8.61), and (8.63). Appropriate generalizations of these inequalities can be obtained for arbitrary a and c when we replace the condition (8.17) by the condition (8.71), and thus one can verify in a straight-forward manner that the desired statement is an immediate consequence of the proof of Theorem 8.2. $\qquad \square$

Reversibility and equivariance in center manifolds

We show in this chapter that for a *nonautonomous* differential equation (in the presence of a *nonuniform* exponential trichotomy; see Chapter 8), the (time) reversibility and equivariance of the associated semiflow descends respectively to the reversibility and equivariance in any center manifold. We note that time-reversal symmetries are among the fundamental symmetries in many "physical" systems, both in classical and quantum mechanics. This is due to the fact that many Hamiltonian systems are reversible (see [53] for many examples). In spite of the crucial differences between reversible and equivariant dynamical systems, the techniques that are useful in any of the two contexts usually carry over to the other one. This will be apparent along the exposition. We follow closely [14].

9.1 Reversibility for nonautonomous equations

9.1.1 The notion of reversibility

We introduce here the notion of reversible (nonautonomous) differential equation. Let X be a Banach space. Consider a continuous function $L\colon \mathbb{R} \times X \to X$ such that $v' = L(t, v)$ has unique and global solutions, and a (Fréchet) differentiable map $S\colon \mathbb{R} \times X \to X$.

Definition 9.1. *We say that the equation $v' = L(t, v)$ is reversible with respect to S if*

$$L(-t, S(t, v)) + \frac{\partial S}{\partial v}(t, v)L(t, v) = -\frac{\partial S}{\partial t}(t, v) \tag{9.1}$$

for every $t \in \mathbb{R}$ and $v \in X$. We also say that the equation $v' = L(t, v)$ is reversible *if it is reversible with respect to some map S.*

We now present a characterization of reversibility, in terms of the solutions of $v' = L(t, v)$. This characterization can in fact be seen as the main justification for the above notion of reversible nonautonomous equation. For each $s \in \mathbb{R}$ and $v_s \in X$ we denote by $\Phi(t, s)(v_s)$ the unique solution of $v' = L(t, v)$ with $v(s) = v_s$. We recall that by hypothesis $\Phi(t, s)$ is defined for all $t, s \in \mathbb{R}$.

Proposition 9.2. *The equation $v' = L(t, v)$ is reversible with respect to the map S if and only if*

$$\Phi(\tau, -t)(S(t, v)) = S(-\tau, \Phi(-\tau, t)(v)) \quad \text{for any } t, \tau \in \mathbb{R} \text{ and } v \in X. \quad (9.2)$$

Proof. We first assume that (9.2) holds. We have

$$\frac{\partial(\Phi(\tau, \pm t)(v))}{\partial \tau} = L(\tau, \Phi(\tau, \pm t)(v)) \quad (9.3)$$

for every $v \in X$. By (9.3), taking derivatives with respect to τ in (9.2),

$$L(\tau, \Phi(\tau, t)(S(t, v))) = -\frac{\partial S}{\partial t}(-\tau, \Phi(-\tau, t)(v))$$
$$- \frac{\partial S}{\partial v}(-\tau, \Phi(-\tau, t)(v))L(-\tau, \Phi(-\tau, t)(v)).$$

Using again (9.2) and setting $w = \Phi(-\tau, t)(v)$ we obtain

$$L(\tau, S(-\tau, w)) = -\frac{\partial S}{\partial t}(\tau, w) - \frac{\partial S}{\partial v}(-\tau, w)L(-\tau, w),$$

thus yielding (9.1).

We now assume that (9.1) holds. Given $t \in \mathbb{R}$ and $v \in X$ we set $z(t) = S(-t, \Phi(-t, s)(v))$. Taking derivatives with respect to t and using (9.3) we obtain

$$z'(t) = -\frac{\partial S}{\partial t}S(-t, \Phi(-t, -s)(v)) - \frac{\partial S}{\partial v}(-t, \Phi(-t, s)(v))L(-t, \Phi(-t, s)(v)).$$

Using now (9.1),

$$z'(t) = L(t, S(-t, \Phi(-t, s)(v))) + \frac{\partial S}{\partial v}(-t, \Phi(-t, s)(v))L(-t, \Phi(-t, s)(v))$$
$$+ \frac{\partial S}{\partial v}(-t, \Phi(-t, s)(v))L(-t, \Phi(-t, s)(v))$$
$$= L(t, S(-t, \Phi(-t, s)(v))) = L(t, z(t)).$$

Thus, $z(t)$ satisfies the initial value problem $z' = L(t, z)$, $z(-s) = S(s, v)$. Since $\Phi(t, -s)(S(s, v))$ is also a solution of this problem, the uniqueness shows that $z(t) = \Phi(t, -s)(S(s, v))$ for any $v \in X$ and $t, s \in \mathbb{R}$, and (9.2) holds. $\square$

In the particular case when the operators $S_t = S(t, \cdot)\colon X \to X$ are linear the following statement establishes several basic properties of reversible systems.

Proposition 9.3. *Assume that L is Fréchet differentiable in v. If the equation $v' = L(t, v)$ is reversible with respect to the map S, and S_t is linear for each $t \in \mathbb{R}$, then the following properties hold:*

1. *the linear equation $v' = A(t)v$ is reversible with respect to S;*
2. *if $S_0^2 = \mathrm{Id}$, then $S_{-t} \circ S_t = \mathrm{Id}$ for every $t \in \mathbb{R}$.*

Proof. Taking derivatives in (9.1) with respect to v, and using the linearity of S_t we obtain

$$\frac{\partial L}{\partial v}(-t, S_t v)S_t + S_t \frac{\partial L}{\partial v}(t, v) = -\frac{\partial S}{\partial t}(t, \cdot).$$

Setting $v = 0$ yields

$$A(-t)S_t + S_t A(t) = -\frac{\partial S}{\partial t}(t, \cdot), \tag{9.4}$$

and this establishes the first property (recall that S_t is linear).

Assume now that $S_0^2 = \mathrm{Id}$. Let $v(t)$ be a solution of the equation $v' = L(t, v)$, that is, $v'(t) = L(t, v(t))$. Using (9.1) we obtain

$$\begin{aligned}
\frac{d((S_{-t} \circ S_t)v(t))}{dt} &= -\frac{\partial S}{\partial t}(-t, S_t v(t)) + S_{-t}\frac{\partial S}{\partial t}(t, v(t)) + (S_{-t} \circ S_t)v'(t) \\
&= L(t, S_{-t}(S_t v(t))) + S_{-t}L(-t, S_t v(t)) \\
&\quad - S_{-t}[L(-t, S_t v(t)) + S_t L(t, v(t))] + (S_{-t} \circ S_t)L(t, v(t)) \\
&= L(t, (S_{-t} \circ S_t)v(t)),
\end{aligned}$$

and $(S_{-t} \circ S_t)v(l)$ is also a solution of $v' = L(t, v)$. Since $S_0^2 = \mathrm{Id}$, we have $(S_{-t} \circ S_t)v(t)|_{t=0} = v(0)$, and by the uniqueness of solutions we conclude that $(S_{-t} \circ S_t)v(t) = v(t)$ for every $t \in \mathbb{R}$. Therefore, $S_{-t} \circ S_t = \mathrm{Id}$ for every $t \in \mathbb{R}$. $\qquad\square$

9.1.2 Relation with the autonomous case

We show here that the notion of reversibility in Section 9.1.1 is a natural extension of the notion of reversibility in the autonomous setting.

Let $L\colon X \to X$ be a continuous function in the Banach space X, such that the equation $v' = L(v)$ generates a flow $(\varphi_t)_{t \in \mathbb{R}}$ in X. We say that $v' = L(v)$ is *reversible* with respect to a map $T\colon X \to X$ if

$$L \circ T = -T' \circ L. \tag{9.5}$$

We note that if the function $L(t, v)$ in Definition 9.1 does not depend on t, then taking $S_t = T$ for all $t \in \mathbb{R}$ the identity (9.1) becomes (9.5).

As in the general nonautonomous case there is a characterization of reversibility of autonomous equations in terms of the solutions of $v' = L(v)$. The following statement is an immediate consequence of Proposition 9.2, and the fact that in the autonomous case $\Phi(t, \tau) = \varphi_{t-\tau}$ for every $t, \tau \in \mathbb{R}$.

Proposition 9.4. *The equation* $v' = L(v)$ *is reversible with respect to the map* T *if and only if* $\varphi_t \circ T = T \circ \varphi_{-t}$ *for every* $t \in \mathbb{R}$.

The following result shows that the notion of reversibility in Section 9.1.1 is a natural extension of the notion of reversibility in the autonomous setting.

Proposition 9.5. *The equation* $v' = L(t, v)$ *is reversible with respect to the map* $S: \mathbb{R} \times X \to X$ *if and only if the autonomous equation*

$$t' = 1, \quad v' = L(t, v) \tag{9.6}$$

is reversible with respect to the map $T: \mathbb{R} \times X \to \mathbb{R} \times X$ *defined by*

$$T(t, v) = (-t, S_t(v)). \tag{9.7}$$

Proof. Since the equation $v' = L(t, v)$ has unique and global solutions, the autonomous equation in (9.6) defines a flow φ_τ on $\mathbb{R} \times X$, given by

$$\varphi_\tau(t, v) = (t + \tau, \Phi(t + \tau, t)(v)),$$

with Φ as in Proposition 9.2. By Proposition 9.4, the equation (9.6) is reversible with respect to the map T if and only if

$$\varphi_r \circ T = T \circ \varphi_{-r}, \quad r \in \mathbb{R}. \tag{9.8}$$

For every $(t, v) \in \mathbb{R} \times X$, we have

$$(\varphi_r \circ T)(t, v) = \varphi_r(-t, S_t(v)) = (r - t, \Phi(r - t, -t)(S_t(v))),$$

and

$$(T \circ \varphi_{-r})(t, v) = T(t - r, \Phi(t - r, t)(v)) = (r - t, S_{t-r}(\Phi(t - r, t)(v))).$$

Comparing the two identities we conclude that (9.8) holds if and only if (9.2) holds (setting $r - t = \tau$), that is, if and only if the equation $v' = L(t, v)$ is reversible with respect to the map S. This completes the proof. □

As a consequence of Proposition 9.5, the reversibility of a nonautonomous equation can always be reduced to that of an autonomous equation. Notice that $T^2(t, v) = (t, (S_{-t} \circ S_t)(v))$. In view of Proposition 9.3, if L is Fréchet differentiable in v, S_t linear for each $t \in \mathbb{R}$, and $S_0^2 = \mathrm{Id}$, then $T^2 = \mathrm{Id}$, that is, T is an involution.

9.1.3 Nonautonomous reversible equations

We provide here a nontrivial example of a nonautonomous reversible equation, to which we can additionally apply our results in Section 9.2 concerning center manifolds and their reversibility.

Example 9.6. Consider linear transformations

$$S_t = \begin{pmatrix} -1 & 0 & 0 \\ 0 & 0 & b(t) \\ 0 & b(-t)^{-1} & 0 \end{pmatrix}$$

for each $t \in \mathbb{R}$, where

$$b(t) = e^{\varepsilon(\sin t - t\cos t - \cos t - t\sin t)}.$$

Note that $S_0^2 = \mathrm{Id}$ and that $S_t \circ S_{-t} = \mathrm{Id}$ for each $t \in \mathbb{R}$. We define a map $S \colon \mathbb{R} \times \mathbb{R}^3 \to \mathbb{R}^3$ by $S(t, v) = S_t v$. We also consider the map $L \colon \mathbb{R} \times \mathbb{R}^3 \to \mathbb{R}^3$ given by

$$L(t, v) = A(t)v + f(t, v), \tag{9.9}$$

where

$$A(t) = \begin{pmatrix} 0 & 0 & 0 \\ 0 & -\omega - \varepsilon t \sin t & 0 \\ 0 & 0 & \omega + \varepsilon t \cos t \end{pmatrix}, \quad \omega > \varepsilon > 0, \tag{9.10}$$

and, setting $v = (x, y, z)$ and $\beta(t) = e^{-2\varepsilon(\cos t + t \sin t)}$,

$$f(t, v) = \begin{cases} \delta t^{k+1} e^{-12\varepsilon t} \alpha(\|v\|^2)\left(x^2, xy + \beta(t)xz, xy + xz\right), & t \geq 0, \\ -S_{-t} f(-t, S_t v), & t < 0, \end{cases} \tag{9.11}$$

for some C^k function $\alpha \colon \mathbb{R} \to \mathbb{R}$ with $\alpha = 0$ outside $[-1, 1]$.

We want to show that the equation $v' = L(t, v)$ with L as in (9.9) is reversible with respect to S. Since each S_t is linear, the identity (9.1) is equivalent to

$$A(-t)S_t v + f(-t, S_t v) + S_t A(t)v + S_t f(t, v) = -\frac{\partial S}{\partial t}(t, v).$$

Thus, it is sufficient to establish that

$$A(-t)S_t + S_t A(t) = -\frac{\partial S}{\partial t}(t, \cdot). \tag{9.12}$$

and

$$f(-t, S_t v) = -S_t f(t, v). \tag{9.13}$$

The identity (9.12) is equivalent to

$$-(\omega + \varepsilon t \sin t)b(t) + b(t)(\omega + \varepsilon t \cos t) = -b'(t),$$

$$\omega - \varepsilon t \cos t - (\omega + \varepsilon t \sin t) = -\frac{b'(-t)}{b(-t)},$$

and both identities can be verified in a straightforward manner.

We now show that (9.13) holds. By construction, for each $t < 0$ we have

$$S_t f(t, v) = -S_t S_{-t} f(-t, S_t v) = -f(-t, S_t v). \tag{9.14}$$

When $t > 0$, it follows from (9.14) (replacing v by $S_{-t}v$, and then t by $-t$) that $S_{-t} f(-t, S_t v) = -f(t, v)$, and thus,

$$S_t f(t, v) = -S_t S_{-t} f(-t, S_t v) = -f(-t, S_t v).$$

Furthermore, $f(0, v) = 0$ and

$$-S_{-t} f(-t, S_t v)\big|_{t=0^-} = -S_0 f(0, S_0 v) = 0.$$

This establishes (9.13), and thus also (9.1). In particular, f is continuous. To verify that it is of class C^k we note that whenever $j = a + b \le k$ we have

$$\frac{\partial f^j}{\partial v^a \partial t^b}(t, v)\big|_{t=0^+} = 0 \quad \text{and} \quad \frac{\partial^j}{\partial v^a \partial t^b}\big[-S_{-t} f(-t, S_t v)\big]\big|_{t=0^-} = 0.$$

We shall see in Section 9.2 that the equation $v' = L(t, v)$ with L as in (9.9) satisfies the hypotheses of Theorems 8.2 and 9.7. In particular, this allows us to conclude that the equation $v' = L(t, v)$ has a reversible global center manifold of class C^k provided that $\omega > (k + 2)\varepsilon$.

9.2 Reversibility in center manifolds

9.2.1 Formulation of the main result

We formulate here our main result about the reversibility in center manifolds, showing that the reversibility of a given equation, with respect to a map S with S_t linear for each $t \in \mathbb{R}$, always descends to the center manifold. We assume in this section that for some constants $C > 0$ and $\theta \ge 0$ we have

$$\frac{1}{C} e^{-\theta|t|} \le \|S_t^{-1}\|^{-1} \le \|S_t\| \le C e^{\theta|t|}, \quad t \in \mathbb{R}. \tag{9.15}$$

We continue to write

$$\mathcal{V}_s = \{v \in X : (s, v) \in \mathcal{V}\} = \{(\xi, \varphi(s, \xi)) : \xi \in E(s)\}, \tag{9.16}$$

where φ is the function given by Theorem 8.2.

Theorem 9.7. *Under the assumptions of Theorem 8.2, if the equation $v' = A(t)v + f(t, v)$ is reversible with respect to a map S with $S_0^2 = \text{Id}$ and S_t linear for each $t \in \mathbb{R}$, and the constants in (8.2)–(8.3) and (9.15) satisfy*

$$\max\{c, a\} + 2(\gamma + \theta) < \min\{b, d\}, \quad \text{with} \quad \gamma = \max\{a', b', c', d'\}, \quad (9.17)$$

then $S_s(\mathcal{V}_s) = \mathcal{V}_{-s}$ for every $s \in \mathbb{R}$.

The statement in the theorem is equivalent to $T(\mathcal{V}) = \mathcal{V}$ with T as in (9.7).

It follows easily from (9.8) that T takes invariant sets into invariant sets, with respect to the flow defined by

$$t' = 1, \quad v' = A(t)v + f(t, v). \quad (9.18)$$

As explained at the end of Section 9.1.2, when $S_0^2 = \text{Id}$ it follows from Proposition 9.3 that T is an involution. In this case, provided that the map S is of class C^1, the map T takes invariant C^1 manifolds into invariant C^1 manifolds (again with respect to the flow defined by (9.18)). Furthermore, it follows easily from Lemma 9.10 that T takes any invariant C^1 manifold $\mathcal{W}$ satisfying

$$\mathbb{R} \times \{0\} \subset \mathcal{W} \text{ and } T_{(s,0)}\mathcal{W} = \mathbb{R} \times E(s) \text{ for every } s \in \mathbb{R} \quad (9.19)$$

to an invariant manifold with the same property. Note that by Theorem 8.2, the center manifold $\mathcal{W} = \mathcal{V}$ satisfies (9.19). However, *it does not follow from Theorem 8.2 that $\mathcal{V}$ is the unique invariant C^1 manifold with this property*, but only that it is unique among those which are graphs of functions in $\mathcal{X}$. The reason is that in view of the possible exponentials in (9.15) (when $\theta > 0$, as in Example 9.6), the images $S_s(\mathcal{V}_s)$ of the graphs $\mathcal{V}_s$ need not be graphs of functions in the same space $\mathcal{X}$ (although it is easy to verify that $S_s(\mathcal{V}_s)$ is always a graph no matter whether $\theta = 0$ or $\theta > 0$). This prevents us from obtaining the invariance property $T(\mathcal{V}) = \mathcal{V}$ in Theorem 9.7 by using the T-invariance of the class of invariant C^1 manifolds $\mathcal{W}$ satisfying (9.19). Instead we give a proof of Theorem 9.7 based on the "explicit" form in (8.15) for the manifold $\mathcal{V}$ (which thus requires explicitly the setup from Chapter 8. We stress that in the present *nonuniform* context (with nonuniform exponential trichotomies and with $\theta > 0$ in (9.15)) we are not aware of any alternative argument to the proof of Theorem 9.7 given below.

We also would like to emphasize that the difficulty described above (of not being able to deduce that any invariant manifold $\mathcal{W}$ satisfying (9.19) is the center manifold in Theorem 8.2) is unrelated to the fact that we consider *global* center manifolds. Indeed, even if we were considering local center manifolds, since we may have exponentials in (9.15) when $\theta > 0$ (or even if we can make θ arbitrarily close to zero but we cannot make it zero), the images $S_s(\widetilde{\mathcal{V}}_s)$ of local center manifolds $\widetilde{\mathcal{V}}_s$ are not necessarily graphs of functions in the same initial space. Thus, the difficulty is exactly the same with local and global center manifolds (and in particular it is unrelated to the possible nonuniqueness of local center manifolds).

Concerning the hypotheses of Theorem 9.7, the requirement that the maps S_t are linear is presumably technical and has to do with the method of proof of Theorem 8.2 which takes care separately of the central and stable–unstable parts (in a similar manner to that in Lemmas 9.12 and 9.13). We believe that it should be possible to obtain a version of Theorem 9.7 when the maps S_t are not necessarily linear, although we should require a different approach to the proof of Theorem 8.2 (we note that in the nonuniform setting Theorem 8.2 is the only result in the literature establishing a *smooth* center manifold theorem for continuous time, and thus we are not aware of any alternative approach that can be used successfully). Condition (9.15) has the sole purpose to maintain sufficiently separated the spectrum of the nonuniform exponential trichotomy when we apply the linear maps S_t. The separation is given by condition (9.17).

We now show that the equation $v' = L(t, v)$ with L as in (9.9) satisfies the hypotheses of Theorems 8.2 and 9.7.

Example 9.8. We consider again the map $L \colon \mathbb{R} \times \mathbb{R}^3 \to \mathbb{R}^3$ in Example 9.6 (see (9.9)–(9.11)). We want to show that we are in the hypotheses of Theorem 8.2 with

$$D = e^{2\varepsilon}, \quad a = c = 0, \quad b = d = \omega - \varepsilon, \quad \gamma = \max\{a', b', c', d'\} = \varepsilon,$$

provided that $\omega > (k + 2)\varepsilon$.

The evolution operator associated to the linear equation $v' = A(t)v$ with $A(t)$ as in (9.10) is given by

$$T(t, s)v = T(t, s)(x, y, z) = (x, U(t, s)y, V(t, s)z),$$

where

$$U(t, s) = e^{-\omega t + \omega s + \varepsilon(t \cos t - s \cos s - \sin t + \sin s)}$$

and

$$V(t, s) = e^{\omega t - \omega s + \varepsilon(t \sin t - s \sin s + \cos t - \cos s)}.$$

Hence, condition G1 in Section 8.1 is satisfied with any k.

We now show that there exists $D > 0$ such that

$$U(t, s) \le De^{(-\omega + \varepsilon)(t - s) + 2\varepsilon |s|} \quad \text{for } t \ge s, \tag{9.20}$$

and

$$V(s, t) \le De^{(-\omega + \varepsilon)(t - s) + 2\varepsilon |t|} \quad \text{for } t \ge s. \tag{9.21}$$

We first note that

$$U(t, s) = e^{(-\omega + \varepsilon)(t - s) + \varepsilon t(\cos t - 1) - \varepsilon s(\cos s - 1) + \varepsilon(\sin s - \sin t)}. \tag{9.22}$$

For $t \ge s \ge 0$, it follows from (9.22) that

$$U(t, s) \le e^{2\varepsilon} e^{(-\omega + \varepsilon)(t - s) + 2\varepsilon s}.$$

Furthermore, if $t = 2k\pi$ and $s = (2l-1)\pi$ with $k, l \in \mathbb{N}$ and $k \geq l$, then

$$U(t, s) = e^{(-\omega+\varepsilon)(t-s)+2\varepsilon s}.$$

For $t \geq 0 \geq s$ it also follows from (9.22) that $U(t, s) \leq e^{2\varepsilon}e^{(-\omega+\varepsilon)(t-s)}$. Finally, for $s \leq t \leq 0$ it follows from (9.22) that

$$U(t, s) \leq e^{2\varepsilon}e^{(-\omega+\varepsilon)(t-s)+2\varepsilon|t|} \leq e^{2\varepsilon}e^{(-\omega+\varepsilon)(t-s)+2\varepsilon|s|}.$$

This establishes (9.20). To obtain (9.21) we proceed in a similar way. Since

$$V(s, t) = e^{(-\omega+\varepsilon)(t-s)+\varepsilon s(\sin s+1)-\varepsilon t(\sin t+1)+\varepsilon(\cos s-\cos t)}, \tag{9.23}$$

for $t \geq s \geq 0$ we have

$$V(s, t) \leq e^{2\varepsilon}e^{(-\omega+\varepsilon)(t-s)+2\varepsilon|s|} \leq e^{2\varepsilon}e^{(-\omega+\varepsilon)(t-s)+2\varepsilon|t|}.$$

For $t \geq 0 \geq s$ it follows from (9.23) that $V(s, t) \leq e^{2\varepsilon}e^{(-\omega+\varepsilon)(t-s)}$. Finally, for $s \leq t \leq 0$ it also follows from (9.23) that

$$V(s, t) \leq e^{2\varepsilon}e^{(-\omega+\varepsilon)(t-s)+2\varepsilon|t|}.$$

This establishes (9.21). Therefore, the equation $v' = A(t)v$ admits a nonuniform exponential trichotomy with $D = e^{2\varepsilon}$, $a = c = 0$, $b = d = \omega - \varepsilon$, and $\gamma = \varepsilon$.

Finally, we show that condition G2 in Section 8.1 holds. For simplicity we consider only $k = 1$. We already know that $f(t, 0) = 0$ and $(\partial f/\partial v)(t, 0) = 0$. Furthermore, for $t \geq 0$,

$$\frac{\partial f}{\partial v}(t, v) = \delta t^{k+1}e^{-12\varepsilon t}\alpha(\|v\|^2)\begin{pmatrix} 2x & 0 & 0 \\ y+\beta(t)z & x & \beta(t)x \\ y+z & x & x \end{pmatrix}$$
$$+ \delta t^{k+1}e^{-12\varepsilon t}\alpha'(\|v\|^2)\begin{pmatrix} 2x^3 & 0 & 0 \\ x^2(y+\beta(t)z) & xy^2 & \beta(t)xz^2 \\ x^2(y+z) & xy^2 & xz^2 \end{pmatrix}. \tag{9.24}$$

Note that there exists $C > 0$ such that

$$t^{k+1}e^{-12\varepsilon t}e^{-2\varepsilon(\cos t+t\sin t)} \leq Ce^{-11\varepsilon t}e^{2\varepsilon(1+t)} = Ce^{2\varepsilon}e^{-9\varepsilon t}.$$

If follows readily from (9.24) that there exists a constant $\kappa > 0$ such that for every $t \geq 0$ and $u, v \in \mathbb{R}^3$,

$$\left\|\frac{\partial f}{\partial v}(t, v)\right\| \leq \kappa\delta e^{-9\varepsilon t}, \quad \left\|\frac{\partial f}{\partial v}(t, u) - \frac{\partial f}{\partial v}(t, v)\right\| \leq \kappa\delta e^{-9\varepsilon t}\|u-v\|.$$

When $t < 0$ we have

$$\frac{\partial}{\partial v}[-S_{-t}f(-t, S_t v)] = -S_{-t}\frac{\partial f}{\partial v}(-t, S_t v)S_t. \tag{9.25}$$

Since
$$e^{-2\varepsilon(1+|t|)} \le |b(t)| \le e^{2\varepsilon(1+|t|)},$$
we obtain $\|S_t\| \le e^{2\varepsilon(1+|t|)}$ for every $t \in \mathbb{R}$. Therefore, each entry of the matrix in (9.25) is at most
$$\gamma_t(v) := C' e^{4\varepsilon(1+|t|)} \left\| \frac{\partial f}{\partial v}(-t, S_t v) \right\|$$
for some constant $C' > 0$. Since
$$\left\| \frac{\partial f}{\partial v}(-t, S_t v) \right\| = \|S_t v\| \cdot \left\| \frac{\partial f}{\partial v}\left(-t, \frac{S_t v}{\|S_t v\|}\right) \right\| \le \kappa \delta e^{-9\varepsilon|t|} e^{2\varepsilon(1+|t|)},$$
we obtain
$$\gamma_t(v) \le \kappa C' \delta e^{-9\varepsilon|t|} e^{6\varepsilon(1+|t|)} = \kappa C' \delta e^{6\varepsilon - 3\varepsilon|t|}.$$
Therefore, there exists $\kappa' > 0$ such that for every $t < 0$ and $u, v \in \mathbb{R}^3$,
$$\left\| \frac{\partial f}{\partial v}(t, v) \right\| \le \kappa' \delta e^{-3\varepsilon|t|}, \quad \left\| \frac{\partial f}{\partial v}(t, u) - \frac{\partial f}{\partial v}(t, v) \right\| \le \kappa' \delta e^{-3\varepsilon|t|} \|u - v\|.$$

Thus, condition G2 in Section 8.1 is satisfied with $k = 1$. It is easy to verify that for $\omega > (k+2)\varepsilon$ the hypotheses of Theorem 8.2 are satisfied. In addition, we can easily verify that the hypotheses of Theorem 9.7 are satisfied provided that $\omega > (3 + 2\sqrt{2})\varepsilon$ (indeed we can take $\theta = \sqrt{2}\varepsilon$ in (9.15)).

9.2.2 Auxiliary results

We first prove several auxiliary lemmas. We continue to denote by $T(t, s)$ the evolution operator associated to the linear equation $v' = A(t)v$, and by $E(t)$ and $F_i(t)$, $i = 1, 2$, the subspaces in (8.10). Furthermore, the evolution operator can be written in the form (see (8.11))
$$T(t, s) = U(t, s) \oplus V_1(t, s) \oplus V_2(t, s).$$

Lemma 9.9. *For each $t \in \mathbb{R}$ and $w = (x, y, z) \in E(t) \times F_1(t) \times F_2(t)$*
$$\|w\| \le \|x\| + \|y\| + \|z\| \le 3De^{\gamma|t|}\|w\|.$$

Proof. The first inequality is clear. For the second one, since $x = P(t)w$, $y = Q_1(t)w$, and $z = Q_2(t)w$, we have
$$\|x\| + \|y\| + \|z\| \le (\|P(t)\| + \|Q_1(t)\| + \|Q_2(t)\|)\|w\|,$$
and the inequality follows readily from (8.4)–(8.5) (setting $s = t$). □

Lemma 9.10. *For every $t \in \mathbb{R}$ we have*
$$S_t(F_1(t)) = F_2(-t), \quad S_t(F_2(t)) = F_1(-t), \quad S_t(E(t)) = E(-t). \quad (9.26)$$

Proof. We have $(\partial f/\partial v)(t,0) = 0$ (by condition G2 in Section 8.1). Hence, by Proposition 9.3, the linear equation defined by $A(t)$ is also reversible with respect to S. Thus, by Proposition 9.2,

$$S_{-t}T(-t,-s) = T(t,s)S_{-s}, \quad s,t \in \mathbb{R}. \tag{9.27}$$

In particular, for any $v \in F_1(-s)$ we have that

$$S_{-t}V_1(-t,-s)v = S_{-t}T(-t,-s)v = T(t,s)S_{-s}v. \tag{9.28}$$

We now write

$$S_{-s}v - w_2 = w_0 + w_1, \quad w_0 \in E(-s), w_1 \in F_1(-s), w_2 \in F_2(-s).$$

Applying the operator $T(t,s)$ to $S_{-s}v - w_2$ we obtain

$$\|T(t,s)S_{-s}v - V_2(t,s)w_2\| = \|U(t,s)w_0 + V_1(t,s)w_1\|. \tag{9.29}$$

Notice that for an invertible linear operator A we have the inequality $\|Av\| \geq \|A^{-1}\|^{-1}\|v\|$. It follows from Lemma 9.9, and (8.4)–(8.5) that for $t \leq s$,

$$\begin{aligned}
3De^{\gamma|t|}\|U(t,s)w_0 + V_1(t,s)w_1\| &\geq \|U(t,s)w_0\| + \|V_1(t,s)w_1\| \\
&\geq \|U(s,t)\|^{-1}\|w_0\| + \|V_1(s,t)\|^{-1}\|w_1\| \\
&\geq \frac{\|w_0\|}{D}e^{-a(s-t)-\gamma|t|} + \frac{\|w_1\|}{D}e^{d(s-t)-\gamma|t|}.
\end{aligned} \tag{9.30}$$

On the other hand, using (9.28), (9.15), and again (8.4)–(8.5),

$$\begin{aligned}
\|T(t,s)S_{-s}v - V_2(t,s)w_2\| &\leq \|S_{-t}V_1(-t,-s)v\| + \|V_2(s,t)^{-1}w_2\| \\
&\leq CDe^{-d(s-t)+\gamma|s|+\theta|t|}\|v\| + De^{-b(s-t)+\gamma|s|}\|w_2\| \\
&\leq D\max\{C\|v\|, \|w_2\|\}e^{\gamma|s|+\theta|t|-\min\{b,d\}(s-t)}.
\end{aligned} \tag{9.31}$$

By (9.17) we have that $a + 2\gamma + 2\theta < \min\{b,d\}$. Together with (9.30) this implies that when $w_0 + w_1 \neq 0$,

$$\lim_{t\to-\infty} e^{(2\theta-\min\{b,d\})t}\|U(t,s)w_0 + V_1(t,s)w_1\| = \infty. \tag{9.32}$$

On the other hand, (9.31) yields

$$\lim_{t\to-\infty} e^{(2\theta-\min\{b,d\})t}\|T(t,s)S_{-s}v - V_2(t,s)w_2\| = 0. \tag{9.33}$$

By (9.29), (9.32) and (9.33) we have a contradiction unless $w_0 + w_1 = 0$. Thus, $S_{-s}v = w_2$ for every $s \in \mathbb{R}$. This shows that

$$S_s(F_1(s)) \subset F_2(-s), \quad s \in \mathbb{R}. \tag{9.34}$$

We prove in a similar manner that $S_s(F_2(s)) \subset F_1(-s)$ for every $s \in \mathbb{R}$. By (9.27), for any $v \in F_2(-s)$ we have that

$$S_{-t}V_2(-t,-s)v = S_{-t}T(-t,-s)v = T(t,s)S_{-s}v. \qquad (9.35)$$

We now write

$$S_{-s}v - w_1 = w_0 + w_2, \quad w_0 \in E(-s), w_1 \in F_1(-s), w_2 \in F_2(-s).$$

Then

$$\|T(t,s)S_{-s}v - V_1(t,s)w_1\| = \|U(t,s)w_0 + V_2(t,s)w_2\|. \qquad (9.36)$$

By Lemma 9.9, and (8.4)–(8.5), for $t \geq s$ we have

$$\begin{aligned}
3De^{\gamma|t|}\|U(t,s)w_0 + V_2(t,s)w_2\| &\geq \|U(t,s)w_0\| + \|V_2(t,s)w_2\| \\
&\geq \|U(t,s)^{-1}\|^{-1}\|w_0\| + \|V_2(t,s)^{-1}\|^{-1}\|w_2\| \\
&\geq \frac{\|w_0\|}{D}e^{-c(t-s)-\gamma|t|} + \frac{\|w_2\|}{D}e^{b(t-s)-\gamma|t|},
\end{aligned} \qquad (9.37)$$

and using (9.35) and (9.15),

$$\begin{aligned}
\|T(t,s)S_{-s}v - V_1(t,s)w_1\| &\leq \|S_{-t}V_2(-s,-t)^{-1}v\| + \|V_1(t,s)w_1\| \\
&\leq CDe^{-b(t-s)+\gamma|s|+\theta|t|}\|v\| + De^{-d(t-s)+\gamma|s|}\|w_1\| \\
&\leq D\max\{C\|v\|,\|w_1\|\}e^{\gamma|s|+\theta|t|-\min\{d,b\}(t-s)}.
\end{aligned} \qquad (9.38)$$

By (9.17) we have that $c + 2\gamma + 2\theta < \min\{d,b\}$. It follows from (9.37) that when $w_0 + w_2 \neq 0$,

$$\lim_{t\to+\infty} e^{-(2\theta-\min\{d,b\})t}\|U(t,s)w_0 + V_2(t,s)w_2\| = \infty. \qquad (9.39)$$

On the other hand, by (9.38),

$$\lim_{t\to+\infty} e^{-(2\theta-\min\{d,b\})t}\|T(t,s)S_{-s}v - V_1(t,s)w_1\| = 0. \qquad (9.40)$$

By (9.36), (9.39) and (9.40) we have a contradiction unless $w_0 + w_2 = 0$. Therefore,

$$S_s(F_2(s)) \subset F_1(-s), \quad s \in \mathbb{R}. \qquad (9.41)$$

Note that from (9.34) and (9.41) we have

$$S_s(F_1(s) \oplus F_2(s)) \subset F_1(-s) \oplus F_2(-s), \quad s \in \mathbb{R}. \qquad (9.42)$$

But since the operator S_s is invertible, with inverse S_{-s} (see Proposition 9.3), we have

$$F_1(s) \oplus F_2(s) \subset S_{-s}(F_1(-s) \oplus F_2(-s)) \subset F_1(s) \oplus F_2(s),$$

using (9.42) with s replaced by $-s$. Hence,

$$S_s(F_1(s) \oplus F_2(s)) = F_1(-s) \oplus F_2(-s), \quad s \in \mathbb{R}.$$

It follows from (9.34) and (9.41) that

$$S_s(F_1(s)) = F_2(-s) \quad \text{and} \quad S_s(F_2(s)) = F_1(-s), \quad s \in \mathbb{R}.$$

Similarly, to prove the third identity in (9.26), it is sufficient to show that $S_s(E(s)) \subset E(-s)$ for every $s \in \mathbb{R}$. We use a similar argument to the above ones. By (9.27), for any $v \in E(-s)$,

$$S_{-t}U(-t, -s)v = S_{-t}T(-t, -s)v = T(t, s)S_{-s}v. \tag{9.43}$$

We now write

$$S_{-s}v - w_0 - w_1 = w_2, \quad w_0 \in E(-s), w_1 \in F_1(-s), w_2 \in F_2(-s). \tag{9.44}$$

Applying the operator $T(t, s)$ to this identity we obtain

$$\|T(t, s)S_{-s}v - U(t, s)w_0 - V_1(t, s)w_1\| = \|V_2(t, s)w_2\|. \tag{9.45}$$

Proceeding as in (9.38), taking into account that $\max\{-d, c, a\} = \max\{c, a\}$, and using (9.15) and (9.43), we find that for $t \geq s$,

$$\begin{aligned} &\|T(t, s)S_{-s}v - U(t, s)w_0 - V_1(t, s)w_1\| \\ &\leq D \max\{\|w_0\|, C\|v\|, \|w_1\|\} e^{\gamma|s| + \theta|t| + \max\{c,a\}(t-s)}. \end{aligned} \tag{9.46}$$

By (9.17) we have that $\max\{c, a\} + 2\gamma + 2\theta < b$. Together with (9.37) this implies that when $w_2 \neq 0$,

$$\lim_{t \to +\infty} e^{-(2\theta + \max\{c,a\})t} \|V_2(t, s)w_2\| = \infty.$$

On the other hand, (9.46) implies that

$$\lim_{t \to +\infty} e^{-(2\theta + \max\{c,a\})t} \|T(t, s)S_{-s}v - U(t, s)w_0 - V_1(t, s)w_1\| = 0.$$

By (9.45) we have a contradiction unless $w_2 = 0$. We can thus rewrite (9.44) as

$$S_{-s}v - w_0 = w_1, \quad w_0 \in E(-s), w_1 \in F_1(-s). \tag{9.47}$$

Then

$$\|T(t, s)S_{-s}v - U(t, s)w_0\| = \|V_1(t, s)w_1\|, \tag{9.48}$$

and proceeding as in (9.46), for $t \leq s$ we have

$$\|T(t, s)S_{-s}v - U(t, s)w_0\| \leq D \max\{\|w_0\|, C\|v\|\} e^{\gamma|s| + \theta|t| + \max\{c,a\}(s-t)}. \tag{9.49}$$

By (9.17) we have that $\max\{c, a\} + 2\gamma + 2\theta < d$. Together with (9.30) this implies that when $w_1 \neq 0$,

$$\lim_{t \to -\infty} e^{(2\theta + \max\{c,a\})t} \|V_1(t,s)w_1\| = \infty.$$

On the other hand, (9.49) implies that

$$\lim_{t \to -\infty} e^{(2\theta + \max\{c,a\})t} \|T(t,s)S_{-s}v - U(t,s)w_0\| = 0.$$

By (9.48) we have again a contradiction unless $w_1 = 0$. But by (9.47) this implies that $S_s(E(s)) \subset E(-s)$ for every $s \in \mathbb{R}$. □

Lemma 9.11. *For every $t \in \mathbb{R}$ we have*

$$S_{-t}V_1(-t,s) = V_2(t,-s)S_s \quad on \ F_1(s),$$
$$S_{-t}V_2(-t,s) = V_1(t,-s)S_s \quad on \ F_2(s),$$
$$S_{-t}U(-t,s) = U(t,-s)S_s \quad on \ E(s).$$

Proof. By Lemma 9.10, if $v \in F_1(s)$ then $S_s v \in F_2(-s)$ and hence, by (9.27),

$$S_{-t}V_1(-t,s)v = S_{-t}T(-t,s)v = T(t,-s)S_s v = V_2(t,-s)S_s v.$$

This establishes the first identity in the lemma. The other identities can be readily obtained in a similar manner. □

We now use the same notation as in Chapter 8 (see Section 8.3.1). In particular, for a fixed s we consider the spaces $\mathcal{B}_+$ and $\mathcal{B}_-$. We define the space $\mathcal{B}_s$ of continuous functions $x \colon \mathbb{R} \times E(s) \to X$ such that

$$x|[s, +\infty) \times E(s) \in \mathcal{B}_+ \quad \text{and} \quad x|(-\infty, s] \times E(s) \in \mathcal{B}_-.$$

By Propositions 8.5 and 8.6, $\mathcal{B}_s$ is a complete metric space with the norm

$$\|x\|_s := \max\left\{\|x|[s, +\infty) \times E(s)\|_+, \|x|(-\infty, s] \times E(s)\|_-\right\},$$

where $\|\cdot\|_+$ and $\|\cdot\|_-$ denote respectively the norms in $\mathcal{B}_+$ and $\mathcal{B}_-$.

Given $s \in \mathbb{R}$ and $\varphi \in X$ (see Section 8.2 for the definition of the space X), we define the operator

$$(J_s x)(t, \xi) = U(t,s)\xi + \int_s^t U(t,\tau)f(\tau, x(\tau,\xi), \varphi(\tau, x(\tau,\xi))) \, d\tau$$

for each $x \in \mathcal{B}_s$ and $(t, \xi) \in \mathbb{R} \times E(s)$. It follows from the proof of Lemma 8.12 that $J_s(\mathcal{B}_s) \subset \mathcal{B}_s$ and $J_s \colon \mathcal{B}_s \to \mathcal{B}_s$ is a contraction. We note that the unique fixed point $x_s \in \mathcal{B}_s$ of J_s is the function x_φ in Lemma 8.12.

Lemma 9.12. *Given $\varphi \in X$ and $s \in \mathbb{R}$, if $S_t\varphi(t, \xi) = \varphi(-t, S_t\xi)$ for every $(t, \xi) \in \mathbb{R} \times E(s)$ then*

$$S_t x_s(t, \xi) = x_{-s}(-t, S_s\xi) \quad \text{for every } (t, \xi) \in \mathbb{R} \times E(s). \tag{9.50}$$

Proof. We consider the space $\mathcal{B}_s \times \mathcal{B}_{-s}$ with the norm

$$\|(x, y)\| = \max\{\|x\|_s, \|y\|_{-s}\},$$

and we define the operator $J = (J_s, J_{-s})$ in $\mathcal{B}_s \times \mathcal{B}_{-s}$. Clearly, J is a contraction in the complete metric space $\mathcal{B}_s \times \mathcal{B}_{-s}$. The unique fixed point of J is the pair (x_s, x_{-s}).

We now consider the subset $\mathcal{C}$ of functions $(x, y) \in \mathcal{B}_s \times \mathcal{B}_{-s}$ such that

$$S_t x(t, \xi) = y(-t, S_s \xi) \text{ for every } (t, \xi) \in \mathbb{R} \times E(s).$$

It is straightforward to verify that $\mathcal{C}$ is closed in $\mathcal{B}_s \times \mathcal{B}_{-s}$. Furthermore, $\mathcal{C} \neq \varnothing$ since $(0, 0) \in \mathcal{C}$. The statement in (9.50) will follow immediately after showing that $J(\mathcal{C}) \subset \mathcal{C}$.

We have

$$S_t(J_s x)(t, \xi) = S_t U(t, s)\xi + S_t(\bar{J}_s x)(t, \xi), \tag{9.51}$$

where

$$(\bar{J}_s x)(t, \xi) = \int_s^t U(t, \tau) f(\tau, x(\tau, \xi), \varphi(\tau, x(\tau, \xi))) \, d\tau.$$

By Lemma 9.11,

$$S_t U(t, s)\xi = U(-t, -s) S_s \xi \tag{9.52}$$

and

$$\begin{aligned} S_t(\bar{J}_s x)(t, \xi) &= \int_s^t S_t U(t, \tau) f(\tau, x(\tau, \xi), \varphi(\tau, x(\tau, \xi))) \, d\tau \\ &= \int_s^t U(-t, -\tau) S_\tau f(\tau, x(\tau, \xi), \varphi(\tau, x(\tau, \xi))) \, d\tau. \end{aligned}$$

Since the equation is reversible and each operator S_t is linear, for every $t \in \mathbb{R}$ and $v \in X$ we have

$$A(-t) S_t v + f(-t, S_t v) + S_t A(t) v + S_t f(t, v) = -\frac{\partial S}{\partial t}(t, v).$$

It follows from (9.4) that for every $t \in \mathbb{R}$ and $v \in X$,

$$f(-t, S_t v) + S_t f(t, v) = 0. \tag{9.53}$$

Using (9.53) and the hypothesis on the function φ, for $(x, y) \in \mathcal{C}$ we have

$$\begin{aligned} S_t(\bar{J}_s x)(t, \xi) &= -\int_s^t U(-t, -\tau) f(-\tau, S_\tau x(\tau, \xi), \varphi(-\tau, S_\tau x(\tau, \xi))) \, d\tau \\ &= -\int_s^t U(-t, -\tau) f(-\tau, y(-\tau, S_s \xi), \varphi(-\tau, y(-\tau, S_s \xi))) \, d\tau \\ &= \int_{-s}^{-t} U(-t, r) f(r, y(r, S_s \xi), \varphi(r, y(r, S_s \xi))) \, dr \\ &= (\bar{J}_{-s} y)(-t, S_s \xi), \end{aligned}$$

with the change of variables $\tau = -r$. By (9.51) and (9.52) this implies that

$$S_t(J_s x)(t, \xi) = U(-t, -s)S_s \xi + (\bar{J}_{-s} y)(-t, S_s \xi) = (J_{-s} y)(-t, S_s \xi).$$

Therefore $(J_s x, J_{-s} y) \in \mathcal{C}$ whenever $(x, y) \in \mathcal{C}$, and $J(\mathcal{C}) \subset \mathcal{C}$.   □

9.2.3 Proof of the reversibility

Lemma 9.13. *The unique function* $\varphi = (\varphi_1, \varphi_2) \in X$ *in Theorem 8.2 satisfies*

$$S_t \varphi_1(t, x) = \varphi_2(-t, S_t x), \quad S_t \varphi_2(t, x) = \varphi_1(-t, S_t x) \qquad (9.54)$$

for every $(t, x) \in \mathbb{R} \times E(t)$.

Proof. By Proposition 8.4, X is a complete metric space with the norm in (8.24). The unique function $\varphi \in X$ in Theorem 8.2 is obtained as the fixed point of the contraction map $\Phi \colon X \to X$ in (8.60). We now consider the subset $\mathcal{Y}$ of X formed by the functions $\varphi = (\varphi_1, \varphi_2) \in X$ satisfying (9.54) for every $(t, \xi) \in \mathbb{R} \times E(t)$. It is straightforward to verify that $\mathcal{Y}$ is closed in X. We will show that the contraction map Φ satisfies $\Phi(\mathcal{Y}) \subset \mathcal{Y}$. It then follows immediately that the unique function $\varphi \in X$ in Theorem 8.2 is also in $\mathcal{Y}$.

We must prove that for every $(s, \xi) \in \mathbb{R} \times E(s)$,

$$S_s(\Phi\varphi)_1(s, \xi) = (\Phi\varphi)_2(-s, S_s \xi), \quad S_s(\Phi\varphi)_2(s, \xi) = (\Phi\varphi)_1(-s, S_s \xi) \quad (9.55)$$

whenever $\varphi \in \mathcal{Y}$. We only prove the first identity in (9.55), since the other one can be obtained in a similar manner. By (8.60), Lemma 9.11, and (9.53) we have

$$S_s(\Phi\varphi)_1(s, \xi) = \int_{-\infty}^{s} S_s V_1(s, \tau) f(\tau, x_s(\tau, \xi), \varphi(\tau, x_s(\tau, \xi))) \, d\tau$$

$$= \int_{-\infty}^{s} V_2(-s, -\tau) S_\tau f(\tau, x_s(\tau, \xi), \varphi(\tau, x_s(\tau, \xi))) \, d\tau$$

$$= -\int_{-\infty}^{s} V_2(-s, -\tau) f(-\tau, S_\tau x_s(\tau, \xi), S_\tau \varphi(\tau, x_s(\tau, \xi))) \, d\tau.$$

Take $\varphi \in \mathcal{Y}$. By (9.54), we have

$$S_\tau \varphi(\tau, \xi) = S_\tau((\varphi_1 + \varphi_2)(\tau, \xi)) = (\varphi_2 + \varphi_1)(-\tau, S_\tau \xi) = \varphi(-\tau, S_\tau \xi). \quad (9.56)$$

It follows from Lemma 9.12 that $S_s(\Phi\varphi)_1(s, \xi)$ is given by

$$-\int_{-\infty}^{s} V_2(-s, -\tau) f(-\tau, x_{-s}(-\tau, S_s \xi), \varphi(-\tau, S_\tau x_s(\tau, \xi))) \, d\tau$$

$$= -\int_{-\infty}^{s} V_2(-s, -\tau) f(-\tau, x_{-s}(-\tau, S_s \xi), \varphi(-\tau, x_{-s}(-\tau, S_s \xi))) \, d\tau.$$

Making the change of variables $\tau = -r$ we obtain

$$S_s(\Phi\varphi)_1(s,\xi) = \int_{+\infty}^{-s} V_2(-s,r)f(r,x_{-s}(r,S_s\xi),\varphi(r,x_{-s}(r,S_s\xi)))\,dr$$
$$= (\Phi\varphi)_2(-s,S_s\xi).$$

This establishes the first identity in (9.55). The other identity can be obtained in a similar manner. □

Proof of Theorem 9.7. Since S_s is linear, it follows from (9.56) that

$$S_s(\xi,\varphi(s,\xi)) = (S_s\xi,\varphi(-s,S_s\xi)).$$

By Lemma 9.12 we have $S_s(E(s)) = E(-s)$ and thus,

$$S_s(\mathcal{V}_s) = \{(\eta,\varphi(-s,\eta)) : \eta \in S_s(E(s))\} = \mathcal{V}_{-s}.$$

This completes the proof. □

9.3 Equivariance for nonautonomous equations

We introduce here the notion of equivariant (nonautonomous) differential equation. The approach is similar to the one in Section 9.1.1 and thus we make the presentation brief. We continue to consider a Banach space X and a continuous function $L\colon \mathbb{R} \times X \to X$ such that $v' = L(t,v)$ has unique and global solutions.

Definition 9.14. *We say that the equation $v' = L(t,v)$ is equivariant with respect to the (Fréchet) differentiable map $S\colon \mathbb{R} \times X \to X$ if*

$$L(t,S_tv) - \frac{\partial S}{\partial v}(t,v)L(t,v) = \frac{\partial S}{\partial t}(t,v) \tag{9.57}$$

for every $t \in \mathbb{R}$ and $v \in X$. We also say that the equation $v' = L(t,v)$ is equivariant if it is equivariant with respect to some map S.

We can also provide characterizations of equivariance, in terms of the solutions of the equation $v' = L(t,v)$. We continue to denote by $\Phi(t,s)(v_s)$ the unique solution of $v' = L(t,v)$ with $v(s) = v_s$. The proof of the following statement essentially repeats arguments in the proofs of Propositions 9.2 and 9.5, and thus it is omitted.

Proposition 9.15. *For a map S, the following statements are equivalent:*

1. the equation $v' = L(t,v)$ is equivariant with respect to S;
2. $\Phi(\tau,t)(S(t,v)) = S(\tau,\Phi(\tau,t)(v))$ for any t, $\tau \in \mathbb{R}$ and $v \in X$;

3. *the autonomous equation* $t' = 1$, $v' = L(t, v)$ *is equivariant with respect to the map* $T \colon \mathbb{R} \times X \to \mathbb{R} \times X$ *defined by*

$$T(t, v) = (t, S(t, v)).$$

The proposition shows that the above notion of equivariance is a natural generalization of the notion of equivariance in the autonomous setting. Indeed, let $L \colon X \to X$ be a continuous function such that the equation $v' = L(v)$ defines a flow $(\varphi_t)_{t \in \mathbb{R}}$ in X. The equation $v' = L(v)$ is called *equivariant* if there is a map $S \colon X \to X$ such that

$$L \circ S = S' \circ L. \qquad (9.58)$$

One can easily show that this happens if and only if $\varphi_t \circ S = S \circ \varphi_t$ for every $t \in \mathbb{R}$. Similarly to what happens in the case of reversibility, if in (9.57) the function $L(t, v)$ does not depend on t, then taking $S_t = S$ for all $t \in \mathbb{R}$ the identity (9.57) becomes (9.58).

9.4 Equivariance in center manifolds

The following statement is a counterpart of Theorem 9.7 concerning equivariance in center manifolds. We use the same notation as in Section 9.2. In particular, we consider the sets $\mathcal{V}_s$ in (9.16).

Theorem 9.16. *Under the assumptions of Theorem 8.2, if the equation $v' = A(t)v + f(t, v)$ is equivariant with respect to a map S with $S_0^2 = \mathrm{Id}$ and S_t linear for each $t \in \mathbb{R}$, and the constants in (3.3) and (9.15) satisfy*

$$a + 2(\gamma + \theta) < b \quad and \quad c + 2(\gamma + \theta) < d, \quad with \quad \gamma = \max\{a', b', c', d'\},$$

then $S_s(\mathcal{V}_s) = \mathcal{V}_s$ for every $s \in \mathbb{R}$.

Proof. The proof is a simple modification of the proof of Theorem 9.7, and thus we only describe the necessary modifications. Following closely the proofs of Lemmas 9.10 and 9.11, we readily obtain the next statement.

Lemma 9.17. *For every $t, s \in \mathbb{R}$ we have*

$$S_t(F_i(t)) = F_i(t), \quad i = 1, 2, \quad S_t(E(t)) = E(t),$$

and

$$S_t V_i(t, s) = V_i(t, s) S_s \ on \ F_i(s), \ i = 1, 2, \quad S_t U(t, s) = U(t, s) S_s \ on \ E(s).$$

In a similar manner to that for (9.53), we can use (9.57) to show that for any $t \in \mathbb{R}$ and $v \in X$,

$$f(t, S_t v) - S_t f(t, v) = 0. \qquad (9.59)$$

Imitating the proofs of Lemmas 9.12 and 9.13, now using Lemma 9.17 and (9.59) instead of Lemmas 9.10 and 9.11 and (9.53), we obtain the following.

Lemma 9.18. *The unique function* $\varphi = (\varphi_1, \varphi_2) \in \mathcal{X}$ *in Theorem 8.2 satisfies*

$$S_t \varphi_1(t, x) = \varphi_1(t, S_t x), \quad S_t \varphi_2(t, x) = \varphi_2(t, S_t x)$$

for every $(t, x) \in \mathbb{R} \times E(t)$. *Furthermore, given* $s \in \mathbb{R}$,

$$S_t x_s(t, \xi) = x_s(t, S_s \xi) \text{ for every } (t, \xi) \in \mathbb{R} \times E(s).$$

We proceed with the proof of the theorem. It follows from Lemma 9.18 that

$$S_s(\xi, \varphi(s, \xi)) = (S_s \xi, \varphi(s, S_s \xi)).$$

On the other hand, by Lemma 9.17 we have $S_s(E(s)) = E(s)$ and thus,

$$S_s(\mathcal{V}_s) = \{(\eta, \varphi(s, \eta)) : \eta \in S_s(E(s))\} = \mathcal{V}_s.$$

This completes the proof. $\qquad\qquad\qquad\qquad\qquad\qquad\qquad\qquad\qquad$ $\square$

Part IV

Lyapunov regularity and stability theory

Lyapunov regularity and exponential dichotomies

We show in this chapter that *any* linear equation $v' = A(t)v$, with $A(t)$ in block form with blocks corresponding to the stable and center-unstable components, admits a strong nonuniform exponential dichotomy. While the extra exponentials in the notion of nonuniform exponential dichotomy substantially complicate the study of invariant manifolds in former chapters, we are able to obtain fairly general results at the expense of a careful control of the nonuniformity. In particular, we showed that if the equation $v' = A(t)v$ has a nonuniform exponential dichotomy with *sufficiently small nonuniformity* (when compared to the Lyapunov exponents), then with mild assumptions on the perturbation f there exist stable and unstable manifolds for the nonlinear equation $v' = A(t)v + f(t, v)$. We note that we do not need the nonuniformity to be zero, only sufficiently small. Therefore, it is important to estimate in quantitative terms how much a nonuniform exponential dichotomy can deviate from a uniform one. Fortunately, there exists a device, introduced by Lyapunov, that allows one to measure this deviation. It is the so-called notion of regularity (see Section 10.1 for the definition), introduced by Lyapunov in his doctoral thesis [57] (the expression is his own), which nowadays seems unfortunately apparently overlooked in the theory of differential equations. We emphasize that we only consider finite-dimensional spaces in this chapter. The infinite-dimensional case is considered in Chapter 11. The material in this chapter is based in [13], which in its turn is inspired in [1]. See [16] for a related study in the case of the discrete time.

10.1 Lyapunov exponents and regularity

Consider a continuous function $A \colon \mathbb{R}_0^+ \to M_n(\mathbb{R})$, where $M_n(\mathbb{R})$ is the set of $n \times n$ matrices with real entries. We assume that

$$\limsup_{t \to +\infty} \frac{1}{t} \log^+ \|A(t)\| = 0, \tag{10.1}$$

where $\log^+ x = \max\{0, \log x\}$ and $\|A(t)\|$ denotes the operator norm. Consider the initial value problem

$$v' = A(t)v, \quad v(0) = v_0, \tag{10.2}$$

with $v_0 \in \mathbb{R}^n$. It can easily be shown, for example using Gronwall's lemma, that this problem has a unique solution and that the solution is global in the future.

Definition 10.1. *We define the* Lyapunov exponent $\lambda \colon \mathbb{R}^n \to \mathbb{R} \cup \{-\infty\}$ *associated with the equation in* (10.2) *by*

$$\lambda(v_0) = \limsup_{t \to +\infty} \frac{1}{t} \log\|v(t)\|, \tag{10.3}$$

where $v(t)$ *is the solution of* (10.2) *(we use the convention that* $\log 0 = -\infty$*).*

Proposition 10.2. *The following properties hold:*

1. $\lambda(\alpha v) = \lambda(v)$ *for each* $v \in \mathbb{R}^n$ *and* $\alpha \in \mathbb{R} \setminus \{0\}$*;*
2. $\lambda(v + w) \le \max\{\lambda(v), \lambda(w)\}$ *for each* $v, w \in \mathbb{R}^n$*;*
3. $\lambda(v + w) = \max\{\lambda(v), \lambda(w)\}$ *for each* $v, w \in \mathbb{R}^n$ *with* $\lambda(v) \ne \lambda(w)$*;*
4. *if for some* $v_1, \ldots, v_m \in \mathbb{R}^n \setminus \{0\}$ *the numbers* $\lambda(v_1), \ldots, \lambda(v_m)$ *are distinct, then the vectors* $v_1, \ldots, v_m$ *are linearly independent.*

Proof. The first three properties follow immediately from the definition. For property 4, assume that on the contrary there exist constants $\alpha_1, \ldots, \alpha_m$ not all zero such that $\sum_{i=1}^m \alpha_i v_i = 0$. It follows from the above properties that

$$-\infty = \lambda\left(\sum_{i=1}^m \alpha_i v_i\right) = \max\{\lambda(v_i) : i = 1, \ldots, m \text{ with } \alpha_i \ne 0\} \ne -\infty.$$

This contradiction yields the desired result. $\qquad\square$

By the last property in Proposition 10.2, the function λ takes at most $p \le n$ distinct values on $\mathbb{R}^n \setminus \{0\}$, say

$$-\infty \le \lambda_1 < \cdots < \lambda_p. \tag{10.4}$$

Moreover, by the first two properties in Proposition 10.2, for each $i = 1, \ldots, p$ the set

$$E_i = \{v_0 \in \mathbb{R}^n : \lambda(v_0) \le \lambda_i\} \tag{10.5}$$

is a linear space. Note that $\lambda(v_0) > \lambda_i$ for every $v_0 \in \mathbb{R}^n \setminus E_i$.

We want to introduce the classical notion of Lyapunov regularity. For this we need to consider the initial value problem of the adjoint equation

$$w' = -A(t)^* w, \quad w(0) = w_0, \tag{10.6}$$

with $w_0 \in \mathbb{R}^n$, where $A(t)^*$ denotes the transpose of $A(t)$. We also consider the associated *Lyapunov exponent* $\mu \colon \mathbb{R}^n \to \mathbb{R} \cup \{-\infty\}$ defined by

$$\mu(w_0) = \limsup_{t \to +\infty} \frac{1}{t} \log \|w(t)\|, \qquad (10.7)$$

where $w(t)$ is the solution of (10.6). Again by Proposition 10.2, the function μ can take at most $q \le n$ distinct values on $\mathbb{R}^n \setminus \{0\}$, say

$$-\infty \le \mu_q < \cdots < \mu_1, \qquad (10.8)$$

and for each $i = 1, \ldots, q$ the set

$$F_i = \{w_0 \in \mathbb{R}^n : \mu(w_0) \le \mu_i\} \qquad (10.9)$$

is a linear space.

Denote by $\langle \cdot, \cdot \rangle$ the standard inner product in $\mathbb{R}^n$. We recall that two bases $v_1, \ldots, v_n$ and $w_1, \ldots, w_n$ of $\mathbb{R}^n$ are said to be *dual* if $\langle v_i, w_j \rangle = \delta_{ij}$ for every i and j, where δ_{ij} is the Kronecker symbol.

Definition 10.3. *We define the* regularity coefficient *of the pair of Lyapunov exponents λ and μ by*

$$\gamma(\lambda, \mu) = \min \max\{\lambda(v_i) + \mu(w_i) : 1 \le i \le n\}, \qquad (10.10)$$

where the minimum is taken over all dual bases $v_1, \ldots, v_n$ and $w_1, \ldots, w_n$ of the space $\mathbb{R}^n$.

We note that since λ and μ take only finitely many values (see (10.4) and (10.8)), the minimum in (10.10) is indeed well-defined.

We establish an additional property implying that $\gamma(\lambda, \mu) \ge 0$.

Proposition 10.4. *If $v_1, \ldots, v_n$ and $w_1, \ldots, w_n$ are dual bases of $\mathbb{R}^n$, then $\lambda(v_i) + \mu(w_i) \ge 0$ for every $i = 1, \ldots, n$.*

Proof. Let $v(t)$ be a solution of $v' = A(t)v$, and $w(t)$ a solution of $w' = -A(t)^*w$. We have

$$\frac{d}{dt}\langle v(t), w(t) \rangle = \langle A(t)v(t), w(t) \rangle + \langle v(t), -A(t)^*w(t) \rangle$$
$$= \langle A(t)v(t), w(t) \rangle - \langle A(t)v(t), w(t) \rangle = 0,$$

and hence,

$$\langle v(t), w(t) \rangle = \langle v(0), w(0) \rangle \text{ for every } t \ge 0.$$

For each i, let $v_i(t)$ be the unique solution of (10.2) with $v_0 = v_i$, and $w_i(t)$ the unique solution of (10.6) with $w_0 = w_i$. We obtain

$$\|v_i(t)\| \cdot \|w_i(t)\| \ge 1$$

for every $t \ge 0$, and hence, $\lambda(v_i) + \mu(w_i) \ge 0$ for every i. $\qquad \square$

We now introduce the important concept of regularity.

Definition 10.5. *We say that the equation in* (10.2) *is* (Lyapunov) *regular if* $\gamma(\lambda, \mu) = 0$.

We note that the notion of regularity can also be expressed solely using the equation in (10.2) (without the need for the adjoint equation in (10.6)). Namely, it is shown in Theorem 10.19 that the equation in (10.2) is Lyapunov regular if and only if

$$\lim_{t \to +\infty} \frac{1}{t} \int_0^t \operatorname{tr} A(\tau) \, d\tau = \sum_{i=1}^p (\dim E_i - \dim E_{i-1}) \lambda_i, \qquad (10.11)$$

with $E_0 = \{0\}$. However, it is sometimes more convenient to use instead the above description of regularity involving the adjoint equation.

10.2 Existence of nonuniform exponential dichotomies

We assume in this section that there is at least one negative Lyapunov exponent, that is, for some $1 \le k \le p$,

$$-\infty \le \lambda_1 < \cdots < \lambda_k < 0 \le \lambda_{k+1} < \cdots < \lambda_p. \qquad (10.12)$$

We emphasize that λ_{k+1} may be zero. Furthermore, we assume that there is a subspace $F \subset \mathbb{R}^n$ such that $E = E_k$ (see (10.12)) and F give a decomposition $\mathbb{R}^n = E \times F$, with respect to which $A(t)$ has the block form in (2.19). We also consider the Lyapunov exponents associated to the blocks $B(t)$ and $C(t)$ in (2.19), that is, to the pair $x' = B(t)x$ and $x' = -B(t)^*x$, as well as to the pair $y' = C(t)y$ and $y' = -C(t)^*y$. The corresponding regularity coefficients are

$$\gamma_U = \gamma(\lambda|E, \mu|E) \quad \text{and} \quad \gamma_V = \gamma(\lambda|F, \mu|F). \qquad (10.13)$$

The following statement shows that *any* linear differential equations as above defines a nonuniform exponential dichotomy. Furthermore, the six numbers in (2.6) can be related to the Lyapunov exponents and to the regularity coefficients.

Theorem 10.6 ([13]). *Assume that the matrix $A(t)$ has the block form in* (2.19) *for every $t \ge 0$, and that the equation $v' = A(t)v$ has at least one negative Lyapunov exponent. Then, for each $\varepsilon > 0$, the equation $v' = A(t)v$ admits a strong nonuniform exponential dichotomy in $\mathbb{R}^+$ with*

$$\underline{a} = \lambda_1 + \varepsilon, \quad \overline{a} = \lambda_k + \varepsilon, \quad a = \gamma_U + 2\varepsilon, \qquad (10.14)$$

$$\underline{b} = \lambda_{k+1} + \varepsilon, \quad \overline{b} = \lambda_p + \varepsilon, \quad b = \gamma_V + 2\varepsilon. \qquad (10.15)$$

Proof. Recall the evolution operators $U(t, s)$ and $V(t, s)$ given by (2.16).

We write $U(t, s)$ in the form $U(t, s) = X(t)X(s)^{-1}$, where $X(t)$ is a fundamental solution matrix of the equation $x' = B(t)x$. We also consider the matrix $Z(t) = [X(t)^*]^{-1}$. Taking derivatives in the identity

$$X(t)X(t)^{-1} = X(t)Z(t)^* = \mathrm{Id}$$

we obtain

$$X'(t)X(t)^{-1} + X(t)Z'(t)^* = 0,$$

and thus

$$X(t)Z'(t)^* = -B(t)X(t)X(t)^{-1} = -B(t).$$

Therefore,

$$Z'(t)^* = -X(t)^{-1}B(t) = -Z(t)^*B(t)$$

and hence, $Z'(t) = -B(t)^*Z(t)$. Since $Z(t)$ is nonsingular for every $t \geq 0$ (since this happens with $X(t)$), its columns form a basis for the space of solutions of the equation $z' = -B(t)^*z$. Thus $Z(t)$ is a fundamental solution matrix of $z' = -B(t)^*z$. Let now $x_1(t), \ldots, x_\ell(t)$ be the columns of $X(t)$, and $z_1(t), \ldots, z_\ell(t)$ the columns of $Z(t)$, where $\ell = \dim E$. For each $j = 1, \ldots, \ell$ we set

$$\alpha_j = \lambda(x_j(0)) \quad \text{and} \quad \beta_j = \mu(z_j(0)),$$

where λ and μ are the Lyapunov exponents in (10.3) and (10.7). Given $\varepsilon > 0$ there is a constant $d_1 = d_1(\varepsilon) \geq 1$ such that for each $j = 1, \ldots, \ell$ and $t \geq 0$,

$$\|x_j(t)\| \leq d_1\, e^{(\alpha_j+\varepsilon)t} \quad \text{and} \quad \|z_j(t)\| \leq d_1\, e^{(\beta_j+\varepsilon)t}.$$

It follows from the identity $Z(t)^*X(t) = \mathrm{Id}$ that $\langle x_i(t), z_j(t)\rangle = \delta_{ij}$ for every i and j. Eventually rechoosing the matrix $X(t)$ we can thus assume that

$$\max\{\alpha_j + \beta_j : j = 1, \ldots, \ell\} = \gamma_U,$$

since when we vary $X(t)$ the maximum in (10.10) can only take a finite number of values (recall that the Lyapunov exponents λ and μ take also a finite number of values). The entries of the matrix $U(t, s) = X(t)Z(s)^*$ are

$$u_{ik}(t, s) = \sum_{j=1}^{\ell} x_{ij}(t)z_{kj}(s),$$

where $x_{ij}(t)$ is the i-th coordinate of $x_j(t)$, and $z_{kj}(s)$ is the k-th coordinate of $z_j(s)$. Therefore,

$$
\begin{aligned}
|u_{ik}(t, s)| &\leq \sum_{j=1}^{\ell} |x_{ij}(t)| \cdot |z_{kj}(s)| \leq \sum_{j=1}^{\ell} \|x_j(t)\| \cdot \|z_j(s)\| \\
&\leq \sum_{j=1}^{\ell} d_1^2 e^{(\alpha_j+\varepsilon)t+(\beta_j+\varepsilon)s} = \sum_{j=1}^{\ell} d_1^2 e^{(\alpha_j+\varepsilon)(t-s)+(\alpha_j+\beta_j+2\varepsilon)s} \\
&\leq \ell d_1^2 e^{(\lambda_k+\varepsilon)(t-s)+(\gamma_U+2\varepsilon)s}.
\end{aligned}
\tag{10.16}
$$

Taking $v = \sum_{k=1}^{\ell} \alpha_k e_k$ with $\|v\|^2 = \sum_{k=1}^{\ell} \alpha_k^2 = 1$, where $e_1, \ldots, e_\ell$ is the canonical (orthonormal) basis of $\mathbb{R}^\ell$, we obtain

$$
\|U(t,s)v\|^2 = \left\| \sum_{i=1}^{\ell} \sum_{k=1}^{\ell} \alpha_k u_{ik}(t,s) e_i \right\|^2
$$

$$
= \sum_{i=1}^{\ell} \left(\sum_{k=1}^{\ell} \alpha_k u_{ik}(t,s) \right)^2 \leq \sum_{i=1}^{\ell} \left(\sum_{k=1}^{\ell} \alpha_k^2 \sum_{k=1}^{\ell} u_{ik}(t,s)^2 \right) \quad (10.17)
$$

$$
= \sum_{i=1}^{\ell} \sum_{k=1}^{\ell} u_{ik}(t,s)^2.
$$

Therefore, writing $D_1 = D_1(\varepsilon) = \ell^2 d_1^2$, we conclude that

$$
\|U(t,s)\| \leq \left(\sum_{i=1}^{\ell} \sum_{k=1}^{\ell} u_{ik}(t,s)^2 \right)^{1/2} \leq D_1 e^{(\lambda_k + \varepsilon)(t-s) + (\gamma_U + 2\varepsilon)s}. \quad (10.18)
$$

This establishes the first inequality in (2.7). Similarly, the entries of the matrix $U(t,s)^{-1} = X(s)Z(t)^*$ are

$$
w_{ik}(t,s) = u_{ik}(s,t) = \sum_{j=1}^{\ell} x_{ij}(s) z_{kj}(t).
$$

Therefore, using (10.16),

$$
|w_{ik}(t,s)| \leq \sum_{j=1}^{\ell} |x_{ij}(s)| \cdot |z_{kj}(t)| \leq \sum_{j=1}^{\ell} d_1^2 e^{(\alpha_j + \varepsilon)s + (\beta_j + \varepsilon)t}
$$

$$
= \sum_{j=1}^{\ell} d_1^2 e^{-(\alpha_j + \varepsilon)(t-s) + (\alpha_j + \beta_j + 2\varepsilon)t}
$$

$$
\leq \ell d_1^2 e^{-(\lambda_1 + \varepsilon)(t-s) + (\gamma_U + 2\varepsilon)t}.
$$

Arguing as in (10.17) and (10.18) we obtain the second inequality in (2.7).
We now consider the operator

$$
V(t,s)^{-1} = (Y(t)Y(s)^{-1})^{-1} = Y(s)Y(t)^{-1},
$$

where $Y(t)$ is a fundamental solution matrix of the equation $w' = -C(t)^*w$. Let $W(t) = [Y(t)^*]^{-1}$. We proceed in a similar manner to that for the operator $U(t,s)$. Namely, starting by taking derivatives in the identity

$$
Y(t)Y(t)^{-1} = Y(t)W(t)^* = \mathrm{Id},
$$

we can show that $W'(t) = -C(t)^*W(t)$, and thus $W(t)$ is a fundamental solution matrix for the equation $w' = -C(t)^*w$. Let now $y_1(t), \ldots, y_{n-\ell}(t)$

be the columns of $Y(t)$, and $w_1(t)$, ..., $w_{n-\ell}(t)$ the columns of $W(t)$ (notice that $n - \ell = \dim F$). For each $j = 1, \ldots, n - \ell$ we set

$$\eta_j = \lambda(y_j(0)) \quad \text{and} \quad \zeta_j = \mu(w_j(0)).$$

Given $\varepsilon > 0$ there exists a constant $d_2 = d_2(\varepsilon) \geq 1$ such that

$$\|y_j(t)\| \leq d_2 e^{(\eta_j + \varepsilon)t} \quad \text{and} \quad \|w_j(t)\| \leq d_2 e^{(\zeta_j + \varepsilon)t}$$

for each $j = 1, \ldots, n - \ell$ and $t \geq 0$. It follows from the identity $W(t)^* Y(t) = \mathrm{Id}$ that $\langle y_i(t), w_j(t) \rangle = \delta_{ij}$ for every i and j, and thus, eventually rechoosing the matrix $Y(t)$ we can assume that

$$\max\{\eta_j + \zeta_j : j = 1, \ldots, n - \ell\} = \gamma_V.$$

The entries of the matrix $Y(s)Y(t)^{-1} = Y(s)W(t)^*$ are

$$v_{ik}(s,t) = \sum_{j=1}^{n-\ell} y_{ij}(s) w_{kj}(t),$$

where $y_{ij}(s)$ is the i-th coordinate of $y_j(s)$, and $w_{kj}(t)$ is the k-th coordinate of $w_j(t)$. We obtain

$$|v_{ik}(s,t)| \leq \sum_{j=1}^{n-\ell} |y_{ij}(s)| \cdot |w_{kj}(t)| \leq \sum_{j=1}^{n-\ell} \|y_j(s)\| \cdot \|w_j(t)\|$$

$$\leq \sum_{j=1}^{n-\ell} d_2^2 e^{(\eta_j + \varepsilon)s + (\zeta_j + \varepsilon)t} = \sum_{j=1}^{n-\ell} d_2^2 e^{-(\eta_j + \varepsilon)(t-s) + (\eta_j + \zeta_j + 2\varepsilon)t}$$

$$\leq (n - \ell) d_2^2 e^{-(\lambda_{k+1} + \varepsilon)(t-s) + (\gamma_V + 2\varepsilon)t},$$

and thus, in a similar manner to that in (10.17) and (10.18), setting $D_2 = (n - \ell)^2 d_2^2$, we conclude that

$$\|V(t,s)^{-1}\| \leq \left(\sum_{i=1}^{n-\ell} \sum_{k=1}^{n-\ell} v_{ik}(s,t)^2 \right)^{1/2} \leq D_2 e^{-(\lambda_{k+1} + \varepsilon)(t-s) + (\gamma_V + 2\varepsilon)t}.$$

This gives the third inequality in (2.7). The last inequality can be obtained in an analogous manner. $\square$

We note that although the proof of the theorem provides explicit values for possible constants $\underline{a}$, $\overline{a}$, a, $\underline{b}$, $\overline{b}$, b, these are not necessarily optimal. In particular, the proof does not discard the possibility of having the limit case $\varepsilon = 0$ in (10.14)–(10.15).

One can easily obtain an entirely analogous statement to that in Theorem 10.6 by assuming that there is at least one positive Lyapunov exponent (instead of at least one negative Lyapunov exponent).

10.3 Bounds for the regularity coefficient

Following the discussion in the introduction to the chapter, it is of considerable interest to obtain sharp lower and upper bounds for the regularity coefficient $\gamma(\lambda, \mu)$ (see Section 10.1 for the definition), if possible expressed solely in terms of the matrices $A(t)$ (in particular without the need to know any explicit information about the solutions of the linear equation in (10.2)). This is the objective of this section. We continue to assume that $A \colon \mathbb{R}_0^+ \to M_n(\mathbb{R})$ is a continuous function satisfying (10.1).

10.3.1 Lower bound

We first obtain a lower bound for the regularity coefficient.

Theorem 10.7. *The regularity coefficient satisfies*

$$\gamma(\lambda, \mu) \geq \frac{1}{n} \left(\limsup_{t \to +\infty} \frac{1}{t} \int_0^t \operatorname{tr} A(\tau)\, d\tau - \liminf_{t \to +\infty} \frac{1}{t} \int_0^t \operatorname{tr} A(\tau)\, d\tau \right).$$

Proof. Let $v_1, \ldots, v_n$ be a basis of $\mathbb{R}^n$, and for each i let $v_i(t)$ be the solution of (10.2) with $v_0 = v_i$. Then the vectors $v_1(t), \ldots, v_n(t)$ are the columns of a fundamental solution matrix $X(t)$ of the equation $v' = A(t)v$. It is well known that for every $t \geq 0$,

$$\frac{\det X(t)}{\det X(0)} = \exp \int_0^t \operatorname{tr} A(\tau)\, d\tau. \tag{10.19}$$

Furthermore, $|\det X(t)| \leq \prod_{i=1}^n \|v_i(t)\|$. Therefore,

$$\limsup_{t \to +\infty} \frac{1}{t} \int_0^t \operatorname{tr} A(\tau)\, d\tau \leq \sum_{i=1}^n \lambda(v_i). \tag{10.20}$$

Let now $w_1, \ldots, w_n$ be another basis of $\mathbb{R}^n$. For each i, we denote by $w_i(t)$ the solutions of (10.6) with $w_0 = w_i$. Proceeding in a similar manner, we obtain

$$\liminf_{t \to +\infty} \frac{1}{t} \int_0^t \operatorname{tr} A(\tau)\, d\tau = -\limsup_{t \to +\infty} \frac{1}{t} \int_0^t \operatorname{tr}(-A(\tau)^*)\, d\tau \geq -\sum_{i=1}^n \mu(w_i). \tag{10.21}$$

Therefore,

$$\limsup_{t \to +\infty} \frac{1}{t} \int_0^t \operatorname{tr} A(\tau)\, d\tau - \liminf_{t \to +\infty} \frac{1}{t} \int_0^t \operatorname{tr} A(\tau)\, d\tau \leq \sum_{i=1}^n (\lambda(v_i) + \mu(w_i)).$$

We now require, in addition, that the bases $v_1, \ldots, v_n$ and $w_1, \ldots, w_n$ satisfy $\langle v_i, w_j \rangle = \delta_{ij}$ for each i and j, and that the minimum in (10.10) is attained at this pair, that is,

$$\gamma(\lambda, \mu) = \max\{\lambda(v_i) + \mu(w_i) : 1 \le i \le n\}.$$

We then obtain

$$\sum_{i=1}^{n} (\lambda(v_i) + \mu(w_i)) \le n \max\{\lambda(v_i) + \mu(w_i) : 1 \le i \le n\} = n\,\gamma(\lambda, \mu). \quad (10.22)$$

The desired result follows immediately from (10.21) and (10.22).   □

We now present a nontrivial geometric consequence of Theorem 10.7. Let $v_1, \ldots, v_n$ be a basis of $\mathbb{R}^n$. We denote by $\Gamma_n(t)$ the n-volume of the parallelepiped defined by the vectors $v_1(t), \ldots, v_n(t)$, where $v_i(t)$ is the solution of (10.2) with $v_0 = v_i$, for $i = 1, \ldots, n$. Thus, as in (10.19),

$$\Gamma_n(t)/\Gamma_n(0) = \exp \int_0^t \operatorname{tr} A(\tau)\, d\tau,$$

and it follows immediately from Theorem 10.7 that

$$\limsup_{t \to +\infty} \frac{1}{t} \log \Gamma_n(t) - \liminf_{t \to +\infty} \frac{1}{t} \log \Gamma_n(t) \le n\,\gamma(\lambda, \mu).$$

This inequality shows that the regularity coefficient measures the deviation from the existence of an exponential growth rate of n-volumes defined by solutions of the differential equation. In particular, when the equation in (10.2) is regular, there exists the exponential growth rate

$$\lim_{t \to +\infty} \frac{1}{t} \log \Gamma_n(t) = \lim_{t \to +\infty} \frac{1}{t} \int_0^t \operatorname{tr} A(\tau)\, d\tau \quad (10.23)$$

(and in particular the limit in the left-hand side of (10.23) is independent of the basis). We refer to Section 10.4 for more general statements. See Chapter 11 for a related discussion in the infinite-dimensional setting.

10.3.2 Upper bound in the triangular case

We now obtain an upper bound for the regularity coefficient. We start with triangular matrices (either all lower triangular or all upper triangular) in which case the results can be written more explicitly. The reduction to the triangular case is performed in Section 10.3.3.

For each $k = 1, \ldots, n$, consider the numbers

$$\underline{\alpha}_k = \liminf_{t \to +\infty} \frac{1}{t} \int_0^t a_k(\tau)\, d\tau \quad \text{and} \quad \overline{\alpha}_k = \limsup_{t \to +\infty} \frac{1}{t} \int_0^t a_k(\tau)\, d\tau,$$

where $a_1(t), \ldots, a_n(t)$ are the entries in the diagonal of $A(t)$. The upper bound for the regularity coefficient $\gamma(\lambda, \mu)$ is expressed in terms of these numbers.

Theorem 10.8. *If $A(t)$ is upper triangular for every $t \geq 0$, then the regularity coefficient satisfies*

$$\gamma(\lambda, \mu) \leq \sum_{k=1}^{n} (\overline{\alpha}_k - \underline{\alpha}_k). \tag{10.24}$$

Proof. We denote by $a_{ij}(t)$ the entries of the matrix $A(t)$ for each i and j. In particular, $a_k(t) = a_{kk}(t)$ for each k. We first establish two auxiliary results. For each $i = 1, \ldots, n$ and $t \geq 0$, set

$$z_{ij}(t) = \begin{cases} 0 & \text{if } i > j \\ e^{\int_0^t a_{ii}(\tau)\, d\tau} & \text{if } i = j, \\ \int_{h_{ij}}^t \sum_{k=i+1}^{j} a_{ik}(s) z_{kj}(s) e^{\int_s^t a_{ii}(\tau)\, d\tau}\, ds & \text{if } i < j \end{cases} \tag{10.25}$$

for some constants h_{ij} to be chosen later. One can easily verify by direct substitution that the columns of the $n \times n$ matrix $Z(t) = (z_{ij}(t))$ form a basis for the space of solutions of $z' = A(t)z$. The columns of $Z(t)$ are $z_j(t) = (z_{1j}(t), \ldots, z_{nj}(t))$ for $j = 1, \ldots, n$. Given $i, j = 1, \ldots, n$, we write

$$\lambda(z_{ij}) = \limsup_{t \to +\infty} \frac{1}{t} \log |z_{ij}(t)|. \tag{10.26}$$

Lemma 10.9. *For every $i, j = 1, \ldots, n$, we have $\lambda(z_{ii}) = \overline{\alpha}_i$ and*

$$\lambda(z_{ij}) \leq \overline{\alpha}_j + \sum_{m=i}^{j-1} (\overline{\alpha}_m - \underline{\alpha}_m). \tag{10.27}$$

Proof. We show that the elements of each column $z_j(t)$ of $Z(t)$ satisfy (10.27) by choosing appropriate constants h_{ij}. Clearly, $\lambda(z_{ii}) = \overline{\alpha}_i$ for $i = 1, \ldots, n$. We now proceed by backward induction on i. Namely, for a given $i < n$, assume that

$$\lambda(z_{kj}) \leq \overline{\alpha}_j + \sum_{m=k}^{j-1} (\overline{\alpha}_m - \underline{\alpha}_m) \text{ whenever } i + 1 \leq k \leq j. \tag{10.28}$$

We want to prove that for $j \geq i + 1$,

$$\lambda(z_{ij}) \leq \overline{\alpha}_j + \sum_{m=i}^{j-1} (\overline{\alpha}_m - \underline{\alpha}_m).$$

By (10.1) and (10.28), for each $\varepsilon > 0$ there exists $D > 0$ such that

$$|a_{ik}(t)| \leq De^{t\varepsilon}, \quad e^{-\int_0^t a_{ii}(\tau)\, d\tau} \leq De^{(-\underline{\alpha}_i + \varepsilon)t},$$

$$|z_{kj}(t)| \leq De^{(\overline{\alpha}_j + \sum_{m=i+1}^{j-1} (\overline{\alpha}_m - \underline{\alpha}_m) + \varepsilon)t}$$

for every $t \geq 0$ and $i+1 \leq k \leq j$. Therefore,

$$\lambda(z_{ij}) \leq \limsup_{t \to +\infty} \frac{1}{t} \log \left(e^{\int_0^t a_{ii}(\tau)\,d\tau} \left| \int_{h_{ij}}^t \sum_{k=i+1}^j |a_{ik}(s)z_{kj}(s)|\, e^{-\int_0^s a_{ii}(\tau)\,d\tau}\, ds \right| \right)$$

$$\leq \overline{\alpha}_i + \limsup_{t \to +\infty} \frac{1}{t} \log \left| \int_{h_{ij}}^t D^3 \sum_{k=i+1}^j e^{(\overline{\alpha}_j + \sum_{m=k}^{j-1}(\overline{\alpha}_m - \underline{\alpha}_m) - \underline{\alpha}_i + 3\varepsilon)s}\, ds \right|$$

$$\leq \overline{\alpha}_i + \limsup_{t \to +\infty} \frac{1}{t} \log \left| \int_{h_{ij}}^t D^3 n e^{(\overline{\alpha}_j + \sum_{m=i+1}^{j-1}(\overline{\alpha}_m - \underline{\alpha}_m) - \underline{\alpha}_i + 3\varepsilon)s}\, ds \right|.$$

Set

$$c_{ij} = \overline{\alpha}_j - \underline{\alpha}_i + \sum_{m=i+1}^{j-1} (\overline{\alpha}_m - \underline{\alpha}_m). \tag{10.29}$$

For every $i < j$, let

$$h_{ij} = \begin{cases} 0 & \text{if } c_{ij} \geq 0 \\ +\infty & \text{if } c_{ij} < 0. \end{cases} \tag{10.30}$$

Then, if $c_{ij} \geq 0$ we have

$$\lambda(z_{ij}) \leq \overline{\alpha}_i + \limsup_{t \to +\infty} \frac{1}{t} \log \frac{D^3 n (e^{(c_{ij}+3\varepsilon)t} - 1)}{c_{ij} + 3\varepsilon}$$

and, if $c_{ij} < 0$ we have

$$\lambda(z_{ij}) \leq \overline{\alpha}_i + \limsup_{t \to +\infty} \frac{1}{t} \log \frac{D^3 n e^{(c_{ij}+3\varepsilon)t}}{|c_{ij} + 3\varepsilon|}.$$

Therefore, in both cases,

$$\lambda(z_{ij}) \leq \overline{\alpha}_i + c_{ij} + 3\varepsilon = \overline{\alpha}_j + \sum_{m=i}^{j-1} (\overline{\alpha}_m - \underline{\alpha}_m) + 3\varepsilon.$$

Since ε is arbitrary, we conclude that (10.27) holds for every $j \geq i+1$ (and thus, for every $j \geq i$). This completes the proof of the lemma. $\square$

We now consider the adjoint equation $w' = -A(t)^*w$. For each $i, j = 1, \ldots, n$ and $t \geq 0$, set

$$w_{ij}(t) = \begin{cases} 0 & \text{if } i < j \\ e^{-\int_0^t a_{jj}(\tau)\,d\tau} & \text{if } i = j, \\ -\int_{h_{ji}}^t \sum_{k=j}^{i-1} a_{ki}(s) w_{kj}(s) e^{-\int_s^t a_{ii}(\tau)\,d\tau}\, ds & \text{if } i > j \end{cases} \tag{10.31}$$

using the constants in (10.30) (these constants are also used in (10.25)). Again, one can easily verify by direct substitution that the columns of the $n \times n$

matrix $W(t) = (w_{ij}(t))$ form a basis for the space of solutions $w' = -A(t)^*w$. The columns of $W(t)$ are $w_j(t) = (w_{1j}(t), \ldots, w_{nj}(t))$ for $j = 1, \ldots, n$. Given $i, j = 1, \ldots, n$, we write

$$\mu(w_{ij}) = \limsup_{t \to +\infty} \frac{1}{t} \log |w_{ij}(t)|.$$

Lemma 10.10. *For every $i, j = 1, \ldots, n$, we have $\mu(w_{jj}) = -\underline{\alpha}_j$ and*

$$\mu(w_{ij}) \le -\underline{\alpha}_j + \sum_{k=j+1}^{i} (\overline{\alpha}_k - \underline{\alpha}_k). \tag{10.32}$$

Proof. We proceed in a similar manner to that in the proof of Lemma 10.9 to show that the elements of each column $w_j(t)$ of $W(t)$ satisfy (10.32). Clearly, $\mu(w_{jj}) = -\underline{\alpha}_j$ for $j = 1, \ldots, n$. We now proceed by induction on i. Namely, for a given $i > 1$, assume that

$$\mu(w_{kj}) \le -\underline{\alpha}_j + \sum_{m=j+1}^{k} (\overline{\alpha}_m - \underline{\alpha}_m) \text{ whenever } j \le k \le i - 1. \tag{10.33}$$

We want to prove that for $j \le i - 1$,

$$\mu(w_{ij}) \le -\underline{\alpha}_j + \sum_{m=j+1}^{i} (\overline{\alpha}_m - \underline{\alpha}_m).$$

It follows from (10.1) and (10.33) that given $\varepsilon > 0$ there exists $D > 0$ such that

$$|a_{ki}(t)| \le De^{t\varepsilon}, \quad e^{\int_0^t a_{ii}(\tau)\,d\tau} \le De^{(\overline{\alpha}_i + \varepsilon)t},$$

$$|w_{kj}(t)| \le De^{(-\underline{\alpha}_j + \sum_{m=j+1}^{i-1}(\overline{\alpha}_m - \underline{\alpha}_m) + \varepsilon)t}$$

for every $t \ge 0$ and $j \le k \le i - 1$. Therefore,

$$\mu(w_{ij}) \le \limsup_{t \to +\infty} \frac{1}{t} \log \left(e^{-\int_0^t a_{ii}(\tau)\,d\tau} \left| \int_{h_{ji}}^{t} \sum_{k=j}^{i-1} |a_{ki}(s) w_{kj}(s)| e^{\int_0^s a_{ii}(\tau)\,d\tau}\,ds \right| \right)$$

$$\le -\underline{\alpha}_i + \limsup_{t \to +\infty} \frac{1}{t} \log \left| \int_{h_{ji}}^{t} D^3 \sum_{k=j}^{i-1} e^{(-\underline{\alpha}_j + \sum_{m=j+1}^{k}(\overline{\alpha}_m - \underline{\alpha}_m) + \overline{\alpha}_i + 3\varepsilon)s}\,ds \right|$$

$$\le -\underline{\alpha}_i + \limsup_{t \to +\infty} \frac{1}{t} \log \left| \int_{h_{ji}}^{t} D^3 n e^{(c_{ji} + 3\varepsilon)s}\,ds \right|,$$

using the constants in (10.29). Then, if $c_{ji} \ge 0$ we have

$$\mu(w_{ij}) \le -\underline{\alpha}_i + \limsup_{t \to +\infty} \frac{1}{t} \log \frac{D^3 n (e^{(c_{ji} + 3\varepsilon)t} - 1)}{c_{ji} + 3\varepsilon}$$

and, if $c_{ji} < 0$ we have

$$\mu(w_{ij}) \leq -\underline{\alpha}_i + \limsup_{t \to +\infty} \frac{1}{t} \log \frac{D^3 n e^{(c_{ji}+3\varepsilon)t}}{|c_{ji} + 3\varepsilon|}.$$

Therefore, in both cases,

$$\mu(w_{ij}) \leq -\underline{\alpha}_i + c_{ji} + 3\varepsilon = -\underline{\alpha}_j + \sum_{m=j+1}^{i} (\overline{\alpha}_m - \underline{\alpha}_m) + 3\varepsilon.$$

Since ε is arbitrary, we conclude that (10.32) holds for every $j \leq i - 1$ (and thus, for every $j \leq i$). This completes the proof of the lemma. $\square$

We now proceed with the proof of Theorem 10.8. It follows from Lemmas 10.9 and 10.10 that

$$\lambda(z_j) = \max\{\lambda(z_{ij}) : i = 1, \ldots, n\} \leq \overline{\alpha}_j + \sum_{m=1}^{j-1} (\overline{\alpha}_m - \underline{\alpha}_m),$$

and

$$\mu(w_j) = \max\{\mu(w_{ij}) : i = 1, \ldots, n\} \leq -\underline{\alpha}_j + \sum_{m=j+1}^{n} (\overline{\alpha}_m - \underline{\alpha}_m).$$

In particular,

$$\lambda(z_i) + \mu(w_i) \leq \sum_{k=1}^{n} (\overline{\alpha}_k - \underline{\alpha}_k) \quad \text{for every } i = 1, \ldots, n. \tag{10.34}$$

Therefore, in view of the definition of the regularity coefficient $\gamma(\lambda, \mu)$, to prove Theorem 10.8 it is enough to show that the bases $(z_1, \ldots, z_n)$ and $(w_1, \ldots, w_n)$ are dual. First we note that

$$\frac{d}{dt} \langle z_i(t), w_j(t) \rangle = \langle B(t)z_i(t), w_j(t) \rangle + \langle z_i(t), -B(t)^* w_j(t) \rangle$$
$$= \langle B(t)z_i(t), w_j(t) \rangle - \langle B(t)z_i(t), w_j(t) \rangle = 0,$$

and hence,

$$\langle z_i(t), w_j(t) \rangle = \langle z_i(0), w_j(0) \rangle \text{ for every } t \geq 0.$$

Clearly $\langle z_i(0), w_j(0) \rangle = 0$ for every $i < j$. Furthermore,

$$\langle z_i(0), w_i(0) \rangle = \sum_{j=1}^{n} z_{ji}(0) w_{ji}(0)$$

$$= \sum_{j \leq i-1} z_{ji}(0) w_{ji}(0) + z_{ii}(0) w_{ii}(0) + \sum_{j \geq i+1}^{n} z_{ji}(0) w_{ji}(0)$$

$$= z_{ii}(0) w_{ii}(0) = 1$$

for every $i = 1, \ldots, n$. We now fix $i > j$ and $t \geq 0$. We have

$$\langle z_i(t), w_j(t) \rangle = \sum_{k=j}^{i} z_{ki}(t) w_{kj}(t)$$

$$= z_{ji}(t) w_{jj}(t) + z_{ii}(t) w_{ij}(t) + \sum_{k=j+1}^{i-1} z_{ki}(t) w_{kj}(t). \qquad (10.35)$$

Note that

$$c_{ji} = c_{ki} + c_{jk} \qquad (10.36)$$

for $k = j + 1, \ldots, i - 1$ (see (10.29)). We consider two cases:

1. If $c_{ji} \geq 0$, then $h_{ji} = 0$ (see (10.30)). By (10.36), for every k such that $j + 1 \leq k \leq i - 1$ we have either $c_{ki} \geq 0$ or $c_{jk} \geq 0$. By (10.30), we have $h_{ki} = 0$ or $h_{jk} = 0$, and thus either $z_{ki}(0) = 0$ or $w_{kj}(0) = 0$ (by direct substitution of $t = 0$ in (10.25)). Furthermore, again since $h_{ji} = 0$, it follows from (10.25) and (10.31) that $z_{ji}(0) = w_{ij}(0)$. Hence, evaluating (10.35) at $t = 0$ we find that all terms in the sum are zero, and thus $\langle z_i, w_j \rangle = 0$.
2. If $c_{ji} < 0$, then $h_{ji} = +\infty$ (see (10.30)). By (10.36) and (10.30), for every k such that $j + 1 \leq k \leq i - 1$ we have either $h_{ki} = +\infty$ or $h_{jk} = +\infty$, and thus $z_{ki}(+\infty) = 0$ or $w_{kj}(+\infty) = 0$. Hence, evaluating (10.35) at $t = +\infty$ we find that all terms in the sum are zero, and thus $\langle z_i, w_j \rangle = 0$.

We conclude that $\langle z_i, w_j \rangle = \delta_{ij}$ for every i and j. The theorem follows from (10.34) and the definition of $\gamma(\lambda, \mu)$. $\qquad \square$

We note that the case of lower triangular matrices can be treated in a similar manner. In fact, with the same notation, the inequality in (10.24) also holds in the case of lower triangular matrices.

10.3.3 Reduction to the triangular case

The following result shows that we can always assume, without loss of generality, that the matrices $A(t)$ in (10.2) are upper triangular for every t, with respect to the canonical basis $e_1, \ldots, e_n$ of $\mathbb{R}^n$. By first performing this reduction for a given $A(t)$, we can then obtain an upper bound for the regularity coefficient $\gamma(\lambda, \mu)$ by applying Theorem 10.8 to the upper triangular matrix function $B(t)$ obtained in the following statement.

Theorem 10.11. *For a continuous function $A \colon \mathbb{R}_0^+ \to M_n(\mathbb{R})$, there exist continuous functions $B \colon \mathbb{R}_0^+ \to M_n(\mathbb{R})$ and $U \colon \mathbb{R}_0^+ \to M_n(\mathbb{R})$ such that:*

1. *$B(t)$ is upper triangular and $\|B(t)\| \leq 2n\|A(t)\|$ for each $t \geq 0$;*
2. *U is differentiable, $U(0) = \mathrm{Id}$, and for each $t \geq 0$, $U(t)$ is unitary and*

$$B(t) = U(t)^{-1} A(t) U(t) - U(t)^{-1} U'(t); \qquad (10.37)$$

3. the initial value problem in (10.2) *is equivalent to*

$$x' = B(t)x, \quad x(0) = v_0, \tag{10.38}$$

and the solutions $v(t)$ of (10.2) *and $u(t)$ of* (10.38) *satisfy $v(t) = U(t)x(t)$.*

Furthermore, if the function A satisfies (10.1), *then*

$$\limsup_{t \to +\infty} \frac{1}{t} \log^+ \|B(t)\| = 0. \tag{10.39}$$

Proof. We apply the Gram–Schmidt orthogonalization procedure to the vectors $v_1(t)$, ..., $v_n(t)$, where $v_i(t)$ is the solution of (10.2) with $v_0 = e_i$. In this manner, we obtain functions $u_1(t)$, ..., $u_n(t)$ such that:

1. $\langle u_i(t), u_j(t) \rangle = \delta_{ij}$ for each i and j;
2. each function $u_k(t)$ is a linear combination of $v_1(t)$, ..., $v_k(t)$.

Note that each $v_k(t)$ is also a linear combination of $u_1(t)$, ..., $u_k(t)$, and thus

$$\langle v_i(t), u_j(t) \rangle = 0 \text{ for } i < j. \tag{10.40}$$

Given $t \geq 0$ we consider the linear operator $U(t)$ such that $U(t)e_i = u_i(t)$ for each i. Clearly, $U(t)$ is unitary for each t, $U(0) = \mathrm{Id}$, and $t \mapsto U(t)$ is differentiable with $U'(t)e_i = u_i'(t)$ for each i. Set now $x(t) = U(t)^{-1}v(t)$. We obtain

$$v'(t) = U'(t)x(t) + U(t)x'(t) = A(t)v(t) = A(t)U(t)x(t), \tag{10.41}$$

and thus $x'(t) = B(t)x(t)$ with $B(t)$ given by (10.37). Clearly, the function B is continuous.

Given $t \geq 0$, let now $V(t)$ be the operator such that $V(t)e_i = v_i(t)$ for each i, and set $X(t) = U(t)^{-1}V(t)$. Since $U(t)$ is unitary, it follows from (10.40) that

$$0 = \langle v_i(t), u_j(t) \rangle - \langle V(t)e_i, U(t)e_j \rangle - \langle X(t)e_i, e_j \rangle \text{ for } i < j. \tag{10.42}$$

Therefore $X(t)$ is upper triangular, and the same happens with $X'(t)$. Proceeding in a similar manner to that in (10.41) but now with $V(t) = U(t)X(t)$ we obtain $X'(t) = B(t)X(t)$ for $t \geq 0$. Thus, $B(t) = X'(t)X(t)^{-1}$ and $B(t)$ is upper triangular.

Since $U(t)$ is unitary, it follows from (10.37) that

$$B(t) + B(t)^* = U(t)^*(A(t) + A(t)^*)U(t) - (U(t)^*U'(t) + U'(t)^*U(t))$$

$$= U(t)^*(A(t) + A(t)^*)U(t) - \frac{d}{dt}(U(t)^*U(t))$$

$$= U(t)^*(A(t) + A(t)^*)U(t). \tag{10.43}$$

For each $i, j = 1, \ldots, n$ and $t \geq 0$, we now write $b_{ij}(t) = \langle B(t)e_i, e_j \rangle$ and $\tilde{a}_{ij}(t) = \langle A(t)u_i(t), u_j(t) \rangle$. Since $B(t)$ is upper triangular, it follows from (10.43) that

$$b_{ii}(t) = \tilde{a}_{ii}(t) \quad \text{and} \quad b_{ij}(t) = \tilde{a}_{ij}(t) + \tilde{a}_{ji}(t) \tag{10.44}$$

whenever $i \neq j$. Since $U(t)$ is unitary, the vectors $u_1(t) = U(t)e_1, \ldots, u_n(t) = U(t)e_n$ form an orthonormal basis of $\mathbb{R}^n$, and thus

$$\|A(t)\| \geq \|A(t)u_i(t)\| = \left\| \sum_{j=1}^{n} \langle Au_i(t), u_j(t) \rangle u_j(t) \right\|$$

$$= \left(\sum_{j=1}^{n} \tilde{a}_{ji}(t)^2 \right)^{1/2} \geq |\tilde{a}_{ij}(t)|$$

for every i and j. It follows from (10.44) that

$$|b_{ij}(t)| \leq 2\|A(t)\| \text{ for every } i \text{ and } j.$$

Thus, given $v = \sum_{i=1}^{n} \alpha_i e_i$ with $\|v\| = (\sum_{i=1}^{n} \alpha_i^2)^{1/2} = 1$, we obtain

$$\|B(t)v\|^2 = \left\| \sum_{i=1}^{n} \sum_{j=1}^{n} \alpha_i \langle B(t)e_i, e_j \rangle e_j \right\|^2$$

$$= \sum_{j=1}^{n} \left(\sum_{i=j}^{n} \alpha_i b_{ij}(t) \right)^2 \leq \sum_{j=1}^{n} \left(\sum_{i=j}^{n} \alpha_i^2 \sum_{i=j}^{n} b_{ij}(t)^2 \right)$$

$$\leq \sum_{j=1}^{n} \sum_{i=j}^{n} b_{ij}(t)^2 \leq 4n^2 \|A(t)\|^2.$$

This shows that $\|B(t)\| \leq 2n\|A(t)\|$, and the property (10.39) follows immediately from (10.1). For the last statement in the theorem it remains to observe that $U(t)$ is invertible for each t, and that $v(0) = x(0) = v_0$ since $U(0) = \text{Id}$. This establishes the theorem. □

Theorem 10.11 implies that, in what respects to the study of Lyapunov regularity, there is no loss of generality in replacing the initial value problem in (10.2) by the corresponding problem with the associated triangular matrix function $B(t)$.

Theorem 10.12. *The equation $v' = A(t)v$ is Lyapunov regular if and only if the equation $x' = B(t)x$ is Lyapunov regular.*

Proof. By Theorem 10.11, the initial value problem in (10.2) is equivalent to (10.38), with the solutions $v(t)$ of (10.2) and $x(t)$ of (10.38) related by $v(t) = U(t)x(t)$ with $U(t)$ unitary for each $t \geq 0$ (and with $U(0) = \text{Id}$). Similarly, the initial value problem in (10.6) is equivalent to

$$y' = -B(t)^* y, \quad y(0) = w_0, \tag{10.45}$$

with the solutions $w(t)$ of (10.6) and $y(t)$ of (10.45) related by $w(t) = U(t)y(t)$ using the same operator $U(t)$. Indeed, using (10.37),

$$
\begin{aligned}
(U(t)y(t))' &= U'(t)y(t) + U(t)y'(t) \\
&= [U'(t)U(t)^{-1} - U(t)B(t)^*U(t)^{-1}]U(t)y(t) \\
&= [-A(t)^* + U'(t)U(t)^{-1} + U(t)U'(t)^*]U(t)y(t) \\
&= \left[-A(t)^* + \frac{d}{dt}(U(t)U(t)^*) \right] U(t)y(t) = -A(t)^*U(t)y(t).
\end{aligned}
$$
$$(10.46)$$

Since $U(t)$ is unitary for each t, the Lyapunov exponents associated with the equations in (10.38) and (10.45), that is,

$$ x' = B(t)x \quad \text{and} \quad y' = -B(t)^*y, $$

coincide, respectively, with the Lyapunov exponents λ and μ associated with the equations in (10.2) and (10.6), that is,

$$ v' = A(t)v \quad \text{and} \quad w' = -A(t)^*w. $$

Therefore, the regularity coefficient of the new pair of equations (10.38) and (10.45) is the same as the regularity coefficient of the pair of equations (10.2) and (10.6). This yields the desired result.   □

Using the characterization of Lyapunov regularity in (10.11) (and proven in Theorem 10.19) one could give an alternative proof of Theorem 10.12 simply using the equations $v' = A(t)v$ and $x' = B(t)x$. Namely, observe first that since $U(t)$ is unitary for each t, the numbers λ_i and $\dim E_i$ in (10.11) are the same for both equations. Furthermore, it follows from (10.37) that

$$
\begin{aligned}
\operatorname{tr} B(t) &= \operatorname{tr}(U(t)^{-1}A(t)U(t)) - \operatorname{tr}(U(t)^{-1}U'(t)) \\
&= \operatorname{tr} A(t) - \operatorname{tr}(U(t)^{-1}U'(t)).
\end{aligned}
$$

Since $U(t)$ is unitary,

$$
\begin{aligned}
0 = (U(t)^*U(t))' &= U'(t)^*U(t) + U(t)^*U'(t) \\
&= (U(t)^*U'(t))^* + U(t)^{-1}U'(t)
\end{aligned}
$$

and hence, $U(t)^{-1}U'(t) = -(U(t)^{-1}U'(t))^*$. Being skew-hermitian the matrix $U(t)^{-1}U'(t)$ has zero trace, and thus $\operatorname{tr} B(t) = \operatorname{tr} A(t)$. In particular, we can always replace A by B in the left-hand side of (10.11).

10.4 Characterizations of regularity

We first introduce a new coefficient that also measures the regularity. Recall the numbers λ_i and μ_i in (10.4) and (10.8). We also consider the values

$$\lambda_1' \leq \cdots \leq \lambda_n'$$

of the Lyapunov exponent λ on $\mathbb{R}^n \setminus \{0\}$ counted with multiplicities, that are obtained by repeating each value λ_i a number of times equal to the difference $\dim E_i - \dim E_{i-1}$ (see (10.5) for the definition of E_i), with $E_0 = \{0\}$. Analogously, we consider the values

$$\mu_1' \geq \cdots \geq \mu_n'$$

of the Lyapunov exponent μ on $\mathbb{R}^n \setminus \{0\}$ counted with multiplicities.

Definition 10.13. *We say that a basis $v_1, \ldots, v_n$ of $\mathbb{R}^n$ is* normal *for the filtration by subspaces*

$$E_1 \subset E_2 \subset \cdots \subset E_p = \mathbb{R}^n$$

if for each $i = 1, \ldots, p$ there exists a basis of E_i composed of vectors in $\{v_1, \ldots, v_n\}$. When $v_1, \ldots, v_n$ is a normal basis for the filtration of subspaces in (10.5) we also say that it is normal *for the Lyapunov exponent λ (or simply* normal *when it is clear from the context to which exponent we are referring to).*

We shall refer to dual bases $v_1, \ldots, v_n$ and $w_1, \ldots, w_n$ of $\mathbb{R}^n$ which are normal respectively for the Lyapunov exponents λ and μ, that is, respectively for the filtration by subspaces

$$E_1 \subset \cdots \subset E_p = \mathbb{R}^n \quad \text{and} \quad F_q \subset \cdots \subset F_1 = \mathbb{R}^n \tag{10.47}$$

in (10.5) and (10.9) as *dual normal bases*.

Proposition 10.14. *There exist dual normal bases $v_1, \ldots, v_n$ and $w_1, \ldots, w_n$ of the space $\mathbb{R}^n$.*

Proof. Let $v_1', \ldots, v_n'$ be a basis of $\mathbb{R}^n$ with

$$\lambda(v_1') \leq \cdots \leq \lambda(v_n'), \tag{10.48}$$

which is normal for the first family of subspaces in (10.47). We consider another filtration by subspaces

$$E_1' \subset E_2' \subset \cdots \subset E_q' = \mathbb{R}^n. \tag{10.49}$$

It is easy to see that there exists a nonsingular upper triangular matrix C (in the basis $v_1', \ldots, v_n'$) such that the new basis $v_1 = Cv_1', \ldots, v_n = Cv_n'$ is normal for the filtration in (10.49). On the other hand, in view of (10.48) and since C is upper triangular, the new basis $v_1, \ldots, v_n$ continues to be normal for the first family of subspaces in (10.47). We now consider the particular case of $E_j' = F_j^\perp$ with $j = 1, \ldots, q$. Then, $v_1, \ldots, v_n$ is a basis of $\mathbb{R}^n$ which is normal simultaneously for the families of subspaces

$$E_1 \subset \cdots \subset E_p = \mathbb{R}^n \quad \text{and} \quad F_1^\perp \subset \cdots \subset F_q^\perp = \mathbb{R}^n$$

Then the (unique) dual basis $w_1, \ldots, w_n$ of $\mathbb{R}^n$ is normal for the family of subspaces F_j with $j = 1, \ldots, q$. $\qquad\square$

Definition 10.15. *We define the* Perron coefficient *of the pair of Lyapunov exponents λ and μ by*

$$\pi(\lambda, \mu) = \max\{\lambda_i' + \mu_i' : 1 \le i \le n\}.$$

The Perron coefficient is related with the regularity coefficient as follows.

Theorem 10.16. *We have*

$$0 \le \pi(\lambda, \mu) \le \gamma(\lambda, \mu) \le n\, \pi(\lambda, \mu).$$

Proof. Consider dual bases $v_1, \ldots, v_n$ and $w_1, \ldots, w_n$ of $\mathbb{R}^n$. Without loss of generality we may assume that $\lambda(v_1) \le \cdots \lambda(v_n)$. Let now σ be a permutation of $\{1, \cdots, n\}$ such that the numbers $\bar\mu_{\sigma(i)} = \mu(w_i)$ satisfy $\bar\mu_1 \ge \cdots \ge \bar\mu_n$.

Lemma 10.17. *We have $\lambda(v_i) \ge \lambda_i'$ and $\bar\mu_i \ge \mu_i'$ for $i = 1, \ldots, n$.*

Proof of the lemma. We consider only the exponent λ, since the argument for μ is entirely similar. Since λ_1 is the minimal value of λ on $\mathbb{R}^n \setminus \{0\}$ we have $\lambda(v_i) \ge \lambda_1 = \lambda_i'$ for $1 \le i \le n$. Note that $\lambda(v_{n_1+1}) > \lambda_1$ where $n_i = \dim E_i$ for each i. Indeed, otherwise we would have $v_1, \ldots, v_{n_1+1} \in E_1$ and

$$n_1 = \dim E_1 \ge \dim \operatorname{span}\{v_1, \ldots, v_{n+1}\} = n_1 + 1.$$

Therefore, $\lambda(v_i) \ge \lambda_2 = \lambda_i'$ for $i = n_1 + 1, \ldots, n_2$. Repeating the same argument finitely many times we find that $\lambda(v_i) \ge \lambda_i'$ for $i = 1, \ldots, n$. $\qquad\square$

We now proceed with the proof of the theorem. Fix an integer i such that $1 \le i \le n$. If $i \ge \sigma(i)$, then $\bar\mu_{\sigma(i)} \ge \bar\mu_i$ and

$$\max\{\lambda(v_i) + \mu(w_i) : 1 \le i \le n\} \ge \lambda(v_i) + \bar\mu_{\sigma(i)} \ge \lambda(v_i) + \bar\mu_i.$$

If $i < \sigma(i)$, then there exists $k > i$ such that $i \ge \sigma(k)$. Otherwise, we would have $\sigma(i+1), \ldots, \sigma(n) \ge i + 1$, and hence $\sigma(i) \le i$. It follows that

$$\max\{\lambda(v_i) + \mu(w_i) : 1 \le i \le n\} \ge \lambda(v_k) + \bar\mu_{\sigma(k)} \ge \lambda(v_i) + \bar\mu_i.$$

Therefore, using Lemma 10.17 we obtain

$$\max\{\lambda(v_i) + \mu(w_i) : 1 \le i \le n\} \ge \max\{\lambda(v_i) + \bar\mu_i : 1 \le i \le n\}$$
$$\ge \max\{\lambda_i' + \mu_i' : 1 \le i \le n\} = \pi(\lambda, \mu).$$

Hence, $\gamma(\lambda, \mu) \ge \pi(\lambda, \mu)$.

On the other hand, by Proposition 10.14 we can consider dual normal bases $v_1, \ldots, v_n$ and $w_1, \ldots, w_n$ of $\mathbb{R}^n$, and hence for which the numbers $\lambda(v_i)$ and $\mu(w_i)$ are respectively the values of the Lyapunov exponents λ and μ on $\mathbb{R}^n \setminus \{0\}$, counted with multiplicities, although possibly not ordered. By Proposition 10.4 we have $\lambda(v_i) + \mu(w_i) \geq 0$ for every i. Therefore,

$$0 \leq \lambda(v_i) + \mu(w_i) \leq \sum_{i=1}^{n}(\lambda(v_i) + \mu(w_i)) = \sum_{i=1}^{n}(\lambda_i' + \mu_i') \leq n\,\pi(\lambda, \mu). \quad (10.50)$$

Hence, $0 \leq \gamma(\lambda, \mu) \leq n\,\pi(\lambda, \mu)$. In particular, $\pi(\lambda, \mu) \geq 0$. This completes the proof of the theorem. □

Theorem 10.16 allows us to give several characterizations of regularity.

Theorem 10.18. *The following properties are equivalent:*

1. $\gamma(\lambda, \mu) = 0$;
2. $\pi(\lambda, \mu) = 0$;
3. *given dual normal bases* $v_1, \ldots, v_n$ *and* $w_1, \ldots, w_n$ *of* $\mathbb{R}^n$ *we have*

$$\lambda(v_i) + \mu(w_i) = 0 \text{ for } i = 1, \ldots, n; \quad (10.51)$$

4. $\lambda_i' + \mu_i' = 0$ *for* $i = 1, \ldots, n$.

Proof. The equivalence of the first two properties is immediate from Theorem 10.16. Let $v_1, \ldots, v_n$ and $w_1, \ldots, w_n$ be dual normal bases. It follows from (10.50) that property 2 implies property 3. Furthermore, it follows from the definition of Perron coefficient and Theorem 10.16 that

$$\lambda_i' + \mu_i' \leq 0 \text{ for } i = 1, \ldots, n.$$

In view of (10.50) we find that if property 2 holds then $\sum_{i=1}^{n}(\lambda_i' + \mu_i') = 0$, and thus

$$\lambda_i' + \mu_i' = 0 \text{ for } i = 1, \ldots, n,$$

that is, property 4 holds. On the other hand, clearly property 4 implies property 2. □

We also present alternative characterizations of regularity expressed in terms of the existence of exponential growth rates of volumes. These characterizations have the advantage of being expressed solely using the equation in (10.2) (without the need for the adjoint equation in (10.6)).

Given vectors $v_1, \ldots, v_m \in \mathbb{R}^n$ we denote the m-volume defined by these vectors by $\Gamma(v_1, \ldots, v_m)$. It is equal to $|\det K|^{1/2}$, where K is the $m \times m$ matrix with entries $k_{ij} = \langle v_i, v_j \rangle$ for each i and j.

Theorem 10.19. *The following properties are equivalent:*

1. $\gamma(\lambda, \mu) = 0$;

2.

$$\lim_{t\to+\infty}\frac{1}{t}\int_0^t \operatorname{tr} A(\tau)\,d\tau = \sum_{i=1}^{p}(\dim E_i - \dim E_{i-1})\lambda_i; \tag{10.52}$$

3. *for any normal basis* $v_1,\ldots,v_n$ *of* $\mathbb{R}^n$ *and any* $m \leq n$ *there exists the limit*

$$\lim_{t\to+\infty}\frac{1}{t}\log \Gamma(v_1(t),\ldots,v_m(t)),$$

where each $v_i(t)$ *is the solution of* (10.2) *with* $v_i(0) = v_i$.

Proof. Assume first that $\gamma(\lambda,\mu) = 0$. By Theorem 10.7, we have

$$\limsup_{t\to+\infty}\frac{1}{t}\int_0^t \operatorname{tr} A(\tau)\,d\tau = \liminf_{t\to+\infty}\frac{1}{t}\int_0^t \operatorname{tr} A(\tau)\,d\tau,$$

and there exists the limit in (10.52). On the other hand, by Proposition 10.14 and Theorem 10.18 there exist dual normal bases $v_1,\ldots,v_n$ and $w_1,\ldots,w_n$ of $\mathbb{R}^n$ such that (10.51) holds. By (10.20) and (10.21), and the existence of the limit in (10.52),

$$-\sum_{i=1}^{n}\mu(w_i) \leq \lim_{t\to+\infty}\frac{1}{t}\int_0^t \operatorname{tr} A(\tau)\,d\tau \leq \sum_{i=1}^{n}\lambda(v_i).$$

In view of (10.51) we find that $\sum_{i=1}^{n}\mu(w_i) = -\sum_{i=1}^{n}\lambda(v_i)$ and

$$\lim_{t\to+\infty}\frac{1}{t}\int_0^t \operatorname{tr} A(\tau)\,d\tau = \sum_{i=1}^{n}\lambda(v_i) = \sum_{i=1}^{n}\lambda_i'$$
$$-\sum_{i=1}^{p}(\dim E_i - \dim E_{i-1})\lambda_i.$$

Before continuing we proceed in a similar manner to that in the proof of Theorem 10.11, now for an arbitrary basis $v_1,\ldots,v_n$ of $\mathbb{R}^n$ instead of $e_1,\ldots,e_n$. Namely, we apply the Gram–Schmidt orthogonalization procedure to the vectors $v_1(t),\ldots,v_n(t)$, where $v_i(t)$ is the solution of (10.2) with $v_0 = v_i$ (instead of $v_0 = e_i$). We thus obtain a unitary matrix $U(t)$ such that the vectors $U(t)v_i = u_i(t)$ are those obtained from the orthogonalization procedure; to see that $U(t)$ is unitary note that

$$\langle U(t)^*U(t)v_i, v_j\rangle = \langle U(t)v_i, U(t)v_j\rangle = \langle u_i(t), u_j(t)\rangle = \delta_{ij}.$$

Set $x_i(t) = U(t)^{-1}v_i(t)$ for $i = 1,\ldots,n$. Repeating arguments in the proof of Theorem 10.11 (see (10.42)) we find that the $n \times n$ matrix $X(t)$ with $X(t)v_1 = x_1(t),\ldots,X(t)v_n = x_n(t)$ is upper triangular (with respect to the basis $v_1,\ldots,v_n$). Since $U(t)$ is unitary, we obtain

$$\langle v_i(t), v_j(t)\rangle = \langle U(t)x_i(t), U(t)x_j(t)\rangle = \langle x_i(t), x_j(t)\rangle.$$

Therefore, since $X(t)$ is upper triangular, for each $m \le n$ we have

$$\Gamma(v_1(t), \ldots, v_m(t)) = \Gamma(x_1(t), \ldots, x_m(t)) = \prod_{i=1}^{m} |x_{ii}(t)|, \qquad (10.53)$$

where $x_{ii}(t) = \langle x_i(t), v_i\rangle$ for each i. We also consider the upper triangular matrix $B(t)$ defined by (10.37).

We now assume that property 2 holds. It is well known that

$$\exp \int_0^t \operatorname{tr} A(\tau)\, d\tau = \frac{\Gamma(v_1(t), \ldots, v_n(t))}{\Gamma(v_1, \ldots, v_n)}.$$

Using the notation in (10.26), we clearly have $\lambda(x_i) \ge \lambda(x_{ii})$ for each i. Hence, provided that the basis $v_1, \ldots, v_n$ is normal it follows from property 2 that

$$\lim_{t \to +\infty} \frac{1}{t} \log \Gamma(v_1(t), \ldots, v_n(t)) = \sum_{i=1}^{n} \lambda(x_i) \ge \sum_{i=1}^{n} \lambda(x_{ii}). \qquad (10.54)$$

On the other hand, by (10.53),

$$\lim_{t \to +\infty} \frac{1}{t} \log \Gamma(v_1(t), \ldots, v_n(t)) = \lim_{t \to +\infty} \frac{1}{t} \sum_{i=1}^{n} \log |x_{ii}(t)| \le \sum_{i=1}^{n} \lambda(x_{ii}).$$
$$(10.55)$$

Comparing (10.54) and (10.55) we find that

$$\lim_{t \to +\infty} \frac{1}{t} \sum_{i=1}^{n} \log |x_{ii}(t)| = \sum_{i=1}^{n} \lambda(x_{ii}). \qquad (10.56)$$

In view of the definition of $\lambda(x_{ii})$ (notice that it is a lim sup), using (10.56) one can easily verify that $\lambda(x_{ii})$ is in fact a limit for $i = 1, \ldots, n$, that is,

$$\lambda(x_{ii}) = \lim_{t \to +\infty} \frac{1}{t} \log |x_{ii}(t)|.$$

Indeed, if

$$\underline{c}_i = \liminf_{t \to +\infty} \frac{1}{t} \log |x_{ii}(t)| < \limsup_{t \to +\infty} \frac{1}{t} \log |x_{ii}(t)| = \overline{c}_i$$

for some $i = j$, then choosing a subsequence k_m such that

$$\frac{1}{k_m} \log |x_{jj}(k_m)| \to \underline{c}_j$$

as $m \to +\infty$ we obtain

$$\lim_{t\to+\infty}\frac{1}{t}\sum_{i=1}^{n}\log|x_{ii}(t)| = \lim_{m\to+\infty}\frac{1}{k_m}\sum_{i=1}^{n}\log|x_{ii}(k_m)|$$

$$= \underline{c}_j + \lim_{m\to+\infty}\frac{1}{k_m}\sum_{i\neq j}\log|x_{ii}(k_m)|$$

$$< \overline{c}_j + \sum_{i\neq j}\overline{c}_i = \sum_{i=1}^{n}\overline{c}_i.$$

It follows from (10.53) that, for each $m \leq n$,

$$\lim_{t\to+\infty}\frac{1}{t}\log\Gamma(v_1(t),\dots,v_m(t)) = \sum_{i=1}^{m}\lim_{t\to+\infty}\frac{1}{t}\log|x_{ii}(t)|.$$

In particular, property 3 holds.

We now show that property 3 implies property 1. Note that for each $m \leq n$,

$$\exp\int_0^t\sum_{i=1}^{m}b_{ii}(s)\,ds = \Gamma(x_1(t),\dots,x_m(t)) = \Gamma(v_1(t),\dots,v_m(t)),$$

where $b_{11}(s),\dots,b_{nn}(s)$ are the entries in the diagonal of $B(s)$. By property 3, for $m \leq n$ there exist the limits

$$\lim_{t\to+\infty}\frac{1}{t}\log\Gamma(v_1(t),\dots,v_m(t)).$$

Therefore, there also exist the limits

$$\lim_{t\to+\infty}\frac{1}{t}\int_0^t b_{mm}(s)\,ds = \lim_{t\to+\infty}\frac{1}{t}\log\frac{\Gamma(v_1(t),\dots,v_m(t))}{\Gamma(v_1(t),\dots,v_{m-1}(t))}.$$

It follows from Theorem 10.8 that $\gamma(\lambda,\mu) = 0$. $\square$

10.5 Equations with negative Lyapunov exponents

The purpose of this section is to describe how the results in former chapters can be applied to nonautonomous linear differential equations with nonzero Lyapunov exponents (without loss of generality we only consider negative exponents since the case of positive exponents is analogous). For simplicity of the exposition we only address the existence of invariant manifolds. We emphasize that we only consider finite-dimensional spaces in this section. The approach is based on the fact that essentially all such equations admit a nonuniform exponential dichotomy (see Theorem 10.6).

We emphasize that there are several difficulties presented by a corresponding generalization to the infinite-dimensional setting. In the case of finite-dimensional systems, we can apply the classical Lyapunov–Perron regularity theory to estimate in an effective manner the deviation of a nonuniform exponential dichotomy from the classical notion of uniform exponential dichotomy. On the other hand, the full extent generalization of the present approach to infinite-dimensional systems requires an appropriate development of the regularity theory in this setting. It is however not as developed and presents some additional technical difficulties, essentially due to the fact that a Lyapunov exponent may then take infinitely many values, not to mention that the ambient space may not have a countable basis. See Chapter 11 for the study of Lyapunov regularity in Hilbert spaces.

10.5.1 Lipschitz stable manifolds

We now present the results on the existence of Lipschitz stable manifolds for equations with negative Lyapunov exponents. The manifolds will be obtained as Lipschitz graphs $W \subset \mathbb{R}^{n+1}$ over an open subset of the space $\mathbb{R}_0^+ \times E$ (see (10.5) and (10.12)), where $E = E_k$ is the linear space corresponding to the negative Lyapunov exponents of (10.2).

Consider continuous functions $A \colon \mathbb{R}_0^+ \to M_n(\mathbb{R})$ and $f \colon \mathbb{R}_0^+ \times \mathbb{R}^n \to \mathbb{R}^n$, satisfying the conditions A1–A2 in Section 4.2 with $X = \mathbb{R}^n$, and thus with $B(X) = M_n(\mathbb{R})$. For simplicity, we also assume that there exists a subspace $F \subset \mathbb{R}^n$ such that $A(t)$ has the block form in (2.19) with respect to the direct sum $\mathbb{R}^n = E \oplus F$ (which is independent of time).

As in Section 10.2 we assume that there is at least one negative Lyapunov exponent (see (10.12)). We consider the conditions

$$\lambda_k + \gamma_U + (\gamma_U + \gamma_V)/q < 0 \quad \text{and} \quad \lambda_k + \gamma_V < \lambda_{k+1}, \tag{10.57}$$

with the same constants as in (10.12) and (10.13). Note that both inequalities in (10.57) are automatically satisfied when the regularity coefficients γ_U and γ_V are sufficiently small, and that the first inequality is satisfied for a given $\gamma_U < |\lambda_k|$ provided that q is sufficiently large.

In view of (10.57) we can choose $\varepsilon > 0$ such that

$$\bar{a} + a < 0 \quad \text{and} \quad \bar{a} + b < \underline{b}, \tag{10.58}$$

with the constants given by (10.14)–(10.15). Note that in view of the first inequality in (10.58) we have indeed $\bar{a} < 0$ as required in (2.6). We also consider the constants α and β given by (4.8) and (4.12) with the values of a and b in (10.14)–(10.15).

The following is an immediate consequence of Theorem 4.1 (we use the same notation as in Section 4.2).

Theorem 10.20. *Assume that A1–A2 hold. If the conditions in (10.57) hold, then for each $\varepsilon > 0$ satisfying (10.58) with the constants in (10.14)–(10.15), there exist $\delta > 0$ and a unique function $\varphi \in \mathfrak{X}_\alpha$ such that the set W in (4.11) has the properties in Theorem 4.1.*

We can also formulate a corresponding statement concerning the existence of unstable manifolds. These are analogous to the former ones, and correspond to consider the Lyapunov exponent $\lambda^- : \mathbb{R}^n \to \mathbb{R} \cup \{-\infty\}$ defined by

$$\lambda^-(v_0) = \limsup_{t \to -\infty} \frac{1}{|t|} \log \|v(t)\| \qquad (10.59)$$

where $v(t)$ is the solution of (10.2). It follows from Proposition 10.2 that the function λ^- takes at most $p^- \leq n$ distinct values on $\mathbb{R}^n \setminus \{0\}$, say

$$-\infty \leq \lambda_{p^-}^- < \cdots < \lambda_{k^-+1}^- < 0 \leq \lambda_{k^-}^- < \cdots < \lambda_1^-$$

for some $1 \leq k^- \leq p^-$, assuming that there is at least one negative Lyapunov exponent, with respect to the Lyapunov exponent λ^- in (10.59). We can then proceed in a similar manner to obtain the existence of Lipschitz unstable manifolds as an application of Theorem 4.9.

10.5.2 Smooth stable manifolds

We present here the results on the existence of *smooth* stable manifolds for equations with negative Lyapunov exponents. We now assume that the functions $A : \mathbb{R}_0^+ \to M_n(\mathbb{R})$ and $f : \mathbb{R}_0^+ \times \mathbb{R}^n \to \mathbb{R}^n$ are of class C^1 and that they satisfy the conditions B1–B2 in Section 5.1. We also consider the conditions

$$q\lambda_k + 4\vartheta < \min\{\lambda_k - \lambda_p, (2-q)\vartheta\} \quad \text{and} \quad \lambda_k + \vartheta < \lambda_{k+1}, \qquad (10.60)$$

where $\vartheta = \max\{\gamma_U, \gamma_V\}$. In view of (10.60) we can choose $\varepsilon > 0$ such that

$$q\overline{a} + 4\max\{a, b\} < \overline{a} - \overline{b} \quad \text{and} \quad \overline{a} + (1 + 2/q)\max\{a, b\} < 0, \qquad (10.61)$$

where the constants in (10.61) take the values given by (10.14)–(10.15). We also consider the constants α and γ given respectively by (4.8) and (5.3) again with the values of a and b in (10.14)–(10.15).

The following is an immediate consequence of Theorem 5.1.

Theorem 10.21. *Assume that B1–B2 hold. If the conditions in* (10.60) *hold, then for each* $\varepsilon > 0$ *satisfying* (10.61) *and* $\varrho > 0$, *there exist* $\delta > 0$ *such that for the unique function* $\varphi \in \mathfrak{X}_\alpha$ *given by Theorem 10.20, the set* $\mathcal{V} \subset \mathcal{W}$ *in* (5.4) *is a smooth manifold of class* C^1 *containing the line* $(\varrho, +\infty) \times \{0\}$ *and satisfying* $T_{(s,0)}\mathcal{V} = \mathbb{R} \times E$ *for every* $s > \varrho$.

In a similar manner to that in Section 5.2 we can apply Theorem 10.21 (and the corresponding version for the case of unstable manifolds) to obtain smooth stable and unstable manifolds of nonuniformly hyperbolic trajectories corresponding respectively to the negative and to the positive Lyapunov exponents for the linear variational equation.

10.6 Measure-preserving flows

We show here that from the point of view of ergodic theory *almost all* linear equations in (10.2) obtained as linear variational equations from a measure-preserving flow have a and b in (2.7) as small as desired. We emphasize that no material in this section is required in any other place in the book.

Recall that a (measurable) flow $\Psi_t \colon M \to M$ *preserves* a measure ν on M provided that $\nu(\Psi_t A) = \nu(A)$ for every measurable set $A \subset M$ and every $t \in \mathbb{R}$. We will show how the following result can be obtained from standard results of ergodic theory.

Theorem 10.22. *If F is a vector field on a smooth Riemannian manifold M defining a flow Ψ_t which preserves a finite measure ν on M such that*

$$\int_M \sup_{-1 \le t \le 1} \log^+ \|d_x \Psi_t\| \, d\nu(x) < \infty, \tag{10.62}$$

then for ν-almost every $x \in M$ the evolution operator defined by the linear variational equation

$$v' = A_x(t)v, \quad \text{with } A_x(t) = d_{\Psi_t x} F \tag{10.63}$$

admits a strong nonuniform exponential dichotomy in $\mathbb{R}$ with arbitrarily small constants a and b.

Before giving the proof of Theorem 10.22, we need to recall some material from the theory of nonuniformly hyperbolic dynamical systems. We refer the reader to [1, 2, 3] for detailed expositions.

Consider a flow Ψ_t in a Riemannian manifold M, as in Theorem 10.22. Given $x \in M$ and $v \in T_x M$, we define the *forward Lyapunov exponent* of (x, v) by

$$\lambda^+(x, v) = \limsup_{t \to +\infty} \frac{1}{t} \log \|(d_x \Psi_t)v\|. \tag{10.64}$$

For each $x \in M$, there exist a positive integer $p^+(x) \le n$, a collection of values $\lambda_1^+(x) < \lambda_2^+(x) < \cdots < \lambda_{p^+(x)}^+(x)$, and linear subspaces

$$\{0\} = E_0^+(x) \subset E_1^+(x) \subset \cdots \subset E_{p^+(x)}^+(x) = T_x M$$

such that:

1. $E_i^+(x) = \{v \in T_x M : \lambda^+(x, v) \le \lambda_i^+(x)\}$;
2. if $v \in E_i^+(x) \setminus E_{i-1}^+(x)$, then $\lambda^+(x, v) = \lambda_i^+(x)$.

The collection of linear spaces $E_i^+(x)$ is called the *forward filtration* of $T_x M$. For each i, let

$$k_i^+(x) = \dim E_i^+(x) - \dim E_{i-1}^+(x).$$

One can easily verify that the functions $x \mapsto p^+(x)$, $x \mapsto \lambda_i^+(x)$, and $x \mapsto k_i^+(x)$, for $i = 1, \ldots, p^+(x)$, are invariant under the flow Ψ_t. This is a consequence of the identity

$$\lambda^+(\Psi_s x, (d_x \Psi_s) v) = \lambda^+(x, v).$$

Definition 10.23. *We say that x is a* forward regular point *for the flow Ψ_t if*

$$\lim_{t \to +\infty} \frac{1}{t} \log|\det(d_x \Psi_t)| = \sum_{i=1}^{p^+(x)} \lambda_i^+(x) k_i^+(x).$$

Similarly, given $x \in M$ and $v \in T_x M$, we define the *backward Lyapunov exponent* of (x, v) by

$$\lambda^-(x, v) = \limsup_{t \to -\infty} \frac{1}{|t|} \log \|(d_x \Psi_t) v\|.$$

For each $x \in M$, there exist a positive integer $p^-(x) \le n$, a collection of values $\lambda_1^-(x) > \cdots > \lambda_{p^-(x)}^-(x)$, and linear subspaces

$$T_x M = E_1^-(x) \supset \cdots \supset E_{p^-(x)}^-(x) \supset E_{p^-(x)+1}^-(x) = \{0\}$$

such that:

1. $E_i^-(x) = \{v \in T_x M : \lambda^-(x, v) \le \lambda_i^-(x)\}$;
2. if $v \in E_i^-(x) \setminus E_{i+1}^-(x)$, then $\lambda^-(x, v) = \lambda_i^-(x)$.

The collection of linear spaces $E_i^+(x)$ is called the *backward filtration* of $T_x M$. For each i, let

$$k_i^-(x) = \dim E_i^-(x) - \dim E_{i+1}^-(x).$$

Similarly, we have $\lambda^-(\Psi_s x, (d_x \Psi_s) v) = \lambda^-(x, v)$, and one can easily verify that the functions $x \mapsto p^-(x)$, $x \mapsto \lambda_i^-(x)$, and $x \to k_i^-(x)$, for $i = 1, \ldots, p^-(x)$, are invariant under the flow Ψ_t.

Definition 10.24. *We say that x is a* backward regular point *for the flow Ψ_t if*

$$\lim_{t \to -\infty} \frac{1}{|t|} \log|\det(d_x \Psi_t)| = \sum_{i=1}^{p^-(x)} \lambda_i^-(x) k_i^-(x).$$

Definition 10.25. *We say that the above forward and backward filtrations* comply *at the point $x \in M$ if the following conditions hold:*

1. $p(x) := p^+(x) = p^-(x)$;

2. *there exists a decomposition* $T_x M = \bigoplus_{i=1}^{p(x)} H_i(x)$ *into subspaces* $H_i(x)$
   such that for each $i = 1, \ldots, p(x)$,

$$(d_x \Psi_t) H_i(x) = H_i(\Psi_t x) \text{ for every } t \in \mathbb{R},$$

$$E_i^+(x) = \bigoplus_{j=1}^{i} H_j(x) \quad and \quad E_i^-(x) = \bigoplus_{j=i}^{p(x)} H_j(x);$$

3. $\lambda_i(x) := \lambda_i^+(x) = -\lambda_i^-(x)$ *for each* $i = 1, \ldots, p(x)$;
4. *if* $v \in H_i(x) \setminus \{0\}$ *and* $i = 1, \ldots, p(x)$, *then*

$$\lim_{t \to \pm\infty} \frac{1}{t} \log \sup_{v \in G_i} \|(d_x \Psi_t)v\| = \lim_{t \to \pm\infty} \frac{1}{t} \log \inf_{v \in G_i} \|(d_x \Psi_t)v\| = \lambda_i(x),$$

where $G_i = \{v \in H_i(x) : \|v\| = 1\}$.

We can now introduce the concept of Lyapunov regularity.

Definition 10.26. *We say that a point* $x \in M$ *is* Lyapunov regular *or simply* regular *for the flow* Ψ_t *if the following conditions hold:*

1. x *is forward and backward regular for the flow* Ψ_t;
2. *the forward and backward filtrations comply at* x.

One can easily verify, as a consequence of the invariance of the functions $p^\pm$, $\lambda_i^\pm$, and $k_i^\pm$ under the flow Ψ_t, that a point $x \in M$ is regular if and only if $\Psi_t x$ is regular for every $t \in \mathbb{R}$.

Although the notion of regularity requires a substantial structure, it turns out that at least from the point of view of ergodic theory it is rather common (see [1, 3]).

Theorem 10.27 (Multiplicative Ergodic Theorem for flows). *Let* Ψ_t *be a flow on* M *preserving the finite measure* ν. *If the condition (10.62) holds, then the invariant set of regular points for* Ψ_t *has full* ν-*measure.*

Given $\varepsilon > 0$ and a regular point $x \in M$ for the flow Ψ_t, we introduce an inner product on the tangent space $T_x M$ by

$$\langle u, v \rangle_x' = \int_{\mathbb{R}} \langle (d_x \Psi_t)u, (d_x \Psi_t)v \rangle_{\Psi_t x} e^{-2\lambda_i(x)t - 2\varepsilon|t|} \, dt$$

if $u, v \in H_i(x)$, where $\langle \cdot, \cdot \rangle_y$ denotes the original inner product on $T_y M$. We also set $\langle u, v \rangle_x' = 0$ if $u \in H_i(x)$ and $v \in H_j(x)$ with $i \neq j$. This is called a *Lyapunov inner product*, and the induced norm is called a *Lyapunov norm*. We also say that a linear transformation $C_\varepsilon(x) \colon T_x M \to T_x M$ is a *Lyapunov change of coordinates* if it satisfies

$$\langle u, v \rangle_x = \langle C_\varepsilon(x)u, C_\varepsilon(x)v \rangle_x'.$$

With the help of the Lyapunov inner product one can establish the following statement, that is usually called Oseledets–Pesin Reduction Theorem for flows.

Theorem 10.28 (see [3]). *Let Ψ_t be a flow on M preserving the finite measure ν. Given $\varepsilon > 0$, if x is a regular point for Ψ_t, then for any Lyapunov change of coordinates the following properties hold:*

1. the linear map $\mathcal{A}_\varepsilon(x,t) = C_\varepsilon(\Psi_t x)^{-1}(d_x\Psi_t)C_\varepsilon(x)$ has the block form

$$\mathcal{A}_\varepsilon(x,t) = \begin{pmatrix} \mathcal{A}_\varepsilon^1(x,t) & & \\ & \ddots & \\ & & \mathcal{A}_\varepsilon^{p(x)}(x,t) \end{pmatrix}, \tag{10.65}$$

where $\mathcal{A}_\varepsilon^i(x,t)$ is a $k_i(x) \times k_i(x)$ matrix for $i = 1, \ldots, p(x)$, and the entries of $\mathcal{A}_\varepsilon(x,t)$ are zero elsewhere;
2. each block $\mathcal{A}_\varepsilon^i(x,t)$ satisfies

$$e^{\lambda_i(x)t - \varepsilon|t|} \leq \|\mathcal{A}_\varepsilon^i(x,t)^{-1}\|^{-1} \leq \|\mathcal{A}_\varepsilon^i(x,t)\| \leq e^{\lambda_i(x)t + \varepsilon|t|}. \tag{10.66}$$

We note that the version of Theorem 10.28 in [3] only considers the case of discrete time, although the necessary changes to pass from discrete time to continuous time are straightforward.

We can now establish the announced result with the help of Theorem 10.28.

Proof of Theorem 10.22. Since Ψ_t is the flow defined by the vector field F, for each $x \in M$ the general solution of the linear variational equation in (10.63) is given by $v(t) = (d_x\Psi_t)v_0$, with $v_0 \in T_xM$. In particular, for each $x \in M$ the values of the Lyapunov exponent λ associated with the equation in (10.3) with $A(t) = A_x(t)$, that is, the numbers in (10.4), coincide with the values of the forward Lyapunov exponent $\lambda^+(x,\cdot)$ in (10.64).

Let $x \in M$ be a Lyapunov regular point, and let

$$\lambda_1(x) < \cdots < \lambda_{k(x)}(x) < 0 \leq \lambda_{k(x)+1}(x) < \cdots < \lambda_{p(x)}(x)$$

be the values of the forward Lyapunov exponent at x (which coincide with the symmetric of the values of the backward Lyapunov exponent at x). By Theorem 10.27, the set of regular points has full ν-measure. For any such point it follows from Theorem 10.28 that after any Lyapunov change of coordinates the derivative $d_x\Psi_t$ has the block form in (10.65). Note that the evolution operator associated with (10.63) is given by

$$T(t,s) = d_{\Psi_s x}\Psi_{t-s} = d_x\Psi_t(d_x\Psi_s)^{-1}.$$

In the new coordinates, it follows readily from (10.66) that for any $t, s \in \mathbb{R}$,

$$e^{-\lambda_i(x)s - \varepsilon|s|}e^{\lambda_i(x)t - \varepsilon|t|} \leq \|\mathcal{A}_\varepsilon^i(x,s)\mathcal{A}_\varepsilon^i(x,t)^{-1}\|^{-1}$$
$$\leq \|\mathcal{A}_\varepsilon^i(x,t)\mathcal{A}_\varepsilon^i(x,s)^{-1}\| \leq e^{\lambda_i(x)t + \varepsilon|t|}e^{-\lambda_i(x)s + \varepsilon|s|},$$

which for $t \geq s \geq 0$ we rewrite in the form

$$e^{(\lambda_i(x)+\varepsilon)(t-s)-2\varepsilon t} \le \|\mathcal{A}_\varepsilon^i(x,s)\mathcal{A}_\varepsilon^i(x,t)^{-1}\|^{-1}$$
$$\le \|\mathcal{A}_\varepsilon^i(x,t)\mathcal{A}_\varepsilon^i(x,s)^{-1}\| \le e^{(\lambda_i(x)+\varepsilon)(t-s)+2\varepsilon s}. \quad (10.67)$$

We now put together the blocks corresponding to negative and nonnegative values of the Lyapunov exponent, that is,

$$\begin{pmatrix} \mathcal{A}_\varepsilon^1(x,t) & & \\ & \ddots & \\ & & \mathcal{A}_\varepsilon^{k(x)}(x,t) \end{pmatrix} \quad \text{and} \quad \begin{pmatrix} \mathcal{A}_\varepsilon^{k(x)+1}(x,t) & & \\ & \ddots & \\ & & \mathcal{A}_\varepsilon^{p(x)}(x,t) \end{pmatrix},$$

that we denote respectively by $\mathcal{B}_\varepsilon(x,t)$ and $\mathcal{C}_\varepsilon(x,t)$. It follows readily from (10.67) (and the fact that the functions $k_i(x)$ are independent of t) that for each $t \ge s \ge 0$,

$$e^{(\lambda_1(x)+\varepsilon)(t-s)-2\varepsilon t} \le \|\mathcal{B}_\varepsilon(x,s)\mathcal{B}_\varepsilon(x,t)^{-1}\|^{-1}$$
$$\le \|\mathcal{B}_\varepsilon(x,t)\mathcal{B}_\varepsilon(x,s)^{-1}\| \le e^{(\lambda_{k(x)}(x)+\varepsilon)(t-s)+2\varepsilon s},$$

$$e^{(\lambda_{k(x)+1}(x)+\varepsilon)(t-s)-2\varepsilon t} \le \|\mathcal{C}_\varepsilon(x,s)\mathcal{C}_\varepsilon(x,t)^{-1}\|^{-1}$$
$$\le \|\mathcal{C}_\varepsilon(x,t)\mathcal{C}_\varepsilon(x,s)^{-1}\| \le e^{(\lambda_{p(x)}(x)+\varepsilon)(t-s)+2\varepsilon s}.$$

Thus, for each regular point $x \in M$, the evolution operator with components

$$U(t,s) = \mathcal{B}_\varepsilon(x,t)\mathcal{B}_\varepsilon(x,s)^{-1} \quad \text{and} \quad V(t,s) = \mathcal{C}_\varepsilon(x,t)\mathcal{C}_\varepsilon(x,s)^{-1}$$

defines a strong nonuniform exponential dichotomy in $\mathbb{R}$, with the constants in (2.6) given by

$$\underline{a} = \lambda_1(x) + \varepsilon, \quad \overline{a} = \lambda_{k(x)}(x) + \varepsilon, \quad a = 2\varepsilon,$$

$$\underline{b} = \lambda_{k(x)+1}(x) + \varepsilon, \quad \overline{b} = \lambda_{p(x)}(x) + \varepsilon, \quad b = 2\varepsilon.$$

The desired result follows now from the arbitrariness of ε. $\square$

By Theorem 10.22, from the point of view of ergodic theory, "most" linear equations admit a strong nonuniform exponential dichotomy with arbitrarily small a and b, up to an appropriate Lyapunov change of coordinates. We emphasize that the study of invariant manifolds in Chapters 4–8 does not require these numbers to be zero (or even to be arbitrarily small), but only to be sufficiently small when compared to the Lyapunov exponents.

Lyapunov regularity in Hilbert spaces

The regularity theory presented in Chapter 10 is closely related to the existence of nonuniform exponential dichotomies (see Section 10.2). Unfortunately, it can only be applied to dynamical systems in finite-dimensional spaces. Hence, it is important to develop counterparts of the theory in infinite-dimensional spaces. The main goal of this chapter is precisely to introduce a version of Lyapunov regularity in Hilbert spaces, imitating as much as possible the classical theory introduced by Lyapunov in $\mathbb{R}^n$. We also describe the geometric consequences of regularity, that are related to the existence of exponential growth rates of norms, angles, and volumes determined by the solutions. We shall see in Chapter 12 that this generalization can be used to establish the persistence of the asymptotic stability of solutions of nonlinear equations under sufficiently small perturbations of Lyapunov regular equations, again in the infinite-dimensional setting of Hilbert spaces. The exposition is based in [7].

11.1 The notion of regularity

We introduce in this section the concept of Lyapunov regularity in a separable Hilbert space by closely imitating the corresponding classical notion in $\mathbb{R}^n$ (see Section 10.1).

Let H be a separable real Hilbert space (we can also consider complex Hilbert spaces with minor changes). We denote by $B(H)$ the space of bounded linear operators on H. Let $A\colon \mathbb{R}_0^+ \to B(H)$. We continue to assume that (10.1) holds, and we consider the initial value problem

$$v' = A(t)v, \quad v(0) = v_0, \tag{11.1}$$

with $v_0 \in H$. Under these assumptions, one can easily show that (11.1) has a unique solution $v(t)$ and that this solution is global for positive time. We define the Lyapunov exponent $\lambda\colon H \to \mathbb{R} \cup \{-\infty\}$ for (11.1) by

$$\lambda(v_0) = \limsup_{t \to +\infty} \frac{1}{t} \log\|v(t)\| \tag{11.2}$$

(with the convention that $\log 0 = -\infty$). We also fix an increasing sequence of subspaces $H_1 \subset H_2 \subset \cdots$ of dimension $\dim H_n = n$ for each $n \in \mathbb{N}$, such that the closure of their union is equal to H. By Proposition 10.2 (since H_n is a finite-dimensional vector space), for each $n \in \mathbb{N}$ the function λ restricted to $H_n \setminus \{0\}$ can take at most n values, say

$$-\infty \le \lambda_{1,n} < \cdots < \lambda_{p_n,n} \text{ for some integer } p_n \le n. \tag{11.3}$$

Furthermore, for each $i = 1, \ldots, p_n$ the set

$$E_{i,n} = \{v \in H_n : \lambda(v) \le \lambda_{i,n}\} \tag{11.4}$$

is a linear subspace of H_n. We can also consider the values

$$\lambda'_{1,n} \le \cdots \le \lambda'_{n,n} \tag{11.5}$$

of the Lyapunov exponent λ on $H_n \setminus \{0\}$ counted with multiplicities, obtained by repeating each value $\lambda_{i,n}$ a number of times equal to the difference $\dim E_{i,n} - \dim E_{i-1,n}$ (with $E_{0,n} = \{0\}$).

We consider the initial value problem for the adjoint equation

$$w' = -A(t)^*w, \quad w(0) = w_0, \tag{11.6}$$

with $w_0 \in H$, where $A(t)^*$ denotes the transpose of the operator $A(t)$. We define the Lyapunov exponent $\mu\colon H \to \mathbb{R} \cup \{-\infty\}$ for (11.6) by

$$\mu(w_0) = \limsup_{t \to +\infty} \frac{1}{t} \log\|w(t)\|. \tag{11.7}$$

Again by Proposition 10.2, for each $n \in \mathbb{N}$ the function μ restricted to $H_n \setminus \{0\}$ can take at most n values, say

$$-\infty \le \mu_{q_n,n} < \cdots < \mu_{1,n} \text{ for some integer } q_n \le n. \tag{11.8}$$

Furthermore, for each $i = 1, \ldots, q_n$ the set

$$F_{i,n} = \{w \in H_n : \mu(w) \le \mu_{i,n}\} \tag{11.9}$$

is a linear subspace of H_n. Similarly, we consider the values

$$\mu'_{1,n} \ge \cdots \ge \mu'_{n,n} \tag{11.10}$$

of the Lyapunov exponent μ on $H_n \setminus \{0\}$ counted with multiplicities, obtained by repeating each value $\mu_{i,n}$ a number of times equal to the difference $\dim F_{i,n} - \dim F_{i+1,n}$ (with $F_{n+1,n} = \{0\}$).

We always assume in this chapter that the Lyapunov exponents λ and μ take exactly the countable number of values in

$$\{\lambda'_{i,n} : n \in \mathbb{N}, i = 1, \ldots, n\} \quad \text{and} \quad \{\mu'_{i,n} : n \in \mathbb{N}, i = 1, \ldots, n\}, \quad (11.11)$$

and no other value. We denote respectively by λ'_i and μ'_i, for $i \in \mathbb{N}$, the values of λ and μ on $H \setminus \{0\}$ counted with multiplicities.

We recall that two bases $v_1, \ldots, v_n$ and $w_1, \ldots, w_n$ of H_n are said to be *dual* if $\langle v_i, w_j \rangle = \delta_{ij}$ for every i and j, where δ_{ij} is the Kronecker symbol. Imitating the abstract theory of Lyapunov exponents in finite-dimensional spaces, we introduce the regularity coefficient.

Definition 11.1. *We define the* regularity coefficient *of λ and μ by*

$$\gamma(\lambda, \mu) = \sup\{\gamma_n(\lambda, \mu) : n \in \mathbb{N}\},$$

where

$$\gamma_n(\lambda, \mu) = \min \max\{\lambda(v_i) + \mu(w_i) : 1 \leq i \leq n\}, \quad (11.12)$$

with the minimum taken over all dual bases $v_1, \ldots, v_n$ and $w_1, \ldots, w_n$ of the space H_n.

It follows from Proposition 10.4, applied to the Lyapunov exponents λ and μ restricted to the finite-dimensional vector space H_n, that $\gamma_n(\lambda, \mu) \geq 0$ for each $n \in \mathbb{N}$, and thus $\gamma(\lambda, \mu) \geq 0$.

Definition 11.2. *We say that the equation in (11.1) is* (Lyapunov) regular *if $\gamma(\lambda, \mu) = 0$.*

Note that $\gamma(\lambda, \mu) = 0$ if and only if $\gamma_n(\lambda, \mu) = 0$ for every $n \in \mathbb{N}$.

We refer to Sections 11.3, 11.4, and 11.5 for several alternative characterizations of Lyapunov regularity. We note that in the finite-dimensional case the notion in Definition 11.2 coincides with the classical notion introduced by Lyapunov. When there exists $\delta > 0$ such that

$$-\infty \leq \lambda'_1 \leq \lambda'_2 \leq \cdots < -\delta \quad \text{and} \quad \mu'_1 \geq \mu'_2 \geq \cdots > \delta, \quad (11.13)$$

the Lyapunov regularity of the equation in (11.1) can be shown to imply that (see Theorem 11.7)

$$\lambda'_i + \mu'_i = 0 \text{ for every } i \in \mathbb{N}. \quad (11.14)$$

In view of Theorem 10.18, property (11.14) can be seen as a justification of the notion of regularity in Hilbert spaces given in Definition 11.2 (see also the discussion in Section 11.4).

We would like to clarify why we work with Hilbert spaces instead of Banach spaces. One should be able to proceed with a formal generalization and effect an analogous approach in the case of Banach spaces, namely for operators $A(t)$ in Banach spaces with a Schauder basis. This is the case for example of the spaces $L^p[0,1]$. We note that a Banach space with a Schauder basis must be separable, although not all separable Banach spaces have a Schauder basis, as shown by Enflo in [35]. The present approach in the case of Hilbert spaces starts by considering finite-dimensional subspaces. To effect a generalization for Banach spaces, one can try to study the adjoint equation in the dual space, and consider corresponding finite-dimensional objects for the Banach space and its dual (starting with the subspaces and the associated differential equations), instead of only finite-dimensional objects for the original space. Due to this additional technical complication, the writing would hide the main principles of the approach presented here, while this does not happen in the case of Hilbert spaces (see in particular the proof of Theorem 11.3). Another difficulty is that several norm estimates in the proofs strongly use the fact that we are in a Hilbert space. In the case of Banach spaces it may not possible to establish such strong estimates. Finally, one of the crucial aspects of the classical concept of regularity is the subexponential asymptotic behavior of the *angles* between solutions (see Section 11.5). In the case of Banach spaces we should be able to consider norms of projections instead of angles (this should be compared with the discussion in Section 2.2).

11.2 Upper triangular reduction

We first perform a reduction of an arbitrary function $A(t)$ to an upper triangular function, in the following sense. We fix an orthonormal basis of H by vectors u_1, u_2, ... (recall that H is a separable Hilbert space), such that $H_n = \text{span}\{u_1, \ldots, u_n\}$ for each n, that is, the first n elements of the basis generate H_n. We show that it is always possible to reduce the case of a general function $A(t)$ to that when $A(t)$ is *upper triangular* for each $t \geq 0$, with respect to the fixed basis u_1, u_2, ... of H. This means that

$$\langle A(t)u_i, u_j \rangle = 0 \text{ for each } t \geq 0 \text{ whenever } i < j.$$

Note that the basis is independent of t. The upper triangular reduction is important in the sequel, since it allows us to reduce the study of an infinite-dimensional system to an (infinite) collection of finite-dimensional systems.

Theorem 11.3. *For a continuous function $A \colon \mathbb{R}_0^+ \to B(H)$, there exist continuous functions $B \colon \mathbb{R}_0^+ \to B(H)$ and $U \colon \mathbb{R}_0^+ \to B(H)$ such that:*

1. *$B(t)$ is upper triangular (that is, $\langle B(t)u_i, u_j \rangle = 0$ whenever $i < j$) and $\|B(t)|H_n\| \leq 2n\|A(t)\|$ for each $t \geq 0$ and $n \in \mathbb{N}$;*
2. *U is Fréchet differentiable, $U(0) = \text{Id}$, and for each $t \geq 0$, $U(t)$ is unitary and (10.37) holds;*
3. *the initial value problem (11.1) is equivalent to*

$$x' = B(t)x, \quad x(0) = v_0, \tag{11.15}$$

and the solutions $v(t)$ of (11.1) and $x(t)$ of (11.15) satisfy $v(t) = U(t)x(t)$.

Furthermore, if the function A satisfies (10.1), *then*

$$\limsup_{t \to +\infty} \frac{1}{t} \log^+ \|B(t)|H_n\| = 0 \text{ for each } n \in \mathbb{N}. \tag{11.16}$$

Proof. We follow closely arguments in the proof of Theorem 10.11, although now in the infinite-dimensional setting. Namely, we construct the operator $U(t)$ by applying the Gram–Schmidt orthogonalization procedure to the vectors $v_1(t)$, $v_2(t)$, ..., where $v_i(t)$ is the solution of (11.1) with $v_0 = u_i$ for each $i \geq 1$ (where u_1, u_2, ... is the fixed orthonormal basis of H). In this manner, we obtain functions $u_1(t)$, $u_2(t)$, ... such that:

  1. $\langle u_i(t), u_j(t) \rangle = \delta_{ij}$ for each i and j;
  2. each function $u_k(t)$ is a linear combination of $v_1(t)$, ..., $v_k(t)$.
Given $t \geq 0$ we define the linear operator $U(t) \colon H \to H$ such that $U(t)u_i = u_i(t)$ for each i. Clearly, the operator $U(t)$ is unitary for each t, and $t \mapsto U(t)$ is Fréchet differentiable with

$$U'(t)u_i = u_i'(t) \text{ for each } i.$$

Proceeding as in the proof of Theorem 10.11 we find that $x(t) = U(t)^{-1}v(t)$ is the solution of the initial value problem (11.15) with $B(t)$ given by (10.37), and that $B(t)$ is upper triangular. Clearly, B is a continuous function.

Write for each $i, j \in \mathbb{N}$ and $t \geq 0$,

$$b_{ij}(t) = \langle B(t)u_i, u_j \rangle \quad \text{and} \quad \tilde{a}_{ij}(t) = \langle A(t)u_i(t), u_j(t) \rangle.$$

Since $U(t)$ is unitary, the vectors $u_1(t) = U(t)u_1$, $u_2(t) = U(t)u_2$, ... form an orthonormal basis of H, and thus

$$\|A(t)\| \geq \|A(t)u_i(t)\| = \left\| \sum_{j=1}^{\infty} \langle Au_i(t), u_j(t) \rangle u_j(t) \right\|$$

$$= \left(\sum_{j=1}^{\infty} \tilde{a}_{ji}(t)^2 \right)^{1/2} \geq |\tilde{a}_{ij}(t)|$$

for every i and j. It follows from (10.44) that $|b_{ij}(t)| \leq 2\|A(t)\|$ for every i and j. Given $v = \sum_{i=1}^{n} \alpha_i u_i \in H_n$ with $\|v\| = (\sum_{i=1}^{n} \alpha_i^2)^{1/2} = 1$, we obtain

$$\|B(t)v\|^2 = \left\| \sum_{i=1}^{n} \sum_{j=1}^{n} \alpha_i \langle B(t)u_i, u_j \rangle u_j \right\|^2$$

$$= \sum_{j=1}^{n} \left(\sum_{i=j}^{n} \alpha_i b_{ij}(t) \right)^2 \leq \sum_{j=1}^{n} \left(\sum_{i=j}^{n} \alpha_i^2 \sum_{i=j}^{n} b_{ij}(t)^2 \right) \tag{11.17}$$

$$\leq \sum_{j=1}^{n} \sum_{i=j}^{n} b_{ij}(t)^2 \leq 4n^2 \|A(t)\|^2.$$

Therefore, $\|B(t)|H_n\| \leq 2n\|A(t)\|$, and the property (11.16) follows immediately from (10.1). For the last statement in the theorem it remains to observe that $U(t)$ is invertible for each t, and that $v(0) = x(0) = v_0$ since $U(0) = \mathrm{Id}$. This establishes the theorem.   □

The advantage of considering upper triangular systems is that we can consider finite-dimensional systems in $H_n = \mathrm{span}\{u_1, \ldots, u_n\}$ given by

$$y_n' = B_n(t)y_n, \text{ with } B_n(t) = B(t)|H_n \text{ and } y_n(0) = v_0|H_n, \qquad (11.18)$$

since for each n the space H_n is invariant under solutions of (11.15). We can obtain the solution of (11.15) in the form $y(t) = \lim_{n\to\infty} y_n(t)$. The property (11.16) ensures that for each $n \in \mathbb{N}$ the initial value problem in (11.18) has a unique and global solution. In this manner the initial value problem (11.1) becomes essentially a finite-dimensional problem.

11.3 Regularity coefficient and Perron coefficient

We use the same notation as in Section 11.1. In particular, we consider the values

$$\lambda_{1,n}' \leq \cdots \leq \lambda_{n,n}' \quad \text{and} \quad \mu_{1,n}' \geq \cdots \geq \mu_{n,n}' \qquad (11.19)$$

respectively of the Lyapunov exponents λ and μ on $H_n \setminus \{0\}$ counted with multiplicities (see (11.5) and (11.10)). We imitate once more the abstract theory of Lyapunov exponents in finite-dimensional spaces.

Definition 11.4. *We define the* Perron coefficient *of λ and μ by*

$$\pi(\lambda, \mu) = \sup\{\lambda_i' + \mu_i' : i \in \mathbb{N}\}.$$

We also consider for each $n \in \mathbb{N}$ the number

$$\pi_n(\lambda, \mu) = \max\{\lambda_{i,n}' + \mu_{i,n}' : i = 1, \ldots, n\}.$$

In the abstract theory of Lyapunov exponents in finite-dimensional spaces the numbers $\gamma_n(\lambda, \mu)$ (see (11.12)) and $\pi_n(\lambda, \mu)$ are called respectively the *regularity coefficient* and the *Perron coefficient* of λ and μ (see Definitions 10.3 and 10.15).

The following theorem establishes some relations between the regularity coefficients and the Perron coefficients.

Theorem 11.5. *For each $n \in \mathbb{N}$,*

$$0 \leq \pi_n(\lambda, \mu) \leq \gamma_n(\lambda, \mu) \leq n\,\pi_n(\lambda, \mu). \qquad (11.20)$$

In addition, if there exists $\delta > 0$ such that (11.13) holds, then

$$0 \leq \pi(\lambda, \mu) = \lim_{n\to\infty} \pi_n(\lambda, \mu) \leq \lim_{n\to\infty} \gamma_n(\lambda, \mu) \leq \gamma(\lambda, \mu). \qquad (11.21)$$

Proof. The first statement follows from Theorem 10.16 applied to the Lyapunov exponents λ and μ restricted to the finite-dimensional space H_n.

To show that the sequence $(\pi_n)_n$ with $\pi_n = \pi_n(\lambda, \mu)$ is convergent, note that by the monotonicity in (11.13), given $\varepsilon > 0$ one can choose $k \in \mathbb{N}$ such that

$$\lambda'_i \in (a - \varepsilon, a) \text{ and } \mu'_i \in (b, b + \varepsilon) \text{ for every } i \geq k, \qquad (11.22)$$

where $a = \sup_i \lambda'_i$ and $b = \inf_i \mu'_i$. In particular,

$$a + b - \varepsilon < \lambda'_k + \mu'_k < a + b + \varepsilon. \qquad (11.23)$$

Furthermore, the numbers λ'_i and μ'_i are obtained respectively from collecting the numbers $\lambda'_{j,n}$ and $\mu'_{j,n}$. More precisely, for each $i \in \mathbb{N}$ there exist integers $n, p, q \in \mathbb{N}$, with $p \leq n$ and $q \leq n$, such that

$$\lambda'_i = \lambda'_{p,n} \quad \text{and} \quad \mu'_i = \mu'_{q,n}.$$

We have $i \geq p$ and $i \geq q$, and these inequalities may be strict. However, since the sequence H_n is increasing, for a given integer k, if n is sufficiently large, then all numbers in

$$\lambda'_1 \leq \cdots \leq \lambda'_k \quad \text{and} \quad \mu'_1 \geq \cdots \geq \mu'_k$$

must occur respectively in the two finite sequences in (11.19) (otherwise they would not occur as values of the Lyapunov exponents λ and μ). But due to the monotonicity of the sequences (see (11.13) and (11.19)), we conclude that

$$\lambda'_{i,n} = \lambda'_i \quad \text{and} \quad \mu'_{i,n} = \mu'_i$$

for every $i \leq k$ (and every sufficiently large n). Therefore, in view of (11.22),

$$\max\{c_k, a + b - \varepsilon\} \leq \pi_n \leq \max\{c_k, a + b + \varepsilon\},$$

where $c_k = \max\{\lambda'_i + \mu'_i : 1 \leq i \leq k\}$. By (11.23), we conclude that

$$c_k - 2\varepsilon \leq \max\{c_k, \lambda'_k + \mu'_k - 2\varepsilon\} \leq \pi_n \leq \max\{c_k, \lambda'_k + \mu'_k + 2\varepsilon\} \leq c_k + 2\varepsilon.$$

Letting $k \to \infty$ we obtain $n \to \infty$, and the arbitrariness of ε in the above inequalities implies that the sequence $(\pi_n)_n$ is convergent, with limit $\pi(\lambda, \mu)$. We now show that the sequence $(\gamma_n)_n$ with $\gamma_n = \gamma_n(\lambda, \mu)$ is convergent. For each $n, m \in \mathbb{N}$ we have

$$\begin{aligned}
\gamma_{n+m} &= \min \max\{\lambda(v_i) + \mu(w_i) : 1 \leq i \leq n + m\} \\
&\leq \min \max\{\lambda(v'_i) + \mu(w'_i) : 1 \leq i \leq n + m\},
\end{aligned} \qquad (11.24)$$

where the first minimum is taken over all dual bases $v_1, \ldots, v_{n+m}$ and $w_1, \ldots, w_{n+m}$ of the space H_{n+m}, and the second minimum is taken over all dual bases $v'_1, \ldots, v'_{n+m}$ and $w'_1, \ldots, w'_{n+m}$ of the space H_{n+m} such that

$$\langle v_1', \ldots, v_n' \rangle = \langle w_1', \ldots, w_n' \rangle = H_n,$$

that is, the first n elements of each basis generate H_n. In a similar manner to that for the sequence $(\pi_n)_n$, it follows from the monotonicity in (11.13) that given $\varepsilon > 0$, if n is sufficiently large, then for each $m \in \mathbb{N}$,

$$\lambda(v_{n+i}) \in (a - \varepsilon, a) \text{ and } \mu(w_{n+i}) \in (b, b + \varepsilon) \text{ for every } i \leq m.$$

It follows from (11.24) that $\gamma_{n+m} \leq \max\{\gamma_n, a + b + \varepsilon\}$. Finally, note that for each n sufficiently large there exists $1 \leq i \leq n$ such that $\lambda(v_i) \in (a - \varepsilon, a)$ and $\mu(w_i) \in (b, b + \varepsilon)$. Therefore, for this i we have $\lambda(v_i) + \mu(w_i) > a + b - \varepsilon$, and hence,

$$\gamma_{n+m} \leq \max\{\gamma_n, a + b + \varepsilon\} \leq \max\{\gamma_n, \lambda(v_i) + \mu(w_i) + 2\varepsilon\} \leq \gamma_n + 2\varepsilon.$$

Letting $m \to \infty$ and then $n \to \infty$, we conclude from the arbitrariness of ε that $\limsup_{n\to\infty} \gamma_n \leq \liminf_{n\to\infty} \gamma_n$. The inequalities in (11.21) follow now immediately from (11.20) taking limits when $n \to \infty$. $\square$

11.4 Characterizations of regularity

Using Theorem 11.5 we can provide several characterizations of regularity (see Theorem 11.7). Further characterizations are given in Section 11.4.

We recall that a basis $v_1, \ldots, v_n$ of the space H_n is *normal* for the filtration by subspaces

$$E_1 \subset E_2 \subset \cdots \subset E_p = H_n$$

if for each $i = 1, \ldots, p$ there exists a basis of E_i composed of vectors in $\{v_1, \ldots, v_n\}$. When $v_1, \ldots, v_n$ is a normal basis for the filtration of subspaces $E_{i,n}$ with $i = 1, \ldots, p_n$ (see (11.4)) we also say that it is *normal for the Lyapunov exponent* λ (or simply *normal* when it is clear from the context to which exponent we are referring to).

We shall refer to dual bases $v_1, \ldots, v_n$ and $w_1, \ldots, w_n$ of H_n which are normal respectively for the Lyapunov exponents λ and μ, that is, respectively for the filtration by subspaces

$$E_{1,n} \subset \cdots \subset E_{p_n,n} = H_n \quad \text{and} \quad F_{q_n,n} \subset \cdots \subset F_{1,n} = H_n$$

in (11.4) and (11.9) as *dual normal bases*. The following is an immediate consequence of Proposition 10.14.

Proposition 11.6. *There exist dual normal bases* $v_1, \ldots, v_n$ *and* $w_1, \ldots, w_n$ *of the space* H_n.

We provide several characterizations of Lyapunov regularity in terms of the regularity and Perron coefficients, and in terms of the values of the Lyapunov exponents λ and μ.

Theorem 11.7. *The following properties are equivalent:*

1. *the equation in* (11.1) *is Lyapunov regular, that is,* $\gamma(\lambda, \mu) = 0$;
2. $\gamma_n(\lambda, \mu) = 0$ *for every* $n \in \mathbb{N}$;
3. $\pi_n(\lambda, \mu) = 0$ *for every* $n \in \mathbb{N}$;
4. *for every* $n \in \mathbb{N}$, *given dual normal bases* $v_1, \ldots, v_n$ *and* $w_1, \ldots, w_n$ *of the space* H_n,

$$\lambda(v_i) + \mu(w_i) = 0 \text{ for } i = 1, \ldots, n; \tag{11.25}$$

5. *for every* $n \in \mathbb{N}$,

$$\lambda'_{i,n} + \mu'_{i,n} = 0 \text{ for } i = 1, \ldots, n. \tag{11.26}$$

In addition, if (11.13) *holds for some* $\delta > 0$, *and the equation in* (11.1) *is Lyapunov regular, then* $\pi(\lambda, \mu) = 0$ *and the property* (11.14) *holds.*

Proof. By (11.20), we have $\gamma_n(\lambda, \mu) \geq 0$ for every $n \in \mathbb{N}$, and the equivalence of the first two properties is immediate from the definition of the regularity coefficient. The fact that these are equivalent to the third property follows readily from the inequalities in (11.20).

We proceed in a similar manner to that in the proof of Theorem 10.16. By Proposition 11.6 we can consider dual normal bases $v_1, \ldots, v_n$ and $w_1, \ldots, w_n$ of H_n, and hence the numbers $\lambda(v_i)$ and $\mu(w_i)$ are respectively the values of the Lyapunov exponents λ and μ on $H_n \setminus \{0\}$, counted with multiplicities, although possibly not ordered. By Proposition 10.4 we have $\lambda(v_i) + \mu(w_i) \geq 0$ for every i. Therefore,

$$0 \leq \lambda(v_i) + \mu(w_i) \leq \sum_{i=1}^{n} (\lambda(v_i) + \mu(w_i))$$
$$= \sum_{i=1}^{n} (\lambda'_{i,n} + \mu'_{i,n}) \leq n \, \pi_n(\lambda, \mu). \tag{11.27}$$

If the equation in (11.1) is regular, we have $\pi_n(\lambda, \mu) = 0$ for every $n \in \mathbb{N}$, and thus (11.25) holds. Moreover, by the definition of $\pi_n(\lambda, \mu)$ we have $\lambda'_{i,n} + \mu'_{i,n} \leq 0$ for every i, and in view of (11.27) we conclude that (11.26) follows from property 4. We now show that property 5 yields regularity. Indeed, it follows from (11.26) that $\pi_n(\lambda, \mu) = 0$ for every $n \in \mathbb{N}$, and thus the equation in (11.1) is regular.

For the last statement, observe that using (11.21) we conclude that a regular equation has Perron coefficient $\pi(\lambda, \mu) = 0$. Furthermore, in a similar manner to that in the proof of Theorem 11.5, it follows from the monotonicity in (11.13) that given $k \in \mathbb{N}$, if n is sufficiently large then

$$\lambda'_{i,n} = \lambda'_i \quad \text{and} \quad \mu'_{i,n} = \mu'_i$$

for $i \leq k$. It follows from (11.26) that $\lambda'_i + \mu'_i = 0$ for every $i \leq k$. The desired result follows now from the arbitrariness of k. $\qquad\square$

We also present alternative characterizations of regularity, expressed in terms of the existence of exponential growth rates of finite-dimensional volumes. Given vectors $v_1, \ldots, v_m \in H$ we denote by $\Gamma(v_1, \ldots, v_m)$ the m-volume defined by these vectors (see Section 10.4 for the definition).

With a slight abuse of notation, given $v \in H$ we denote by $v(t)$ the solution of (11.1) with $v(0) = v$. For a given continuous function $A(t)$ we consider also the new function $B(t)$ given by Theorem 11.3 which is upper triangular for each $t \geq 0$.

Theorem 11.8. *The following properties are equivalent:*

1. the equation in (11.1) is Lyapunov regular, that is, $\gamma(\lambda, \mu) = 0$;
2. for each $n \in \mathbb{N}$, and each normal basis $v_1, \ldots, v_n$ of H_n,

$$\lim_{t \to +\infty} \frac{1}{t} \log \Gamma(v_1(t), \ldots, v_n(t)) = \sum_{i=1}^{p_n} \lambda_{i,n} = \sum_{j=1}^{n} \lambda'_{j,n};$$

3. given $n, m \in \mathbb{N}$ with $m \leq n$, and a normal basis $v_1, \ldots, v_n$ of H_n the limit

$$\lim_{t \to +\infty} \frac{1}{t} \log \Gamma(v_1(t), \ldots, v_m(t))$$

exists;
4. for each $n \in \mathbb{N}$,

$$\lim_{t \to +\infty} \frac{1}{t} \int_0^t \operatorname{tr}(B(s)|H_n) \, ds = \sum_{i=1}^{p_n} \lambda_{i,n} = \sum_{j=1}^{n} \lambda'_{j,n}.$$

Proof. We first proceed as in the proof of Theorem 10.12. We recall that by Theorem 11.3 the initial value problem (11.1) is equivalent to

$$x' = B(t)x, \quad x(0) = v_0, \tag{11.28}$$

with the solutions $v(t)$ of (11.1) and $x(t)$ of (11.28) related by $v(t) = U(t)x(t)$ with $U(t)$ unitary for each $t \geq 0$. Similarly, the initial value problem (11.6) is equivalent to

$$y' = -B(t)^* y, \quad y(0) = w_0, \tag{11.29}$$

with the solutions $w(t)$ of (11.6) and $y(t)$ of (11.29) related by $w(t) = U(t)y(t)$ using the same operator $U(t)$ (see (10.46) in the proof of Theorem 10.12). Since the operator $U(t)$ is unitary for each t, the Lyapunov exponents for the equations in (11.28) and (11.29) coincide, respectively, with the Lyapunov exponents λ and μ for the equations in (11.1) and (11.6). We continue to denote respectively by λ and μ the Lyapunov exponents of (11.28) and (11.29). Furthermore, the regularity coefficient of the new pair of equations ((11.28) and (11.29)) is the same at that for the equations (11.1) and (11.6).

In view of the above discussion, the equation in (11.28) is Lyapunov regular if and only if the same happens with (11.1). By Theorem 11.7, these equations are Lyapunov regular if and only if $\gamma_n(\lambda, \mu) = 0$ for every $n \in \mathbb{N}$.

Since $B(t)$ is upper triangular with respect to the basis $u_1, u_2, \ldots$ of H, and for each n the space H_n is spanned by $u_1, \ldots, u_n$, we have $B(t)H_n \subset H_n$ for each $t \geq 0$ and each $n \in \mathbb{N}$. Therefore, we can consider the equation

$$x' = (B(t)|H_n)x$$

in the finite-dimensional vector space H_n. By Theorem 10.19 the following properties are equivalent:

1. $\gamma_n(\lambda, \mu) = 0$;
2.

$$\lim_{t \to +\infty} \frac{1}{t} \int_0^t \mathrm{tr}(B(s)|H_n)\, ds = \sum_{i=1}^{p_n} \lambda_{i,n};$$

3. given $m \leq n$, and a normal basis $v_1, \ldots, v_n$ of H_n the limit

$$\lim_{t \to +\infty} \frac{1}{t} \log \Gamma(x_1(t), \ldots, x_m(t))$$

exists, where each $x_i(t)$ is the solution of (11.28) with $x_i(0) = v_i$. Note that since $U(t)$ is unitary for each t, we have

$$\langle v_i(t), v_j(t) \rangle = \langle U(t)x_i(t), U(t)x_j(t) \rangle = \langle x_i(t), x_j(t) \rangle.$$

Therefore,

$$\Gamma(v_1(t), \ldots, v_m(t)) = \Gamma(x_1(t), \ldots, x_m(t)). \tag{11.30}$$

Furthermore, it is well known in the finite-dimensional setting that

$$\frac{\Gamma(v_1(t), \ldots, v_n(t))}{\Gamma(v_1, \ldots, v_n)} = \exp \int_0^t \mathrm{tr}(B(s)|H_n)\, ds. \tag{11.31}$$

The desired statement can now be easily obtained by putting together the above results. □

11.5 Lower and upper bounds for the coefficients

We obtain here sharp lower and upper bounds for the Perron coefficient and the regularity coefficient, in terms of the upper triangular operator $B(t)$.

We first obtain bounds for the exponential growth rate of volumes.

Theorem 11.9. *For each given $n \in \mathbb{N}$, let $v_1, \ldots, v_n$ be a normal basis of H_n such that $v_1, \ldots, v_{n-1}$ is a normal basis of H_{n-1}. Then we have the following properties:*

1.

$$\limsup_{t\to+\infty} \frac{1}{t} \log \Gamma_n(t) = \limsup_{t\to+\infty} \frac{1}{t} \int_0^t \operatorname{tr}(B(s)|H_n)\, ds \le a_n,$$

$$\liminf_{t\to+\infty} \frac{1}{t} \log \Gamma_n(t) = \liminf_{t\to+\infty} \frac{1}{t} \int_0^t \operatorname{tr}(B(s)|H_n)\, ds \qquad (11.32)$$

$$\ge a_n - n\,\pi_n(\lambda,\mu)$$

$$\ge a_n - n\,\gamma_n(\lambda,\mu),$$

where

$$\Gamma_n(t) = \Gamma(v_1(t), \ldots, v_n(t)) \quad and \quad a_n = \sum_{i=1}^{p_n} \lambda_{i,n} = \sum_{j=1}^{n} \lambda'_{j,n};$$

2.

$$\limsup_{t\to+\infty} \frac{1}{t} \log \rho_n(t) = \limsup_{t\to+\infty} \frac{1}{t} \int_0^t b_{nn}(s)\, ds$$

$$\le \lambda(v_n) + (n-1)\pi_{n-1}(\lambda,\mu)$$

$$\le \lambda(v_n) + (n-1)\gamma_{n-1}(\lambda,\mu), \qquad (11.33)$$

$$\liminf_{t\to+\infty} \frac{1}{t} \log \rho_n(t) = \liminf_{t\to+\infty} \frac{1}{t} \int_0^t b_{nn}(s)\, ds$$

$$\ge \lambda(v_n) - n\pi_n(\lambda,\mu) \ge \lambda(v_n) - n\gamma_n(\lambda,\mu),$$

where $b_{nn}(t) = \langle B(t)u_n, u_n \rangle$, and $\rho_n(t)$ is the distance from $v_n(t)$ to the space $U(t)H_{n-1}$.

Proof. The equalities in (11.32) follow readily from (11.31). For the first inequality, note that $\Gamma_n(t) \le \prod_{j=1}^{n} \|v_j(t)\|$. Since $v_1, \ldots, v_n$ is a normal basis of H_n we obtain

$$\limsup_{t\to+\infty} \frac{1}{t} \log \Gamma_n(t) \le \sum_{j=1}^{n} \lambda(v_j) = a_n.$$

For the remaining inequalities in (11.32), note first that given a basis $w_1, \ldots, w_n$ of H_n we have an analogous identity to (11.31), namely

$$\frac{\Gamma(w_1(t), \ldots, w_n(t))}{\Gamma(w_1, \ldots, w_n)} = \exp \int_0^t \operatorname{tr}(-(B(s)|H_n)^*)\, ds, \qquad (11.34)$$

where $w_1(t), \ldots, w_n(t)$ are the solutions of (11.6) with $w_i(0) = w_i$ for each $i = 1, \ldots, n$. By (11.31) and (11.34),

$$\liminf_{t\to+\infty} \frac{1}{t} \log \Gamma_n(t) = \liminf_{t\to+\infty} \frac{1}{t} \int_0^t \operatorname{tr}(B(s)|H_n) \, ds$$

$$= -\limsup_{t\to+\infty} \frac{1}{t} \int_0^t \operatorname{tr}(-(B(s)|H_n)^*) \, ds$$

$$= -\limsup_{t\to+\infty} \frac{1}{t} \log \Gamma(w_1(t),\ldots,w_n(t)) \geq -\sum_{j=1}^n \mu(w_j).$$

We now assume that $w_1, \ldots, w_n$ is a normal basis (with respect to μ). Then, using (11.20),

$$\liminf_{t\to+\infty} \frac{1}{t} \log \Gamma_n(t) \geq a_n - \sum_{j=1}^n (\lambda(v_j) + \mu(w_j)) = a_n - \sum_{j=1}^n (\lambda'_{j,n} + \mu'_{j,n})$$

$$\geq a_n - n\,\pi_n(\lambda,\mu) \geq a_n - n\,\gamma_n(\lambda,\mu).$$

This completes the proof of the first statement.

For the second statement, we observe that in view of (11.30), since $B(t)$ is upper triangular,

$$\rho_n(t) = \frac{\Gamma_n(t)/\Gamma_n(0)}{\Gamma_{n-1}(t)/\Gamma_{n-1}(0)}, \tag{11.35}$$

and

$$\int_0^t b_{nn}(s) \, ds = \int_0^t \operatorname{tr}(B(s)|H_n) \, ds - \int_0^t \operatorname{tr}(B(s)|H_{n-1}) \, ds$$

$$= \log \frac{\Gamma_n(t)/\Gamma_n(0)}{\Gamma_{n-1}(t)/\Gamma_{n-1}(0)}. \tag{11.36}$$

Thus,

$$\rho_n(t) = \exp \int_0^t b_{nn}(s) \, ds,$$

and we obtain the equalities in (11.33). It follows from (11.35) and (11.32) that

$$\limsup_{t\to+\infty} \frac{1}{t} \log \rho_n(t) \leq \limsup_{t\to+\infty} \frac{1}{t} \log \Gamma_n(t) - \liminf_{t\to+\infty} \frac{1}{t} \log \Gamma_{n-1}(t)$$

$$\leq \sum_{j=1}^n \lambda(v_j) - \sum_{j=1}^{n-1} \lambda(v_j) + (n-1)\pi_{n-1}(\lambda,\mu)$$

$$= \lambda(v_n) + (n-1)\pi_{n-1}(\lambda,\mu)$$

$$\leq \lambda(v_n) + (n-1)\gamma_{n-1}(\lambda,\mu),$$

using (11.20) in the last inequality. Similarly, we obtain

$$\liminf_{t\to+\infty}\frac{1}{t}\log\rho_n(t)\geq\liminf_{t\to+\infty}\frac{1}{t}\log\Gamma_n(t)-\limsup_{t\to+\infty}\frac{1}{t}\log\Gamma_{n-1}(t)$$

$$\geq\sum_{j=1}^{n}\lambda(v_j)-n\pi_n(\lambda,\mu)-\sum_{j=1}^{n-1}\lambda(v_j)$$

$$=\lambda(v_n)-n\pi_n(\lambda,\mu)\geq\lambda(v_n)-n\gamma_n(\lambda,\mu).$$

This completes the proof. □

We note that there always exists bases v_1, ..., v_n as in the statement of Theorem 11.9: given a normal basis v_1, ..., v_{n-1} of H_{n-1}, it is sufficient to select any vector

$$v_n\in(E_{k,n}\setminus E_{k-1,n})\cap(H_n\setminus H_{n-1}),$$

where $k\leq p_n$ is the smallest integer such that $E_{k,n}\cap(H_n\setminus H_{n-1})\neq\varnothing$.

We now obtain the bounds for the Perron coefficient and the regularity coefficient. Set

$$\underline{\beta}_i=\liminf_{t\to+\infty}\frac{1}{t}\int_0^t\langle B(s)u_i,u_i\rangle\,ds\quad\text{and}\quad\overline{\beta}_i=\limsup_{t\to+\infty}\frac{1}{t}\int_0^t\langle B(s)u_i,u_i\rangle\,ds.$$

Theorem 11.10. *If $B(t)$ is the upper triangular operator obtained from $A(t)$ as in Theorem 11.3, then*

$$\sup_{n\geq1}\frac{1}{n^2}\sum_{i=1}^{n}(\overline{\beta}_i-\underline{\beta}_i)\leq\gamma(\lambda,\mu)\leq\sum_{i=1}^{\infty}(\overline{\beta}_i-\underline{\beta}_i).\qquad(11.37)$$

In addition, if (11.13) holds for some $\delta>0$, then

$$\limsup_{n\to\infty}\frac{1}{2n}\sum_{i=1}^{n}\frac{\overline{\beta}_i-\underline{\beta}_i}{i^2}\leq\pi(\lambda,\mu)\leq\sum_{i=1}^{\infty}(\overline{\beta}_i-\underline{\beta}_i).\qquad(11.38)$$

Proof. It follows from Theorem 10.8 that

$$\gamma_n(\lambda,\mu)\leq\sum_{i=1}^{n}(\overline{\beta}_i-\underline{\beta}_i).$$

This readily gives the second inequality in (11.37). Thus, by Theorem 11.5, we also obtain the second inequality in (11.38) provided that (11.13) holds. By Theorem 11.9 (see (11.33)), for each $i\in\mathbb{N}$,

$$\overline{\beta}_i-\underline{\beta}_i\leq(i-1)\gamma_{i-1}(\lambda,\mu)+i\gamma_i(\lambda,\mu)$$
$$\leq(2i-1)\max\{\gamma_i(\lambda,\mu):i=1,\ldots,n\}.\qquad(11.39)$$

Summing over i we obtain

$$\sum_{i=1}^{n}(\overline{\beta}_i - \underline{\beta}_i) \le n^2 \max\{\gamma_i(\lambda, \mu) : i = 1, \dots, n\}.$$

This establishes the first inequality in (11.37). For the first inequality in (11.38), note that by (11.39) and (11.20),

$$\overline{\beta}_i - \underline{\beta}_i \le (i-1)^2 \pi_{i-1}(\lambda, \mu) + i^2 \pi_i(\lambda, \mu)$$
$$\le i^2[\pi_{i-1}(\lambda, \mu) + \pi_i(\lambda, \mu)].$$

Again by Theorem 11.5, we have $\pi(\lambda, \mu) = \lim_{n\to\infty} \pi_n(\lambda, \mu)$ when (11.13) holds, and thus,

$$\pi(\lambda, \mu) = \lim_{n\to\infty} \frac{1}{n} \sum_{i=1}^{n} \pi_i(\lambda, \mu).$$

Therefore,

$$\limsup_{n\to\infty} \frac{1}{2n} \sum_{i=1}^{n} \frac{\overline{\beta}_i - \underline{\beta}_i}{i^2} \le \lim_{n\to\infty} \frac{1}{2n} \sum_{i=1}^{n}[\pi_{i-1}(\lambda, \mu) + \pi_i(\lambda, \mu)] = \pi(\lambda, \mu).$$

This completes the proof. $\square$

By Theorem 11.10, the equation $v' = A(t)v$ is Lyapunov regular if and only if $\underline{\beta}_i = \overline{\beta}_i$ for every $i \in \mathbb{N}$. In fact we can formulate a slightly stronger statement.

Theorem 11.11. *The following properties are equivalent:*

1. the equation in (11.1) is Lyapunov regular, that is, $\gamma(\lambda, \mu) = 0$;
2. for each $n \in \mathbb{N}$, the limit

$$\lim_{t\to+\infty} \frac{1}{t} \int_0^t \operatorname{tr}(B(s)|H_n)\, ds$$

exists;
3. for each $n \in \mathbb{N}$, the limit

$$\lim_{t\to+\infty} \frac{1}{t} \int_0^t \langle B(s)u_n, u_n\rangle\, ds$$

exists.

Proof. The equivalence between the first and third properties is immediate from Theorem 11.10. The equivalence to the second property follows readily from (11.36). $\square$

Stability of nonautonomous equations in Hilbert spaces

We study in this chapter the persistence of the asymptotic stability of the zero solution of a nonautonomous linear equation $v' = A(t)v$ under a perturbation f, that is, for the equation $v' = A(t)v + f(t,v)$. We recall that there are examples, going back to Perron, showing that an arbitrarily small perturbation of an asymptotically stable nonautonomous linear equation may be unstable. In fact it may be exponentially unstable in some directions, even if all Lyapunov exponents of the linear equation are negative. It is of course possible to provide additional assumptions of general nature under which the stability persists. This is the case for example with the assumption of uniform asymptotic stability for the linear equation, although this requirement is dramatically restrictive for a nonautonomous system. Incidentally, this assumption is analogous to the restrictive assumption of existence of a uniform exponential dichotomy for the evolution operator of a nonautonomous equation (instead of a nonuniform exponential dichotomy). It is thus important to look for general assumptions that are substantially weaker than the uniform asymptotic stability, under which one can still establish the persistence of stability in the nonlinear equation under sufficiently small perturbations. This is the case of the Lyapunov regularity (see Chapters 10 and 11). In particular, we show that if the linear equation is *Lyapunov regular*, then for any sufficiently small perturbation f with $f(t, 0) = 0$ for every $t \geq 0$, the zero solution of the perturbed nonlinear equation is asymptotically stable. We follow closely [7].

12.1 Setup

We consider nonlinear perturbations $v' = A(t)v + f(t,v)$ of the linear equation $v' = A(t)v$, and study the persistence of the stability of solutions under sufficiently small perturbations. Without loss of generality, we always assume that the operator $A(t)$ is *upper triangular* for every t with respect to the fixed orthonormal basis $u_1, u_2, \ldots$ of H considered in Section 11.2 (see Theorem 11.3).

We continue to assume in this chapter that the Lyapunov exponents λ and μ (see (11.2) and (11.7)) take exactly the values in (11.11), and no other value.

Consider the initial value problem

$$v' = A(t)v + f(t,v), \quad v(0) = v_0, \tag{12.1}$$

with $v_0 \in H$. We also consider the conditions:

H1. $A \colon \mathbb{R}_0^+ \to \mathcal{B}(H)$ is a continuous function satisfying (10.1) and

$$\langle A(t)u_i, u_j \rangle = 0 \text{ for every } i < j \text{ and every } t \geq 0; \tag{12.2}$$

H2. $f \colon \mathbb{R}_0^+ \times H \to H$ is a continuous function satisfying $f(t,0) = 0$ for all $t \geq 0$, and there exists constants $c, r > 0$ such that

$$\|f(t,u) - f(t,v)\| \leq c\|u - v\|(\|u\|^r + \|v\|^r)$$

for every $t \geq 0$, and $u, v \in H$;

H3. $|\langle v_0, u_n \rangle| < \|v_0\|/a_n$ for every $n \geq 0$, and

$$|\langle f(t,u) - f(t,v), u_n \rangle| \leq \frac{1}{a_n}\|u - v\|(\|u\|^r + \|v\|^r) \tag{12.3}$$

for every $t \geq 0$, $u, v \in H$, and $n \geq 0$, for some positive increasing sequence $(a_n)_n$ that diverges sufficiently fast.

Under the conditions H1–H2, it can easily be shown that the perturbed equation in (12.1) has a unique solution $v(t)$. We note that $v(t) \equiv 0$ is always a solution of (12.1).

A description of the required speed of a_n in (12.3) is given in Section 12.3. We remark that the condition (12.3) corresponds to the requirement that the perturbation is sufficiently small (with respect to some basis). It should be noted that when the perturbation is finite-dimensional, that is, when there exists $n \in \mathbb{N}$ such that $f(t,v) \in H_n$ for every $t \geq 0$ and $v \in H$, then the requirement (12.3) is not needed, since in this case

$$\langle f(t,u) - f(t,v), u_m \rangle = 0$$

for every $m > n$. On the other hand, we emphasize that the perturbations that we consider need not be finite-dimensional.

Consider now the condition

$$r \sup\{\lambda_i' : i \in \mathbb{N}\} + \gamma(\lambda, \mu) < 0, \tag{12.4}$$

where the numbers λ_i' are the values of the Lyapunov exponent λ on H. Since $\gamma(\lambda, \mu) \geq 0$ (see Section 11.1), this implies that

$$\sup\{\lambda_i' : i \in \mathbb{N}\} < 0. \tag{12.5}$$

This property ensures the asymptotic stability of the linear equation in (11.1). We recall from the introduction that the asymptotic stability of (11.1) is not sufficient to ensure the stability of the zero solution of (12.1). In fact, there exist examples for which a small perturbation f makes zero an exponentially *unstable* solution (an explicit example is given in the introduction).

12.2 Stability results

We formulate here the results on the persistence of stability of the zero solution of (11.1) under perturbations. It should be emphasized that the results deal with equations in which the operators $A(t)$ are bounded for every t. This has some drawbacks, since stability questions arise naturally in nonautonomous partial differential equations in which the operators $A(t)$ may be unbounded.

Theorem 12.1. *If conditions H1–H3 and (12.4) hold, then for any positive sequence $(a_n)_n$ diverging sufficiently fast, given $\varepsilon > 0$ sufficiently small there exists a constant $a > 0$ such that any solution of the equation (12.1) with $\|v_0\|$ sufficiently small is global and satisfies*

$$\|v(t)\| \le ae^{(\sup\{\lambda_i':i\in\mathbb{N}\}+\varepsilon)t}\|v_0\| \ \text{for every } t \ge 0. \tag{12.6}$$

Note that $\sup\{\lambda_i' : i \in \mathbb{N}\} + \varepsilon < 0$ for every sufficiently small $\varepsilon > 0$. The proof of Theorem 12.1 and of the remaining results in this section are given in Sections 12.4 and 12.5. The following is an immediate corollary of Theorem 12.1 for regular equations.

Theorem 12.2. *If conditions H1–H3 and (12.5) hold, and the equation in (11.1) is Lyapunov regular, then for any positive sequence $(a_n)_n$ diverging sufficiently fast, given $\varepsilon > 0$ sufficiently small there exists a constant $a > 0$ such that any solution of the equation (12.1) with $\|v_0\|$ sufficiently small is global and satisfies (12.6).*

Theorem 12.1 establishes the persistence of stability of the zero solution allowing a certain degree of nonregularity for the equation in (11.1), that is, it may happen that $\gamma(\lambda, \mu) > 0$. We note that by (12.4) a higher order r of the perturbation f allows a larger regularity coefficient. When $\gamma(\lambda, \mu) > 0$ the angles between distinct solutions may vary with exponential speed, essentially related to $\gamma(\lambda, \mu)$, although this speed is small when compared to the values of the Lyapunov exponent, that is, to $\inf\{|\lambda_i'| : i \in \mathbb{N}\}$. This strongly contrasts to what happens in Theorem 12.2 in which case the regularity assumption forces the angles between distinct solutions to vary at most with subexponential speed. We refer to Section 11.5 for a detailed discussion.

We now formulate an abstract stability result which will be obtained as a consequence of the proof of Theorem 12.1. It is somewhat more explicit about the required speed of a_n in (12.3). We continue to assume that the operator $A(t)$ is upper triangular for every t. Let $X(t)$ be (upper triangular) monodromy operators for the equation $v' = A(t)v$. These are operators such that the solution with $v(0) = v_0$ is given by $v(t) = X(t)X(0)^{-1}v_0$.

Theorem 12.3. *Assume that conditions H1–H3 hold, and that there exist constants $\alpha < 0$ and $\beta > 0$, with $r\alpha + \beta < 0$, and a positive sequence $(c_n)_n$ with $\sum_{k=1}^{\infty} c_k/a_k < \infty$ such that for every $n \in \mathbb{N}$ and $t \ge s \ge 0$,*

$$\|X(t)X(s)^{-1}|H_n\| \le c_n e^{\alpha(t-s)+\beta s}.\tag{12.7}$$

Then there exists a constant $a > 0$ such that any solution of the equation (12.1) with $\|v_0\|$ sufficiently small is global and satisfies

$$\|v(t)\| \le ae^{\alpha t}\|v_0\| \text{ for every } t \ge 0.\tag{12.8}$$

Note that Theorem 12.3 tells us that the required speed for the sequence $(a_n)_n$ is related to norm estimates for the evolution operators $X(t)X(s)^{-1}$ in finite-dimensional spaces (we can set for example $a_n = c_n n^{1+\tau}$ with $\tau > 0$).

The following is another consequence of the proof of Theorem 12.1. It has the advantage of not mentioning the spaces H_n, although at the expense of requiring more from the evolution operators.

Theorem 12.4. *Assume that conditions H1–H2 hold, and that there exist $\alpha < 0$ and $\beta > 0$, with $r\alpha + \beta < 0$, and $C > 0$ such that*

$$\|X(t)X(s)^{-1}\| \le Ce^{\alpha(t-s)+\beta s} \text{ for every } t \ge s \ge 0.$$

Then there exists a constant $a > 0$ such that any solution of the equation (12.1) with $\|v_0\|$ sufficiently small is global and satisfies (12.8).

We also consider the finite-dimensional case. For simplicity we consider the space $H = \mathbb{R}^n$ with the standard inner product. In this case we can obtain the following stronger statement, where $M(\mathbb{R}^n)$ is the set of $n \times n$ matrices with real entries.

Theorem 12.5 ([1, Theorem 1.4.3]). *Assume that:*

1. *$A\colon \mathbb{R}_0^+ \to M(\mathbb{R}^n)$ is a continuous function satisfying (10.1);*
2. *$f\colon \mathbb{R}_0^+ \times \mathbb{R}^n \to \mathbb{R}^n$ is a continuous function satisfying $f(t,0) = 0$ for all $t \ge 0$, and there exist constants c, $r > 0$ such that for every $t \ge 0$, and $u, v \in \mathbb{R}^n$,*

$$\|f(t,u) - f(t,v)\| \le c\|u-v\|(\|u\|^r + \|v\|^r);$$

3. *$r \sup\{\lambda_i' : i = 1,\ldots,n\} + \gamma_n(\lambda,\mu) < 0.$*

Then the solution $v(t) \equiv 0$ of the perturbed equation (12.1) is exponentially stable.

We will obtain Theorem 12.5 as a consequence of the infinite-dimensional version in Theorem 12.1.

12.3 Smallness of the perturbation

We describe here the required speed of the sequence $(a_n)_n$ in (12.3). For each fixed $n \in \mathbb{N}$, we consider dual bases $v_1, \ldots, v_n$ and $w_1, \ldots, w_n$ of H_n such that

$$\max\{\lambda(v_i) + \mu(w_i) : i = 1, \ldots, n\} = \gamma_n(\lambda, \mu) \qquad (12.9)$$

(this is always possible since the minimum in (11.12) attains at most a finite number of values). It follows easily from the definition of the Lyapunov exponents that given $n \in \mathbb{N}$ and $\varepsilon > 0$ there exists a constant $D_{\varepsilon,n} > 0$ such that

$$\|v_i(t)\| \leq D_{\varepsilon,n} e^{(\lambda(v_i)+\varepsilon)t} \quad \text{and} \quad \|w_i(t)\| \leq D_{\varepsilon,n} e^{(\mu(w_i)+\varepsilon)t}, \qquad (12.10)$$

for every $t \geq 0$ and $i = 1, \ldots, n$, where $v_i(t)$ is the solution of (11.1) with $v_0 = v_i$, and $w_i(t)$ is the solution of (11.6) with $w_0 = w_i$ for each i. We assume that the sequence $(a_n)_n$ diverges sufficiently fast so that

$$d := \sum_{k=1}^{\infty} \frac{k^2 D_{\varepsilon,k}^2}{a_k} < \infty \qquad (12.11)$$

for some choice of dual bases $v_1, \ldots, v_n$ and $w_1, \ldots, w_n$ of H_n satisfying (12.9), and of $\varepsilon > 0$ satisfying

$$r(\sup\{\lambda_i' : i \in \mathbb{N}\} + \varepsilon) + \gamma(\lambda, \mu) + 2\varepsilon < 0. \qquad (12.12)$$

Note that in view of (12.4) any sufficiently small $\varepsilon > 0$ satisfies (12.12).

In the particular case of a regular equation, the constants $D_{\varepsilon,n}$ in (12.10) can be made somewhat more explicit. We recall the numbers $\lambda_{i,n}$ and p_n in (11.3), and the numbers $\mu_{i,n}$ and q_n in (11.8).

Proposition 12.6. *When the equation in* (11.1) *is Lyapunov regular, we have* $D_{\varepsilon,n} \leq \max\{c_{\varepsilon,n}, d_{\varepsilon,n}\}$ *with*

$$c_{\varepsilon,n} = \sup_{1 \leq i \leq p_n} \sup \left\{ \sup_{t \geq 0} \frac{\|v(t)\|}{e^{(\lambda_{i,n}+\varepsilon)t}} : v(0) \in E_{i,n} \right\},$$

$$d_{\varepsilon,n} = \sup_{1 \leq i \leq q_n} \sup \left\{ \sup_{t \geq 0} \frac{\|w(t)\|}{e^{(\mu_{i,n}+\varepsilon)t}} : w(0) \in F_{i,n} \right\},$$

where $v(t)$ *is a solution of* (11.1) *and* $w(t)$ *is a solution of* (11.6).

Proof. Due to Proposition 11.6 there exist dual normal bases $v_1, \ldots, v_n$ and $w_1, \ldots, w_n$ of the space H_n. Furthermore, by Theorem 11.7 the regularity implies that $\lambda(v_i) + \mu(w_i) = 0$ for every i, and hence

$$0 \leq \gamma_n(\lambda, \mu) \leq \max\{\lambda(v_i) + \mu(w_i) : i = 1, \ldots, n\} = 0.$$

Therefore, we can consider these bases $v_1, \ldots, v_n$ and $w_1, \ldots, w_n$ when we define $D_{\varepsilon,n}$ by the inequalities (12.10). Since these are normal bases we readily obtain the desired result. $\qquad \square$

In the case of a "uniform" behavior of the Lyapunov exponents, we can be more explicit about the smallness condition on the perturbation f. Namely, assume that for each $\varepsilon > 0$ there exists $C = C(\varepsilon) > 0$ such that

$$\|v(t)\| \le Ce^{(\lambda(v)+\varepsilon)t}\|v(0)\| \quad \text{and} \quad \|w(t)\| \le Ce^{(\mu(v)+\varepsilon)t}\|w(0)\| \qquad (12.13)$$

for every $t \ge 0$ and every $v(0) \in H$, where $v(t)$ is a solution of (11.1) and $w(t)$ is a solution of (11.6). The following is a version of Theorem 12.1 in this particular case.

Theorem 12.7. *Assume that conditions H1–H3, (12.4), and (12.13) hold. If $\sum_{k=1}^{\infty} k^2/a_k < \infty$, then given $\varepsilon > 0$ sufficiently small there exists a constant $a > 0$ such that any solution of the equation (12.1) with $\|v_0\|$ sufficiently small is global and satisfies (12.6).*

Proof. This is an immediate consequence of Theorem 12.1 and of the above description of the required speed of $(a_n)_n$ in (12.3): set $D_{\varepsilon,n} = C$ in (12.11). $\qquad \square$

Alternatively, Theorem 12.7 can be obtained combining Theorem 12.3 with the norm estimates for the evolution operators obtained in Theorem 12.8.

12.4 Norm estimates for the evolution operators

Here we establish crucial estimates for the proofs of the stability results. We use the same notation as in the proof of Theorem 11.3. Namely, let $v_i(t)$ be the solution of (11.1) with $v_0 = u_i$ for each $i \ge 1$. We define an operator $V(t)\colon H \to H$ such that $V(t)u_i = v_i(t)$ for each $i \ge 1$. Then, proceeding as in the proof of Theorem 10.11, we find that the operator $X(t) = U(t)^{-1}V(t)$ (with $U(t)$ as in Theorem 11.3) is upper triangular and satisfies

$$X'(t) = B(t)X(t) \text{ for } t \ge 0, \qquad (12.14)$$

that is, $X(t)$ is a monodromy operator for the equation $x' = B(t)x$. In the following result we obtain bounds on the norm of the evolution operator $X(t)X(s)^{-1}$ restricted to each finite-dimensional space H_n by combining information about the solutions of the equations

$$v' = A(t)v \quad \text{and} \quad w' = -A(t)^*w$$

through the study of the Lyapunov exponents λ and μ. For each $n \in \mathbb{N}$, we fix dual bases $v_1, \ldots, v_n$ and $w_1, \ldots, w_n$ of H_n satisfying (12.9) and (12.10). We recall that $\lambda'_{n,n} = \lambda_{p_n,n}$ (see (11.3) and (11.5)) is the top value of the Lyapunov exponent λ (for the equation (11.1)) on $H_n \setminus \{0\}$.

Theorem 12.8. *For every $n \in \mathbb{N}$, $\varepsilon > 0$, and $t \ge s \ge 0$ we have*

$$\|X(t)X(s)^{-1}|H_n\| \le n^2 D_{\varepsilon,n}^2 e^{(\lambda'_{n,n}+\varepsilon)(t-s)+(\gamma_n(\lambda,\mu)+2\varepsilon)s}.$$

Proof. Consider the operator $Y(t) = [X(t)^{-1}]^*$ for each t. Taking derivatives in the identity

$$X(t)X(t)^{-1} = X(t)Y(t)^* = \mathrm{Id}$$

we obtain

$$X'(t)X(t)^{-1} + X(t)Y'(t)^* = 0.$$

It follows from (12.14) that

$$X(t)Y'(t)^* = -B(t)X(t)X(t)^{-1} = -B(t).$$

Therefore,

$$Y'(t)^* = -X(t)^{-1}B(t) = -Y(t)^*B(t)$$

and hence,

$$Y'(t) = -B(t)^*Y(t). \tag{12.15}$$

By (12.14), the function $x_i(t) = X(t)v_i$ is a solution of $x' = B(t)x$ for each $i = 1, \ldots, n$. Similarly, by (12.15), the function $y_i(t) = Y(t)w_i$ is a solution of $y' = -B(t)^*y$ for each $i = 1, \ldots, n$. Note that

$$x_i(t) = U(t)^{-1}v_i(t) \quad \text{and} \quad y_i(t) = U(t)^{-1}w_i(t), \tag{12.16}$$

where $w_i(t) = [V(t)^{-1}]^*w_i$ for each i. Using (10.37) we obtain

$$\begin{aligned} w_i'(t) &= U'(t)y_i(t) + U(t)y_i'(t) \\ &= [U'(t)U(t)^{-1} - U(t)B(t)^*U(t)^{-1}]w_i(t) \\ &= [-A(t)^* + U'(t)U(t)^{-1} + U(t)U'(t)^*]w_i(t) \\ &= \left[-A(t)^* + \frac{d}{dt}(U(t)U(t)^*)\right]w_i(t) = -A(t)^*w_i(t). \end{aligned}$$

Therefore, $w_i(t)$ is the solution of (11.6) with $w_0 = w_i$ for each i.

Since $U(t)$ is unitary, it follows from (12.10) and (12.16) that

$$\|x_i(t)\| \le D_{\varepsilon,n}e^{(\lambda(v_i)+\varepsilon)t} \quad \text{and} \quad \|y_i(t)\| \le D_{\varepsilon,n}e^{(\mu(w_i)+\varepsilon)t}$$

for every $t \ge 0$ and $i = 1, \ldots, n$. Given i and j such that $1 \le i \le n$ and $1 \le j \le n$ we consider the number

$$a_{ij} = \langle X(t)X(s)^{-1}u_i, u_j \rangle.$$

Since $X(t)$ is upper triangular for every $t \ge 0$, we have $a_{ij} = 0$ for $i < j$. We now consider the case when $i \ge j$. Observe that

$$X(t)X(s)^{-1} = X(t)Y(s)^*$$

for any $t \ge s \ge 0$. Since each operator $X(t)$ leaves invariant the space H_n, and $v_1, \ldots, v_n$ and $w_1, \ldots, w_n$ are dual bases, we obtain

$$a_{ij} = \langle Y(s)^* u_i, X(t)^* u_j \rangle = \sum_{k=1}^{n} \langle Y(s)^* u_i, w_k \rangle \langle v_k, X(t)^* u_j \rangle$$

$$= \sum_{k=1}^{n} \langle u_i, Y(s) w_k \rangle \langle X(t) v_k, u_j \rangle$$

$$= \sum_{k=1}^{n} \langle u_i, y_k(s) \rangle \langle x_k(t), u_j \rangle,$$

and thus, using (12.9),

$$|a_{ij}| \leq \sum_{k=1}^{n} \|y_k(s)\| \cdot \|x_k(t)\|$$

$$\leq \sum_{k=1}^{n} D_{\varepsilon,n}^2 e^{(\lambda(v_k)+\varepsilon)t + (\mu(w_k)+\varepsilon)s}$$

$$= \sum_{k=1}^{n} D_{\varepsilon,n}^2 e^{(\lambda(v_k)+\varepsilon)(t-s) + (\lambda(v_k)+\mu(w_k)+2\varepsilon)s}$$

$$\leq n D_{\varepsilon,n}^2 e^{(\lambda'_{n,n}+\varepsilon)(t-s) + (\gamma_n(\lambda,\mu)+2\varepsilon)s}.$$

We can now proceed in a similar manner to that in the proof of Theorem 11.3 (see (11.17)) to conclude that given $v = \sum_{i=1}^{n} \alpha_i u_i \in H_n$ with $\|v\| = 1$,

$$\|X(t)X(s)^{-1}v\|^2 = \left\| \sum_{i=1}^{n} \sum_{j=1}^{n} \alpha_i \langle X(t)X(s)^{-1} u_i, u_j \rangle u_j \right\|^2$$

$$= \sum_{j=1}^{n} \left(\sum_{i=j}^{n} \alpha_i a_{ij} \right)^2$$

$$\leq \sum_{j=1}^{n} \left(\sum_{i=j}^{n} \alpha_i^2 \sum_{i=j}^{n} a_{ij}^2 \right) \leq \sum_{j=1}^{n} \sum_{i=j}^{n} a_{ij}^2.$$

Therefore,

$$\|X(t)X(s)^{-1}v\| \leq n^2 D_{\varepsilon,n}^2 e^{(\lambda'_{n,n}+\varepsilon)(t-s) + (\gamma_n(\lambda,\mu)+2\varepsilon)s}.$$

This establishes the desired inequality.   □

Note that in Theorem 12.8 the operators $A(t)$ need not be upper triangular. When the operators $X(t)$ are diagonal we can somewhat improve the statement in Theorem 12.8.

Theorem 12.9. *Assume that the operator $X(t)$ is diagonal for every $t \geq 0$. Then for every $n \in \mathbb{N}$, $\varepsilon > 0$, and $t \geq s \geq 0$ we have*

$$\|X(t)X(s)^{-1}|H_n\| \leq D_{\varepsilon,n}^2 e^{(\lambda'_{n,n}+\varepsilon)(t-s) + (\tau_n+2\varepsilon)s},$$

where

$$\tau_n = \max\{\lambda(u_i) + \mu(u_i) : i = 1, \ldots, n\} \geq 0.$$

Proof. We use the same notation as in the proof of Theorem 12.8. Let now $v = \sum_{i=1}^{n} \alpha_i u_i \in H_n$ with $\|v\| = 1$. Using the fact that the operators $Y(s)^*$ and $X(t)^*$ are diagonal, we obtain

$$\|X(t)X(s)^{-1}v\| = \left\| \sum_{i=1}^{n} \alpha_i \langle X(t)X(s)^{-1}u_i, u_i \rangle u_i \right\|$$

$$= \left(\sum_{i=1}^{n} \alpha_i^2 \langle Y(s)^*u_i, X(t)^*u_i \rangle^2 \right)^{1/2}$$

$$\leq \max_{1 \leq i \leq n} |\langle Y(s)^*u_i, X(t)^*u_i \rangle|$$

$$= \max_{1 \leq i \leq n} |\langle Y(s)^*u_i, u_i \rangle \langle u_i, X(t)^*u_i \rangle|.$$

Therefore,

$$\|X(t)X(s)^{-1}|H_n\| \leq \max_{1 \leq i \leq n} |\langle u_i, Y(s)u_i \rangle \langle X(t)u_i, u_i \rangle|$$

$$\leq \max_{1 \leq i \leq n} (\|Y(s)u_i\| \cdot \|X(t)u_i\|)$$

$$\leq D_{\varepsilon,n}^2 \max_{1 \leq i \leq n} e^{(\mu(u_i)+\varepsilon)s + (\lambda(u_i)+\varepsilon)t}$$

$$\leq D_{\varepsilon,n}^2 \max_{1 \leq i \leq n} e^{(\lambda(u_i)+\varepsilon)(t-s) + (\lambda(u_i)+\mu(u_i)+2\varepsilon)s}$$

$$\leq D_{\varepsilon,n}^2 e^{(\lambda'_{n,n}+\varepsilon)(t-s) + (\tau_n+2\varepsilon)s}.$$

The fact that $\tau_n \geq 0$ is an immediate consequence of Proposition 10.4. This completes the proof. $\qquad\square$

12.5 Proofs of the stability results

We use the same notation as in Section 12.4 but now applied to the case when $A(t)$ is upper triangular for every t. In this case we can take $U(t) = \mathrm{Id}$ for every t in Theorem 11.3, and thus we can consider the monodromy operators $X(t) = V(t)$ (the operator $V(t)$ is defined in Section 12.4). We shall always make this choice.

Proof of Theorem 12.1. We denote by $v(t)$ the solution of the initial value problem (12.1). This problem is equivalent to the integral equation

$$v(t) = X(t)v_0 + \int_0^t X(t)X(s)^{-1}f(s, v(s))\, ds. \tag{12.17}$$

Consider the operator

$$(Tv)(t) = X(t)v_0 + \int_0^t X(t)X(s)^{-1}f(s,v(s))\,ds$$

on the space

$$\mathcal{B}_\delta = \{v \colon [0,\infty) \to H \text{ continuous} : \|v(t)\| \le \delta e^{\alpha t} \text{ for every } t \ge 0\},$$

where $\delta > 0$ (to be chosen later), and $\alpha = \sup\{\lambda_i' : i \in \mathbb{N}\} + \varepsilon$ for some $\varepsilon > 0$ such that $\alpha < 0$ (recall that (12.5) is a consequence of (12.4)). We introduce the norm on $\mathcal{B}_\delta$ given by

$$\|v\| = \sup\{\|v(t)\|e^{-\alpha t} : t \ge 0\}.$$

One can easily verify that $\mathcal{B}_\delta$ becomes a complete metric space with respect to the induced distance. Observe now that by Theorem 12.8, for every $n \in \mathbb{N}$, $\varepsilon > 0$, and $t \ge s \ge 0$,

$$\|X(t)X(s)^{-1}|H_n\| \le n^2 D_{\varepsilon,n}^2 e^{(\lambda_{n,n}' + \varepsilon)(t-s) + (\gamma_n(\lambda,\mu) + 2\varepsilon)s}$$
$$\le n^2 D_{\varepsilon,n}^2 e^{\alpha(t-s) + \beta s}, \tag{12.18}$$

where $\beta = \gamma(\lambda,\mu) + 2\varepsilon$. Let $v_1, v_2 \in \mathcal{B}_\delta$. Since $X(t)$ is upper triangular for every t, using (12.18) and condition H3 we obtain

$$\|X(t)X(s)^{-1}(f(s,v_1(s)) - f(s,v_2(s)))\|$$
$$= \left\| X(t)X(s)^{-1} \sum_{k=1}^\infty \langle f(s,v_1(s)) - f(s,v_2(s)), u_k\rangle u_k \right\|$$
$$\le \sum_{k=1}^\infty |\langle f(s,v_1(s)) - f(s,v_2(s)), u_k\rangle| \cdot \|X(t)X(s)^{-1}|H_k\|$$
$$\le \sum_{k=1}^\infty \frac{1}{a_k}\|v_1(s) - v_2(s)\|(\|v_1(s)\|^r + \|v_2(s)\|^r)k^2 D_{\varepsilon,k}^2 e^{\alpha(t-s)+\beta s} \tag{12.19}$$
$$\le \sum_{k=1}^\infty \frac{k^2 D_{\varepsilon,k}^2}{a_k}\|v_1 - v_2\|(\|v_1\|^r + \|v_2\|^r)e^{\alpha t + (r\alpha + \beta)s}$$
$$\le \sum_{k=1}^\infty \frac{2\delta^r k^2 D_{\varepsilon,k}^2}{a_k}\|v_1 - v_2\|e^{\alpha t + (r\alpha+\beta)s}.$$

That is,

$$\|X(t)X(s)^{-1}(f(s,v_1(s)) - f(s,v_2(s)))\| \le 2d\delta^r\|v_1 - v_2\|e^{\alpha t + (r\alpha+\beta)s}, \tag{12.20}$$

where d is the constant in (12.11). We assume that $d < \infty$ for some $\varepsilon > 0$ such that (see (12.12))

$$ r\alpha + \beta = r(\sup\{\lambda_i' : i \in \mathbb{N}\} + \varepsilon) + \gamma(\lambda, \mu) + 2\varepsilon < 0, $$

which is always possible due to (12.4). The assumption $d < \infty$ corresponds to require that the sequence $(a_n)_n$ diverges sufficiently fast. Therefore,

$$ \|(Tv_1)(t) - (Tv_2)(t)\| \leq 2d\delta^r \|v_1 - v_2\| e^{\alpha t} \int_0^t e^{(r\alpha+\beta)s} \, ds $$

$$ \leq 2d\kappa\delta^r \|v_1 - v_2\| e^{\alpha t}, $$

where $\kappa = \int_0^\infty e^{(r\alpha+\beta)s} \, ds$. Hence,

$$ \|Tv_1 - Tv_2\| \leq \theta \|v_1 - v_2\|, \tag{12.21} $$

where $\theta = 2d\kappa\delta^r$. Choose now $\delta \in (0,1)$ such that $\theta < 1$. For each $v_0 \in H$ satisfying condition H3 we obtain in a similar manner, using (12.18) with $s = 0$, that

$$ \|X(t)v_0\| \leq \lim_{n \to \infty} \sum_{k=1}^n |\langle v_0, u_k \rangle| \cdot \|X(t)|H_k\| $$
$$ \leq \sum_{k=1}^\infty \frac{k^2 D_{\varepsilon,k}^2}{a_k} e^{\alpha t} \|v_0\| = d e^{\alpha t} \|v_0\|. \tag{12.22} $$

Note that $X(t)v_0 = (T0)(t)$. Therefore, for each $v \in \mathcal{B}_\delta$, setting $v_1 = v \in \mathcal{B}_\delta$ and $v_2 = 0$ in (12.21), we obtain

$$ \|(Tv)(t)\| e^{-\alpha t} \leq \|X(t)v_0\| + \|Tv - T0\| \leq d\|v_0\| + \theta\delta < \delta $$

provided that v_0 is chosen sufficiently small. Therefore, $T(\mathcal{B}_\delta) \subset \mathcal{B}_\delta$, and the operator T is a contraction on the complete metric space $\mathcal{B}_\delta$. Hence, there exists a unique function $v \in \mathcal{B}_\delta$ which solves (12.17). It remains to establish the stability of the zero solution. For this, set

$$ u(t) = (T0)(t) = X(t)v_0, $$

and observe that the solution $v(t)$ can be obtained by

$$ v(t) = \lim_{n \to +\infty} (T^n 0)(t) = \sum_{k=0}^{+\infty} [(T^{k+1}0)(t) - (T^k 0)(t)]. $$

It follows from (12.21) and (12.22) that

$$ \|v\| \leq \sum_{k=0}^{+\infty} \theta^n \|u\| = \frac{\|u\|}{1-\theta} \leq \frac{d\|v_0\|}{1-\theta}. $$

Therefore,

$$ \|v(t)\| \leq \frac{d\|v_0\|}{1-\theta} e^{\alpha t} \text{ for every } t \geq 0. \tag{12.23} $$

This concludes the proof of the theorem. $\square$

Proof of Theorem 12.3. We can repeat almost verbatim the proof of Theorem 12.1, replacing the inequality (12.18) by the condition (12.7), and the inequalities (12.20) (see also (12.19)) and (12.22) respectively by

$$\|X(t)X(s)^{-1}(f(s,v_1(s)) - f(s,v_2(s)))\| \le 2\eta\delta^r\|v_1 - v_2\|e^{\alpha t + (r\alpha + \beta)s},$$

where $\eta = \sum_{k=1}^{\infty} c_k/a_k < \infty$, and

$$\|X(t)v_0\| \le \eta e^{\alpha t}\|v_0\| \text{ for each } v_0 \in H \text{ satisfying condition H3.}$$

That is, we obtain similar inequalities to those in (12.20) and (12.22), with d replaced by η. It then follows from the proof of Theorem 12.1 (see (12.23)) that choosing $\delta \in (0,1)$ such that

$$\theta := 2\eta\delta^r \int_0^\infty e^{(r\alpha + \beta)s}\, ds < 1,$$

any solution $v(t)$ of the equation (12.1) with $\|v_0\|$ sufficiently small satisfies the estimate (12.8) with $a = \eta/(1-\theta)$. $\square$

Proof of Theorem 12.4. As in the proof of Theorem 12.3 we can repeat almost verbatim the proof of Theorem 12.1, replacing the inequalities (12.20) and (12.22) respectively by

$$\begin{aligned}
&\|X(t)X(s)^{-1}(f(s,v_1(s)) - f(s,v_2(s)))\| \\
&\le \|X(t)X(s)^{-1}\| \cdot \|f(s,v_1(s)) - f(s,v_2(s))\| \\
&\le Ce^{\alpha(t-s)+\beta s}c\|v_1(s) - v_2(s)\|(\|v_1(s)\|^r + \|v_2(s)\|^r) \\
&\le Cc\|v_1 - v_2\|(\|v_1\|^r + \|v_2\|^r)e^{\alpha t + (r\alpha + \beta)s} \\
&\le 2Cc\delta^r\|v_1 - v_2\|e^{\alpha t + (r\alpha + \beta)s},
\end{aligned}$$

and

$$\|X(t)v_0\| \le \|X(t)\| \cdot \|v_0\| \le Ce^{\alpha t}\|v_0\|.$$

We can now proceed in a similar manner to that in the proof of Theorem 12.1 to obtain the desired result. $\square$

Proof of Theorem 12.5. Note that condition H2 is explicitly stated as an hypothesis in the theorem. Furthermore, since in the proof of Theorem 12.1 the series are now replaced by finite sums, we do not need (12.2) or condition H3, and thus in particular any sequence $(a_n)_n$ controlling the smallness of the perturbation. In addition, the third hypothesis in the theorem is equivalent to (12.4). The statement is thus an immediate consequence of Theorem 12.1. $\square$

References

1. L. Barreira and Ya. Pesin, *Lyapunov Exponents and Smooth Ergodic Theory*, University Lecture Series 23, Amer. Math. Soc., 2002.
2. L. Barreira and Ya. Pesin, *Smooth ergodic theory and nonuniformly hyperbolic dynamics*, with appendix by O. Sarig, in Handbook of Dynamical Systems 1B, B. Hasselblatt and A. Katok Eds., Elsevier, 2006, pp. 57–263.
3. L. Barreira and Ya. Pesin, *Nonuniform Hyperbolicity: Dynamics of Systems with Nonzero Lyapunov Exponents*, Encyclopedia of Mathematics and its Applications 115, Cambridge Univ. Press, 2007.
4. L. Barreira and C. Valls, *Center manifolds for nonuniformly partially hyperbolic diffeomorphisms*, J. Math. Pures Appl. **84** (2005), 1693–1715.
5. L. Barreira and C. Valls, *Higher regularity of invariant manifolds for nonautonomous equations*, Nonlinearity **18** (2005), 2373–2390.
6. L. Barreira and C. Valls, *Smoothness of invariant manifolds for nonautonomous equations*, Comm. Math. Phys. **259** (2005), 639–677.
7. L. Barreira and C. Valls, *Stability of nonautonomous differential equations in Hilbert spaces*, J. Differential Equations **217** (2005), 204–248.
8. L. Barreira and C. Valls, *Center manifolds for nonuniformly partially hyperbolic trajectories*, Ergodic Theory Dynam. Systems, **26** (2006), 1707–1732.
9. L. Barreira and C. Valls, *Existence of stable manifolds for nonuniformly hyperbolic C^1 dynamics*, Discrete Contin. Dyn. Syst. **16** (2006), 307–327.
10. L. Barreira and C. Valls, *A Grobman–Hartman theorem for nonuniformly hyperbolic dynamics*, J. Differential Equations **228** (2006), 285–310.
11. L. Barreira and C. Valls, *Smooth invariant manifolds in Banach spaces with nonuniform exponential dichotomy*, J. Funct. Anal. **238** (2006), 118–148.
12. L. Barreira and C. Valls, *Stable manifolds for nonautonomous equations without exponential dichotomy*, J. Differential Equations **221** (2006), 58–90.
13. L. Barreira and C. Valls, *Nonuniform exponential dichotomies and Lyapunov regularity*, J. Dynam. Differential Equations **19** (2007), 215–241.
14. L. Barreira and C. Valls, *Reversibility and equivariance in center manifolds*, Discrete Contin. Dyn. Syst. **18** (2007), 677–699.
15. L. Barreira and C. Valls, *Smooth center manifolds for nonuniformly partially hyperbolic trajectories*, J. Differential Equations **237** (2007), 307–342.
16. L. Barreira and C. Valls, *Stability theory and Lyapunov regularity*, J. Differential Equations **232** (2007), 675–701.

17. L. Barreira and C. Valls, *Conjugacies for linear and nonlinear perturbations of nonuniform behavior*, J. Funct. Anal., to appear.

18. L. Barreira and C. Valls, *Robustness of nonuniform exponential dichotomies in Banach spaces*, preprint.

19. G. Belickiĭ, *Functional equations, and conjugacy of local diffeomorphisms of finite smoothness class*, Functional Anal. Appl. **7** (1973), 268–277.

20. G. Belickiĭ, *Equivalence and normal forms of germs of smooth mappings*, Russian Math. Surveys **33** (1978), 107–177.

21. G. Belickiĭ, *On the Grobman–Hartman theorem in the class C^α*, preprint.

22. J. Carr, *Applications of Centre Manifold Theory*, Applied Mathematical Sciences 35, Springer, 1981.

23. C. Chicone and Yu. Latushkin, *Center manifolds for infinite dimensional nonautonomous differential equations*, J. Differential Equations **141** (1997), 356–399.

24. C. Chicone and Yu. Latushkin, *Evolution Semigroups in Dynamical Systems and Differential Equations*, Mathematical Surveys and Monographs 70, Amer. Math. Soc., 1999.

25. S.-N. Chow and H. Leiva, *Existence and roughness of the exponential dichotomy for skew-product semiflow in Banach spaces*, J. Differential Equations **120** (1995), 429–477.

26. S.-N. Chow, W. Liu and Y. Yi, *Center manifolds for invariant sets*, J. Differential Equations **168** (2000), 355–385.

27. S.-N. Chow, W. Liu and Y. Yi, *Center manifolds for smooth invariant manifolds*, Trans. Amer. Math. Soc. **352** (2000), 5179–5211.

28. S.-N. Chow and K. Lu, *C^k centre unstable manifolds* Proc. Roy. Soc. Edinburgh Sect. A **108** (1988), 303–320.

29. E. Coddington and N. Levinson, *Theory of Ordinary Differential Equations*, McGraw-Hill, 1955.

30. C. Constantine and T. Savits, *A multivariate Faà di Bruno formula with applications*, Trans. Amer. Math. Soc. **348** (1996), 503–520.

31. W. Coppel, *Dichotomies and reducibility*, J. Differential Equations **3** (1967), 500–521.

32. W. Coppel, *Dichotomies in Stability Theory*, Lect. Notes in Math. 629, Springer, 1978.

33. Ju. Dalec′kiĭ and M. Kreĭn, *Stability of Solutions of Differential Equations in Banach Space*, Translations of Mathematical Monographs 43, Amer. Math. Soc., 1974.

34. M. Elbialy, *On sequences of $C_b^{k,\delta}$ maps which converge in the uniform C^0-norm*, Proc. Amer. Math. Soc. **128** (2000), 3285–3290.

35. P. Enflo, *A counterexample to the approximation problem in Banach spaces*, Acta Math. **130** (1973), 309–317.

36. C. Faà di Bruno, *Note sur une nouvelle formule du calcul différentiel*, Quart. J. Math. **1** (1855), 359–360.

37. A. Fathi, M. Herman and J.-C. Yoccoz, *A proof of Pesin's stable manifold theorem*, in Geometric Dynamics (Rio de Janeiro, 1981), J. Palis Ed., Lect. Notes. in Math. 1007, Springer, 1983, pp. 177–215.

38. D. Grobman, *Homeomorphism of systems of differential equations*, Dokl. Akad. Nauk SSSR **128** (1959), 880–881.

39. D. Grobman, *Topological classification of neighborhoods of a singularity in n-space*, Mat. Sb. (N.S.) **56 (98)** (1962), 77–94.

40. M. Guysinsky, B. Hasselblatt and V. Rayskin, *Differentiability of the Hartman-Grobman linearization*, Discrete Contin. Dyn. Syst. **9** (2003), 979–984.

41. J. Hale, *Asymptotic Behavior of Dissipative Systems*, Mathematical Surveys and Monographs 25, Amer. Math. Soc., 1988.

42. J. Hale, L. Magalhães and W. Oliva, *Dynamics in Infinite Dimensions*, Applied Mathematical Sciences 47, Springer, 2002.

43. P. Hartman, *A lemma in the theory of structural stability of differential equations*, Proc. Amer. Math. Soc. **11** (1960), 610–620.

44. P. Hartman, *On local homeomorphisms of Euclidean spaces*, Bol. Soc. Mat. Mexicana (2) **5** (1960), 220–241.

45. P. Hartman, *On the local linearization of differential equations*, Proc. Amer. Math. Soc. **14** (1963), 568–573.

46. D. Henry, *Geometric Theory of Semilinear Parabolic Equations*, Lect. Notes in Math. 840, Springer, 1981.

47. D. Henry, *Exponential dichotomies, the shadowing lemma and homoclinic orbits in Banach spaces*, in Dynamical Phase Transitions (São Paulo, 1994), Resenhas IME-USP **1** (1994), 381–401.

48. A. Katok, *Lyapunov exponents, entropy and periodic orbits for diffeomorphisms*, Inst. Hautes Études Sci. Publ. Math. **51** (1980), 137–173.

49. A. Katok and B. Hasselblatt, *Introduction to the Modern Theory of Dynamical Systems*, with a supplement by A. Katok and L. Mendoza, Encyclopedia of Mathematics and its Applications 54, Cambridge University Press, Cambridge, 1995.

50. A. Katok and L. Mendoza, *Dynamical systems with nonuniformly hyperbolic behavior*, in Introduction to the Modern Theory of Dynamical Systems, A. Katok and B. Hasselblatt, Cambridge Univ. Press, 1995.

51. A. Katok and J.-M. Strelcyn, *Invariant Manifolds, Entropy and Billiards; Smooth Maps with Singularities*, with the collaboration of F. Ledrappier and F. Przytycki, Lect. Notes. in Math. 1222, Springer, 1986.

52. A. Kelley, *The stable, center-stable, center, center-unstable, unstable manifolds*, J. Differential Equations **3** (1967), 546–570.

53. J. Lamb and J. Roberts, *Time-reversal symmetry in dynamical systems: a survey*, in Time-Reversal Symmetry in Dynamical Systems (Coventry, 1996), Phys. D **112** (1998), 1–39.

54. O. Lanford III, *Bifurcation of periodic solutions into invariant tori: the work of Ruelle and Takens*, in Nonlinear Problems in the Physical Sciences and Biology: Proceedings of a Battelle Summer Institute (Seattle, 1972), I. Stakgold, D. Joseph and D. Sattinger Eds., Lect. Notes in Math. 322, Springer, 1973, pp. 159–192.

55. F. Ledrappier and L.-S. Young, *The metric entropy of diffeomorphisms I. Characterization of measures satisfying Pesin's entropy formula*, Ann. of Math. (2) **122** (1985), 509–539.

56. E. Lukacs, *Applications of Faà di Bruno's formula in mathematical statistics*, Amer. Math. Monthly **62** (1955), 340–348.

57. A. Lyapunov, *The General Problem of the Stability of Motion*, Taylor & Francis, 1992.

58. R. Mañé, *Lyapunov exponents and stable manifolds for compact transformations*, in Geometric dynamics (Rio de Janeiro, 1981), J. Palis Ed., Lect. Notes in Math. 1007, Springer, 1983, pp. 522–577.

59. J. Massera and J. Schäffer, *Linear differential equations and functional analysis. I*, Ann. of Math. (2) **67** (1958), 517–573.
60. J. Massera and J. Schäffer, *Linear Differential Equations and Function Spaces*, Pure and Applied Mathematics 21, Academic Press, 1966.
61. P. McSwiggen, *A geometric characterization of smooth linearizability*, Michigan Math. J. **43** (1996), 321–335.
62. A. Mielke, *A reduction principle for nonautonomous systems in infinite-dimensional spaces*, J. Differential Equations **65** (1986), 68–88.
63. J. Moser, *On a theorem of Anosov*, J. Differential Equations **5** (1969), 411–440.
64. R. Naulin and M. Pinto, *Admissible perturbations of exponential dichotomy roughness*, Nonlinear Anal. **31** (1998), 559–571.
65. V. Oseledets, *A multiplicative ergodic theorem. Liapunov characteristic numbers for dynamical systems*, Trans. Moscow Math. Soc. **19** (1968), 197–221.
66. J. Palis, *On the local structure of hyperbolic points in Banach spaces*, An. Acad. Brasil. Ci. **40** (1968), 263–266.
67. K. Palmer, *A generalization of Hartman's linearization theorem*, J. Math. Anal. Appl. **41** (1973), 753–758.
68. K. Palmer, *Exponential dichotomies and transversal homoclinic points*, J. Differential Equations **55** (1984), 225–256.
69. O. Perron, *Die Stabilitätsfrage bei Differentialgleichungen*, Math. Z. **32** (1930), 703–728.
70. Ya. Pesin, *Families of invariant manifolds corresponding to nonzero characteristic exponents*, Math. USSR-Izv. **10** (1976), 1261–1305.
71. Ya. Pesin, *Characteristic Ljapunov exponents, and smooth ergodic theory*, Russian Math. Surveys **32** (1977), 55–114.
72. Ya. Pesin, *Geodesic flows on closed Riemannian manifolds without focal points*, Math. USSR-Izv. **11** (1977), 1195–1228.
73. V. Pliss, *A reduction principle in the theory of stability of motion*, Izv. Akad. Nauk SSSR Ser. Mat. **28** (1964), 1297–1324.
74. V. Pliss and G. Sell, *Robustness of exponential dichotomies in infinite-dimensional dynamical systems*, J. Dynam. Differential Equations **11** (1999), 471–513.
75. L. Popescu, *Exponential dichotomy roughness on Banach spaces*, J. Math. Anal. Appl. **314** (2006), 436–454.
76. C. Pugh, *On a theorem of P. Hartman*, Amer. J. Math. **91** (1969), 363–367.
77. C. Pugh, *The $C^{1+\alpha}$ hypothesis in Pesin theory*, Inst. Hautes Études Sci. Publ. Math. **59** (1984), 143–161.
78. C. Pugh and M. Shub, *Ergodic attractors*, Trans. Amer. Math. Soc. **312** (1989), 1–54.
79. V. Rayskin, *α-Hölder linearization*, J. Differential Equations **147** (1998), 271–284.
80. D. Ruelle, *Ergodic theory of differentiable dynamical systems*, Inst. Hautes Études Sci. Publ. Math. **50** (1979), 27–58.
81. D. Ruelle, *Characteristic exponents and invariant manifolds in Hilbert space*, Ann. of Math. (2) **115** (1982), 243–290.
82. R. Sacker, *Existence of dichotomies and invariant splittings for linear differential systems IV*, J. Differential Equations **27** (1978), 106–137.
83. R. Sacker and G. Sell, *Existence of dichotomies and invariant splittings for linear differential systems I*, J. Differential Equations **15** (1974), 429–458.
84. R. Sacker and G. Sell, *Existence of dichotomies and invariant splittings for linear differential systems II*, J. Differential Equations **22** (1976), 478–496.

85. R. Sacker and G. Sell, *Existence of dichotomies and invariant splittings for linear differential systems III*, J. Differential Equations **22** (1976), 497–522.
86. R. Sacker and G. Sell, *Dichotomies for linear evolutionary equations in Banach spaces*, J. Differential Equations **113** (1994), 17–67.
87. G. Sell, *Smooth linearization near a fixed point*, Amer. J. Math. **107** (1985), 1035–1091.
88. G. Sell and Y. You, *Dynamics of Evolutionary Equations*, Applied Mathematical Sciences 143, Springer, 2002.
89. S. Sternberg, *Local contractions and a theorem of Poincaré*, Amer. J. Math. **79** (1957), 809–824.
90. S. Sternberg, *On the structure of local homeomorphisms of euclidean n-space. II.*, Amer. J. Math. **80** (1958), 623–631.
91. B. Tan, *σ-Hölder continuous linearization near hyperbolic fixed points in* $\mathbb{R}^n$, J. Differential Equations **162** (2000), 251–269.
92. P. Thieullen, *Fibrés dynamiques asymptotiquement compacts. Exposants de Lyapunov. Entropie. Dimension*, Ann. Inst. H. Poincaré. Anal. Non Linéaire **4** (1987), 49–97.
93. A. Vanderbauwhede, *Centre manifolds, normal forms and elementary bifurcations*, in Dynamics Reported 2, Wiley, 1989, pp. 89–169.
94. A. Vanderbauwhede and G. Iooss, *Center manifold theory in infinite dimensions*, in Dynamics Reported (N.S.) 1, Springer, 1992, pp. 125–163.
95. A. Vanderbauwhede and S. van Gils, *Center manifolds and contractions on a scale of Banach spaces*, J. Funct. Anal. **72** (1987), 209–224.
96. S. van Strien, *Smooth linearization of hyperbolic fixed points without resonance conditions*, J. Differential Equations **85** (1990), 66–90.

Index

Lecture Notes in Mathematics

For information about earlier volumes
please contact your bookseller or Springer
LNM Online archive: springerlink.com

Vol. 1781: E. Bolthausen, E. Perkins, A. van der Vaart, Lectures on Probability Theory and Statistics. Ecole d' Eté de Probabilités de Saint-Flour XXIX-1999. Editor: P. Bernard (2002)

Vol. 1782: C.-H. Chu, A. T.-M. Lau, Harmonic Functions on Groups and Fourier Algebras (2002)

Vol. 1783: L. Grüne, Asymptotic Behavior of Dynamical and Control Systems under Perturbation and Discretization (2002)

Vol. 1784: L. H. Eliasson, S. B. Kuksin, S. Marmi, J.-C. Yoccoz, Dynamical Systems and Small Divisors. Cetraro, Italy 1998. Editors: S. Marmi, J.-C. Yoccoz (2002)

Vol. 1785: J. Arias de Reyna, Pointwise Convergence of Fourier Series (2002)

Vol. 1786: S. D. Cutkosky, Monomialization of Morphisms from 3-Folds to Surfaces (2002)

Vol. 1787: S. Caenepeel, G. Militaru, S. Zhu, Frobenius and Separable Functors for Generalized Module Categories and Nonlinear Equations (2002)

Vol. 1788: A. Vasil'ev, Moduli of Families of Curves for Conformal and Quasiconformal Mappings (2002)

Vol. 1789: Y. Sommerhäuser, Yetter-Drinfel'd Hopf algebras over groups of prime order (2002)

Vol. 1790: X. Zhan, Matrix Inequalities (2002)

Vol. 1791: M. Knebusch, D. Zhang, Manis Valuations and Prüfer Extensions I: A new Chapter in Commutative Algebra (2002)

Vol. 1792: D. D. Ang, R. Gorenflo, V. K. Le, D. D. Trong, Moment Theory and Some Inverse Problems in Potential Theory and Heat Conduction (2002)

Vol. 1793: J. Cortés Monforte, Geometric, Control and Numerical Aspects of Nonholonomic Systems (2002)

Vol. 1794: N. Pytheas Fogg, Substitution in Dynamics, Arithmetics and Combinatorics. Editors: V. Berthé, S. Ferenczi, C. Mauduit, A. Siegel (2002)

Vol. 1795: H. Li, Filtered-Graded Transfer in Using Noncommutative Gröbner Bases (2002)

Vol. 1796: J.M. Melenk, hp-Finite Element Methods for Singular Perturbations (2002)

Vol. 1797: B. Schmidt, Characters and Cyclotomic Fields in Finite Geometry (2002)

Vol. 1798: W.M. Oliva, Geometric Mechanics (2002)

Vol. 1799: H. Pajot, Analytic Capacity, Rectifiability, Menger Curvature and the Cauchy Integral (2002)

Vol. 1800: O. Gabber, L. Ramero, Almost Ring Theory (2003)

Vol. 1801: J. Azéma, M. Émery, M. Ledoux, M. Yor (Eds.), Séminaire de Probabilités XXXVI (2003)

Vol. 1802: V. Capasso, E. Merzbach, B. G. Ivanoff, M. Dozzi, R. Dalang, T. Mountford, Topics in Spatial Stochastic Processes. Martina Franca, Italy 2001. Editor: E. Merzbach (2003)

Vol. 1803: G. Dolzmann, Variational Methods for Crystalline Microstructure – Analysis and Computation (2003)

Vol. 1804: I. Cherednik, Ya. Markov, R. Howe, G. Lusztig, Iwahori-Hecke Algebras and their Representation Theory. Martina Franca, Italy 1999. Editors: V. Baldoni, D. Barbasch (2003)

Vol. 1805: F. Cao, Geometric Curve Evolution and Image Processing (2003)

Vol. 1806: H. Broer, I. Hoveijn. G. Lunther, G. Vegter, Bifurcations in Hamiltonian Systems. Computing Singularities by Gröbner Bases (2003)

Vol. 1807: V. D. Milman, G. Schechtman (Eds.), Geometric Aspects of Functional Analysis. Israel Seminar 2000-2002 (2003)

Vol. 1808: W. Schindler, Measures with Symmetry Properties (2003)

Vol. 1809: O. Steinbach, Stability Estimates for Hybrid Coupled Domain Decomposition Methods (2003)

Vol. 1810: J. Wengenroth, Derived Functors in Functional Analysis (2003)

Vol. 1811: J. Stevens, Deformations of Singularities (2003)

Vol. 1812: L. Ambrosio, K. Deckelnick, G. Dziuk, M. Mimura, V. A. Solonnikov, H. M. Soner, Mathematical Aspects of Evolving Interfaces. Madeira, Funchal, Portugal 2000. Editors: P. Colli, J. F. Rodrigues (2003)

Vol. 1813: L. Ambrosio, L. A. Caffarelli, Y. Brenier, G. Buttazzo, C. Villani, Optimal Transportation and its Applications. Martina Franca, Italy 2001. Editors: L. A. Caffarelli, S. Salsa (2003)

Vol. 1814: P. Bank, F. Baudoin, H. Föllmer, L.C.G. Rogers, M. Soner, N. Touzi, Paris-Princeton Lectures on Mathematical Finance 2002 (2003)

Vol. 1815: A. M. Vershik (Ed.), Asymptotic Combinatorics with Applications to Mathematical Physics. St. Petersburg, Russia 2001 (2003)

Vol. 1816: S. Albeverio, W. Schachermayer, M. Talagrand, Lectures on Probability Theory and Statistics. Ecole d'Eté de Probabilités de Saint-Flour XXX-2000. Editor: P. Bernard (2003)

Vol. 1817: E. Koelink, W. Van Assche (Eds.), Orthogonal Polynomials and Special Functions. Leuven 2002 (2003)

Vol. 1818: M. Bildhauer, Convex Variational Problems with Linear, nearly Linear and/or Anisotropic Growth Conditions (2003)

Vol. 1819: D. Masser, Yu. V. Nesterenko, H. P. Schlickewei, W. M. Schmidt, M. Waldschmidt, Diophantine Approximation. Cetraro, Italy 2000. Editors: F. Amoroso, U. Zannier (2003)

Vol. 1820: F. Hiai, H. Kosaki, Means of Hilbert Space Operators (2003)

Vol. 1821: S. Teufel, Adiabatic Perturbation Theory in Quantum Dynamics (2003)

Vol. 1822: S.-N. Chow, R. Conti, R. Johnson, J. Mallet-Paret, R. Nussbaum, Dynamical Systems. Cetraro, Italy 2000. Editors: J. W. Macki, P. Zecca (2003)

Vol. 1823: A. M. Anile, W. Allegretto, C. Ringhofer, Mathematical Problems in Semiconductor Physics. Cetraro, Italy 1998. Editor: A. M. Anile (2003)

Vol. 1824: J. A. Navarro González, J. B. Sancho de Salas, $\mathscr{C}^\infty$ – Differentiable Spaces (2003)

Vol. 1825: J. H. Bramble, A. Cohen, W. Dahmen, Multiscale Problems and Methods in Numerical Simulations, Martina Franca, Italy 2001. Editor: C. Canuto (2003)

Vol. 1826: K. Dohmen, Improved Bonferroni Inequalities via Abstract Tubes. Inequalities and Identities of Inclusion-Exclusion Type. VIII, 113 p, 2003.

Vol. 1827: K. M. Pilgrim, Combinations of Complex Dynamical Systems. IX, 118 p, 2003.

Vol. 1828: D. J. Green, Gröbner Bases and the Computation of Group Cohomology. XII, 138 p, 2003.

Vol. 1829: E. Altman, B. Gaujal, A. Hordijk, Discrete-Event Control of Stochastic Networks: Multimodularity and Regularity. XIV, 313 p, 2003.

Vol. 1830: M. I. Gil', Operator Functions and Localization of Spectra. XIV, 256 p, 2003.

Vol. 1831: A. Connes, J. Cuntz, E. Guentner, N. Higson, J. E. Kaminker, Noncommutative Geometry, Martina Franca, Italy 2002. Editors: S. Doplicher, L. Longo (2004)

Vol. 1832: J. Azéma, M. Émery, M. Ledoux, M. Yor (Eds.), Séminaire de Probabilités XXXVII (2003)

Vol. 1884: N. Hayashi, E.I. Kaikina, P.I. Naumkin, I.A. Shishmarev, Asymptotics for Dissipative Nonlinear Equations (2006)

Vol. 1885: A. Telcs, The Art of Random Walks (2006)

Vol. 1886: S. Takamura, Splitting Deformations of Degenerations of Complex Curves (2006)

Vol. 1887: K. Habermann, L. Habermann, Introduction to Symplectic Dirac Operators (2006)

Vol. 1888: J. van der Hoeven, Transseries and Real Differential Algebra (2006)

Vol. 1889: G. Osipenko, Dynamical Systems, Graphs, and Algorithms (2006)

Vol. 1890: M. Bunge, J. Funk, Singular Coverings of Toposes (2006)

Vol. 1891: J.B. Friedlander, D.R. Heath-Brown, H. Iwaniec, J. Kaczorowski, Analytic Number Theory, Cetraro, Italy, 2002. Editors: A. Perelli, C. Viola (2006)

Vol. 1892: A. Baddeley, I. Bárány, R. Schneider, W. Weil, Stochastic Geometry, Martina Franca, Italy, 2004. Editor: W. Weil (2007)

Vol. 1893: H. Hanßmann, Local and Semi-Local Bifurcations in Hamiltonian Dynamical Systems, Results and Examples (2007)

Vol. 1894: C.W. Groetsch, Stable Approximate Evaluation of Unbounded Operators (2007)

Vol. 1895: L. Molnár, Selected Preserver Problems on Algebraic Structures of Linear Operators and on Function Spaces (2007)

Vol. 1896: P. Massart, Concentration Inequalities and Model Selection, Ecole d'Été de Probabilités de Saint-Flour XXXIII-2003. Editor: J. Picard (2007)

Vol. 1897: R. Doney, Fluctuation Theory for Lévy Processes, Ecole d'Été de Probabilités de Saint-Flour XXXV-2005. Editor: J. Picard (2007)

Vol. 1898: H.R. Beyer, Beyond Partial Differential Equations, On linear and Quasi-Linear Abstract Hyperbolic Evolution Equations (2007)

Vol. 1899: Séminaire de Probabilités XL. Editors: C. Donati-Martin, M. Émery, A. Rouault, C. Stricker (2007)

Vol. 1900: E. Bolthausen, A. Bovier (Eds.), Spin Glasses (2007)

Vol. 1901: O. Wittenberg, Intersections de deux quadriques et pinceaux de courbes de genre 1, Intersections of Two Quadrics and Pencils of Curves of Genus 1 (2007)

Vol. 1902: A. Isaev, Lectures on the Automorphism Groups of Kobayashi-Hyperbolic Manifolds (2007)

Vol. 1903: G. Kresin, V. Maz'ya, Sharp Real-Part Theorems (2007)

Vol. 1904: P. Giesl, Construction of Global Lyapunov Functions Using Radial Basis Functions (2007)

Vol. 1905: C. Prévôt, M. Röckner, A Concise Course on Stochastic Partial Differential Equations (2007)

Vol. 1906: T. Schuster, The Method of Approximate Inverse: Theory and Applications (2007)

Vol. 1907: M. Rasmussen, Attractivity and Bifurcation for Nonautonomous Dynamical Systems (2007)

Vol. 1908: T.J. Lyons, M. Caruana, T. Lévy, Differential Equations Driven by Rough Paths, Ecole d'Été de Probabilités de Saint-Flour XXXIV-2004 (2007)

Vol. 1909: H. Akiyoshi, M. Sakuma, M. Wada, Y. Yamashita, Punctured Torus Groups and 2-Bridge Knot Groups (I) (2007)

Vol. 1910: V.D. Milman, G. Schechtman (Eds.), Geometric Aspects of Functional Analysis. Israel Seminar 2004-2005 (2007)

Vol. 1911: A. Bressan, D. Serre, M. Williams, K. Zumbrun, Hyperbolic Systems of Balance Laws. Lectures given at the C.I.M.E. Summer School held in Cetraro, Italy, July 14–21, 2003. Editor: P. Marcati (2007)

Vol. 1912: V. Berinde, Iterative Approximation of Fixed Points (2007)

Vol. 1913: J.E. Marsden, G. Misiołek, J.-P. Ortega, M. Perlmutter, T.S. Ratiu, Hamiltonian Reduction by Stages (2007)

Vol. 1914: G. Kutyniok, Affine Density in Wavelet Analysis (2007)

Vol. 1915: T. Bıyıkoğlu, J. Leydold, P.F. Stadler, Laplacian Eigenvectors of Graphs. Perron-Frobenius and Faber-Krahn Type Theorems (2007)

Vol. 1916: C. Villani, F. Rezakhanlou, Entropy Methods for the Boltzmann Equation. Editors: F. Golse, S. Olla (2008)

Vol. 1917: I. Veselić, Existence and Regularity Properties of the Integrated Density of States of Random Schrödinger (2008)

Vol. 1918: B. Roberts, R. Schmidt, Local Newforms for GSp(4) (2007)

Vol. 1919: R.A. Carmona, I. Ekeland, A. Kohatsu-Higa, J.-M. Lasry, P.-L. Lions, H. Pham, E. Taflin, Paris-Princeton Lectures on Mathematical Finance 2004. Editors: R.A. Carmona, E. Çinlar, I. Ekeland, E. Jouini, J.A. Scheinkman, N. Touzi (2007)

Vol. 1920: S.N. Evans, Probability and Real Trees. Ecole d'Été de Probabilités de Saint-Flour XXXV-2005 (2008)

Vol. 1921: J.P. Tian, Evolution Algebras and their Applications (2008)

Vol. 1922: A. Friedman (Ed.), Tutorials in Mathematical BioSciences IV. Evolution and Ecology (2008)

Vol. 1923: J.P.N. Bishwal, Parameter Estimation in Stochastic Differential Equations (2008)

Vol. 1924: M. Wilson, Littlewood-Paley Theory and Exponential-Square Integrability (2008)

Vol. 1925: M. du Sautoy, Zeta Functions of Groups and Rings (2008)

Vol. 1926: L. Barreira, V. Claudia, Stability of Nonautonomous Differential Equations (2008)

Recent Reprints and New Editions

Vol. 1618: G. Pisier, Similarity Problems and Completely Bounded Maps. 1995 – 2nd exp. edition (2001)

Vol. 1629: J.D. Moore, Lectures on Seiberg-Witten Invariants. 1997 – 2nd edition (2001)

Vol. 1638: P. Vanhaecke, Integrable Systems in the realm of Algebraic Geometry. 1996 – 2nd edition (2001)

Vol. 1702: J. Ma, J. Yong, Forward-Backward Stochastic Differential Equations and their Applications. 1999 – Corr. 3rd printing (2007)

Vol. 830: J.A. Green, Polynomial Representations of GL_n, with an Appendix on Schensted Correspondence and Littelmann Paths by K. Erdmann, J.A. Green and M. Schocker 1980 – 2nd corr. and augmented edition (2007)

Printed in the United States
By Bookmasters